MOLECULAR NEUROBIOLOGY

Recombinant DNA Approaches

CURRENT TOPICS IN NEUROBIOLOGY

Series Editor

Samuel H. Barondes
*Professor and Chairman, Department of Psychiatry
and Director, Langley Porter Psychiatric Institute
University of California, San Francisco
San Francisco, California*

Cell Culture in the Neurosciences
Edited by Jane E. Bottenstein and Gordon Sato

Molecular Neurobiology: Recombinant DNA Approaches
Edited by Steve Heinemann and James Patrick

Neuroimmunology
Edited by Jeremy Brockes

Neuronal Development
Edited by Nicholas C. Spitzer

Neuronal Recognition
Edited by Samuel H. Barondes

Peptides in Neurobiology
Edited by Harold Gainer

Tissue Culture of the Nervous System
Edited by Gordon Sato

A Continuation Order Plan is available for this series. A continuation order will bring delivery of each new volume immediately upon publication. Volumes are billed only upon actual shipment. For further information please contact the publisher.

MOLECULAR NEUROBIOLOGY

Recombinant DNA Approaches

Edited by
Steve Heinemann
and
James Patrick

The Salk Institute
San Diego, California

PLENUM PRESS • NEW YORK AND LONDON

Library of Congress Cataloging in Publication Data

Molecular neurobiology.

(Current topics in neurobiology)
Includes bibliography and index.
1. Molecular neurobiology. 2. Recombinant DNA. I. Heinemann, Steve. II. Patrick,
James, 1942–   . III. Series. [DNLM: 1. Cloning, Molecular. 2. DNA, Recombinant. 3.
Gene Expression Regulation. 4. Molecular Biology. 5. Nervous System. 6.
Neurobiology. WL 100 M7188]
QP356.2.M65 1987   599′.0188   87-2512
ISBN 978-1-4615-7490-3   ISBN 978-1-4615-7488-0 (eBook)
DOI 10.1007/978-1-4615-7488-0 0

© 1987 Plenum Press, New York
Softcover reprint of the hardcover 1st edition 1987
A Division of Plenum Publishing Corporation
233 Spring Street, New York, N.Y. 10013

Contributors

GIGI ASOULINE

Molecular Neurobiology Laboratory
The Salk Institute
La Jolla, California 92037

MARC BALLIVET

Molecular Neurobiology Laboratory
The Salk Institute
La Jolla, California 92037

JIM BOULTER

Molecular Neurobiology Laboratory
The Salk Institute
La Jolla, California 92037

JOHN CONNOLLY

Molecular Neurobiology Laboratory
The Salk Institute
La Jolla, California 92037

EVAN DENERIS

Molecular Neurobiology Laboratory
The Salk Institute
La Jolla, California 92037

NICK J. DIBB

MRC Laboratory of Molecular Biology
University Postgraduate Medical School
Cambridge CB2 2QH, England

JANET RETTIG EMANUEL

Department of Cell Biology
Yale University School of Medicine
New Haven, Connecticut 06510

GLEN A. EVANS   Cancer Biology Laboratory
                       The Salk Institute
                       San Diego, California 92138

KAREN EVANS   Molecular Neurobiology Laboratory
                       The Salk Institute
                       La Jolla, California 92037

SYLVIA EVANS   Molecular Neurobiology Laboratory
                       The Salk Institute
                       La Jolla, California 92037

LLOYD D. FRICKER   Molecular Pharmacology Department
                       Albert Einstein College of Medicine
                       Bronx, New York 10461

JOHN FORREST   Molecular Neurobiology Laboratory
                       The Salk Institute
                       La Jolla, California 92037

PAUL GARDNER   Molecular Neurobiology Laboratory
                       The Salk Institute
                       La Jolla, California 92037

SUSAN GARETZ   Department of Cell Biology
                       Yale University School of Medicine
                       New Haven, Connecticut 06510

ORA GOLDBERG   Department of Organic Chemistry
                       The Weizmann Institute of Science
                       Rehovot 76100, Israel

DAN GOLDMAN   Molecular Neurobiology Laboratory
                       The Salk Institute
                       La Jolla, California 92037

MARK GRIMES   Institute for Advanced Biomedical
                           Research
                       The Oregon Health Sciences University
                       Portland, Oregon 97201

STEVE HEINEMANN   Molecular Neurobiology Laboratory
                        The Salk Institute
                        La Jolla, California 92037

EDWARD HERBERT   Institute for Advanced Biomedical
                            Research
                        The Oregon Health Sciences University
                        Portland, Oregon 97201

JONATHAN KARN   MRC Laboratory of Molecular Biology
                        University Postgraduate Medical School
                        Cambridge CB2 2QH, England

ABHA KOCHHAR   Molecular Neurobiology Laboratory
                        The Salk Institute
                        La Jolla, California 92037

GREG LEMKE   Molecular Neurobiology Laboratory
                        The Salk Institute for Biological Studies
                        La Jolla, California 92037

ROBERT LEVENSON   Department of Cell Biology
                        Yale University School of Medicine
                        New Haven, Connecticut 06510

DANE LISTON   Institute for Advanced Biomedical
                            Research
                        The Oregon Health Sciences University
                        Portland, Oregon 97201

WALTER LUYTEN   Molecular Neurobiology Laboratory
                        The Salk Institute
                        La Jolla, California 92037

ANNE C. MAHON   Department of Biological Sciences
                        Stanford University
                        Stanford, California 94305

PAM MASON   Molecular Neurobiology Laboratory
                        The Salk Institute
                        La Jolla, California 92037

DAVID M. MILLER   MRC Laboratory of Molecular Biology
                         University Postgraduate Medical School
                         Cambridge CB2 2QH, England

E. JANE MITCHELL   MRC Laboratory of Molecular Biology
                         University Postgraduate Medical School
                         Cambridge CB2 2QH, England

JIM PATRICK   Molecular Neurobiology Laboratory
                         The Salk Institute
                         La Jolla, California 92037

CATHERINE PRODY   Department of Neurobiology
                         The Weizmann Institute of Science
                         Rehovot 76100, Israel

RICHARD H. SCHELLER   Department of Biological Sciences
                         Stanford University
                         Stanford, California 94305

JAY W. SCHNEIDER   Department of Cell Biology
                         Yale University School of Medicine
                         New Haven, Connecticut 06510

HERMONA SOREQ   Department of Neurobiology
                         The Weizmann Institute of Science
                         Rehovot 76100, Israel

DINA ZEVIN-SONKIN   Department of Neurobiology
                         The Weizmann Institute of Science
                         Rehovot 76100, Israel

DOUG TRECO   Molecular Neurobiology Laboratory
                         The Salk Institute
                         La Jolla, California 92037

KEIJI WADA   Molecular Neurobiology Laboratory
                         The Salk Institute
                         La Jolla, California 92037

Preface

This book is a collection of papers describing some of the first attempts to apply the techniques of recombinant DNA and molecular biology to studies of the nervous system. We believe this is an important new direction for brain research that will eventually lead to insights not possible with more traditional approaches. At first glance, the marriage of molecular biology to brain research seems an unlikely one because of the tremendous disparity in the histories of these two disciplines and the problems they face. Molecular biology is by nature a reductionist approach to biology. Molecular biologists have always tried to attack central questions in the most direct approach possible, usually in the most simple system available: a bacterium or a bacterial virus. Important experiments can usually be repeated quickly and cheaply, in many cases by the latest group of graduate students entering the field. The success of molecular biology has been so profound because the result of each important experiment has made the next critical question obvious, and usually answerable, in short order.

Studies of the nervous system have a very different history. First, the human brain is what really interests us and it is the most complex structure that we know in biology. The central question is clear: How do we carry out higher functions such as learning and thinking? However, at present there is no widely accepted and testable theory of learning and no clear path to such a theory. Numerous attempts have been made to find simple brains with interesting properties and, while some insights have been gained from these studies, the search for simple systems has justified the investment of scarce talent and resources into the study of many different species. Thus many experiments are not replicated quickly; sometimes they are forgotten before it is possible to know whether they are true or not. In many cases, the experiments require expensive specialized equipment as well as expertise and are hard to replicate. Thus brain research has not always had the quick feedback it needs to make steady progress.

Despite these problems, however, there have been some great triumphs in neurobiology. The working out of the basis of the action potential and synaptic transmission are impressive examples. More recently progress in understanding the visual system has been encouraging. This progress has induced a new group of investigators trained in cell biology and molecular biology to enter the neurosciences. At first glance, there is reason to be skeptical that the reductionist approach of molecular biology can make important contributions to the study of a system as complex as the brain. But there are some encouraging developments that make this a pessimistic view. Molecular biologists have recently moved into three very different areas involving complex systems: cancer, immunology, and the general problem of development. It is already clear that knowledge of the structure of the genes involved in these complex systems has dramatically changed the way we think about these problems. We believe that molecular biology will make this same kind of contribution to the study of the brain. For example, many plausible theories of brain function postulate that the stimulation of nerve networks results in a stable change in the efficiency of synaptic transmission at individual synapses. This must involve a change in some molecule associated with the synapse. Since we know little about the detailed structure of the molecules present at synapses in the brain, it is difficult to test these ideas. It seems to us that the best hope for studying the molecules important for brain function is through the use of gene cloning and the techniques of molecular biology. The chapters in this book represent a beginning.

Steve Heinemann
Jim Patrick

Contents

3. Molecular Biology of the Neural and Muscle Nicotinic
   Acetylcholine Receptors
   STEVE HEINEMANN, GIGI ASOULINE, MARC BALLIVET, JIM
   BOULTER, JOHN CONNOLLY, EVAN DENERIS, KAREN EVANS,
   SYLVIA EVANS, JOHN FORREST, PAUL GARDNER, DAN GOLDMAN,
   ABHA KOCHHAR, WALTER LUYTEN, PAM MASON, DOUG
   TRECO, KEIJI WADA, AND JIM PATRICK

5. Small Cardioactive Peptides in A and B: Chemical
   Messengers in the *Aplysia* Nervous System
   ANNE C. MAHON AND RICHARD H. SCHELLER

8. Specificity of Prohormone Processing: The Promise of
   Molecular Biology
   LLOYD D. FRICKER, DANE LISTON, MARK GRIMES, AND
   EDWARD HERBERT

1

The Molecular Biology of the Na,K-ATPase and Other Genes Involved in the Ouabain-Resistant Phenotype

ROBERT LEVENSON, JANET RETTIG EMANUEL, SUSAN GARETZ, and JAY W. SCHNEIDER

1. INTRODUCTION

Na,K-ATPase is the membrane-embedded enzyme responsible for the active transport of sodium and potassium ions in most animal cells. This protein, commonly referred to as the sodium pump, plays a central role in mediating the electrical activity of the brain. In neurons, the activity of the sodium pump maintains the ion gradients that provide the driving force for the action potential (Thomas, 1972). The uptake of neurotransmitters and the efflux of calcium from neurons are also coupled to the activity of the ATPase (Iverson and Kelly, 1975). In glia, the sodium pump plays a critical role in potassium uptake during periods of intense neuronal activity (Hertz, 1977).

In all tissues from which Na,K-ATPase has been purified, the en-

ROBERT LEVENSON, JANET RETTIG EMANUEL, SUSAN GARETZ, and JAY W. SCHNEIDER • Department of Cell Biology, Yale University School of Medicine, New Haven, Connecticut 06510.

zyme has been shown to consist of two subunits. The large (α) subunit is a polypeptide of ~100,000 M_r which contains the site for ATP hydrolysis (Cantley, 1981) and the binding site for the cardiac glycoside ouabain, a potent inhibitor of the ATPase (Ruoho and Kyte, 1974). The smaller β-subunit is a glycosylated polypeptide of ~55,000 M_r, which has no known physiological function (Cantley, 1981).

Two separate, biochemically distinct forms of the sodium pump are found in the brain. One sodium pump isoform is found in astrocytes and unmyelinated sympathetic neurons; the other is restricted to the axolemma of myelinated nerve (Sweadner, 1979). The two isoforms from mammalian brain have been separated and their catalytic subunits have both structural and pharmacological differences (Sweadner, 1979; Specht and Sweadner, 1984). It therefore appears likely that the two pump isoforms may play different physiological roles in brain function.

Understanding the structure, function, and regulation of the Na,K-ATPase is a challenging problem. The pump, like other ion-transport proteins, has been difficult to study by conventional protein chemistry techniques. From a general perspective, purification and characterization of the polypeptides responsible for the transport of a particular ion continues to be a technically difficult goal to achieve, especially in situations in which such proteins are present in low abundance. The application of somatic cell genetic and recombinant DNA techniques to studying this class of proteins has the potential to provide new insights into the mechanism and regulation of ion-transport systems.

A central feature of the sodium pump is its sensitivity to the cardiac glycosides. An important clinical application of these drugs is in the treatment of congestive heart failure. It is generally believed that the cardiac glycosides increase the force of myocardial contraction by inhibiting active transport (Sweadner and Goldin, 1980). Biochemical experiments performed by Ruoho and Kyte (1974) suggest that the α-subunit of the pump binds ouabain. Analysis of human cell culture lines has shown that ouabain-resistant mutants exhibit decreased affinity for the drug (Baker, 1976) and that cell lines from different species vary considerably with respect to the level of ouabain that is cytotoxic to wild-type cells (Mankovitz *et al.*, 1974). Because ouabain cytotoxicity appears to be attributable to inhibition of Na^+ and K^+ transport (Robbins and Baker, 1977), it is believed that variations in ouabain sensitivity are due to specific alterations in the amino acid sequence of the ATPase α-subunit ouabain binding site.

Our research has focused on the molecular analysis of the Na,K-ATPase and the underlying basis of cellular resistance to ouabain. This chapter describes the isolation of a cDNA clone that codes for the catalytic (α) subunit of the sodium pump. The availability of this clone has enabled us to address several key aspects of the ATPase. These include (1) the number and organization of ATPase genes, (2) regulation of sodium pump expression, and (3) the relationship between the expression of the ATPase α-subunit and ouabain resistance. We have employed both antibody-screening and DNA-transfer techniques to isolate several different genes that may code for proteins involved in ion transport. The relationship of these genes to the Na,K-ATPase and the phenotype of ouabain resistance is discussed.

2. MOLECULAR CLONING OF THE NA,K-ATPASE CATALYTIC SUBUNIT

Our approach to isolating cDNA coding for the sodium pump catalytic subunit has been to use antibodies directed against the rat sodium pump to screen a rat brain λgt11 cDNA expression library. In this system, originally developed by Young and Davis (1983), a recombinant phage harboring cDNA of interest can be identified by the immunoreactivity of its cDNA/β-galactosidase fusion protein product after expression in *Escherichia coli*. Using this method, we isolated a phage clone, λrb5, which contained a 1200-base pair (bp) cDNA insert encoding a portion of the rat brain Na,K-ATPase α-subunit (Schneider *et al.*, 1985). Direct DNA sequence analysis of λrb5 revealed that a portion of the cDNA insert contained a perfect nucleotide sequence match with the amino acid sequence of a characteristic tryptic digestion fragment of the ATPase α-subunit, the FITC-binding peptide (Farley *et al.*, 1984). This approach provided unequivocal evidence that λrb5 cDNA contains a nucleotide sequence coding for a portion of the rat brain Na,K-ATPase catalytic subunit. A restriction map of rb5 cDNA is shown in Fig. 1. The DNA sequence of the α-subunit FITC site and its flanking region are shown below the map.

3. ORGANIZATION AND EXPRESSION OF THE RAT SODIUM PUMP α-SUBUNIT GENE

Southern blot analysis of EcoR1 digested total rat DNA was carried out using rb5 cDNA as a probe. As shown in Fig. 2A, only one rat DNA fragment of 4.6 kb hybridized with rb5 cDNA. This result suggests that the gene coding for the ATPase α-subunit is present in one or very few copies in the rat genome. Northern analysis of mRNA sequences from

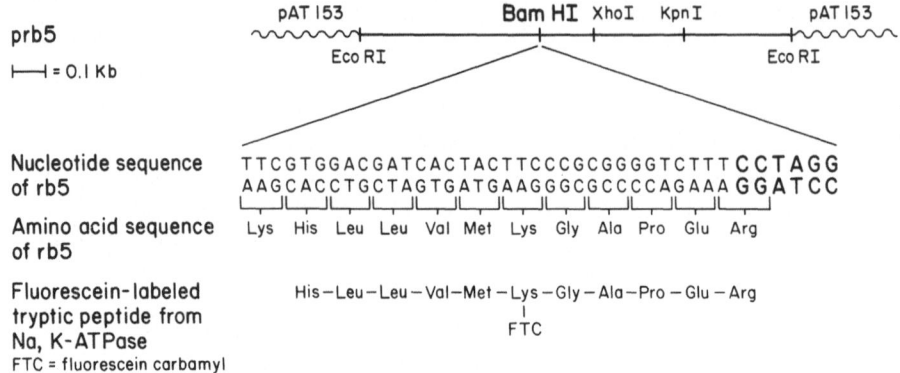

Figure 1. Restriction map and partial DNA sequence of Na,K-ATPase α-subunit cDNA. A restriction enzyme map of λrb5 cDNA was constructed by a series of single and double enzymatic digestions. rb5 cDNA was digested with Bam HI and EcoRI and the EcoRI-Bam HI fragments subcloned in the phage vector M13mp8. DNA sequence analysis was carried out by the dideoxy chain termination method (Sanger *et al.* 1977). A comparison of the rb5 cDNA nucleotide sequence and the FITC peptide amino acid sequence is shown below the restriction map.

various rat tissues demonstrated that a single mRNA species of ~5 kb codes for the catalytic subunit of the sodium pump. As shown in Fig. 2B, the level of α-subunit mRNA varies significantly from tissue to tissue in the rat. Quantitatively more ATPase mRNA is present in kidney and brain than in liver, a situation that corresponds to the level of sodium pump polypeptides expressed in these tissues (Sweadner and Goldin, 1980). These results suggest that transcriptional or post-transcriptional control mechanisms may account for the large fluctuations in ATPase activity known to occur in different tissues.

4.   ORGANIZATION AND EXPRESSION OF THE α-SUBUNIT GENE IN OUABAIN-RESISTANT CELL LINES

The availability of molecular probes for the ATPase has permitted us to initiate experiments designed to analyze the underlying basis of ouabain resistance. Our initial studies focused on a ouabain-resistant human cell line, HeLa C⁺, which was provided to us by Dr. Rich Ash of the University of Utah School of Medicine. HeLa C⁺ cells were derived by UV mutagenesis of HeLa cells and step selection in ouabain. Previous experiments carried out by Ash *et al.* (1984) showed that ouabain resistance in HeLa C⁺ cells occurred in concert with the expression of increased levels of Na,K-ATPase polypeptides, the appearance of minute chromosomes, and decreased ouabain binding. These observations suggested that ouabain resistance in this cell line resulted from amplification and/or mutation of the gene coding for the ATPase α-subunit.

Hybridization analysis on Southern and slot blots demonstrated that HeLa C⁺ DNA sequences reactive with the rb5 ATPase cDNA clone

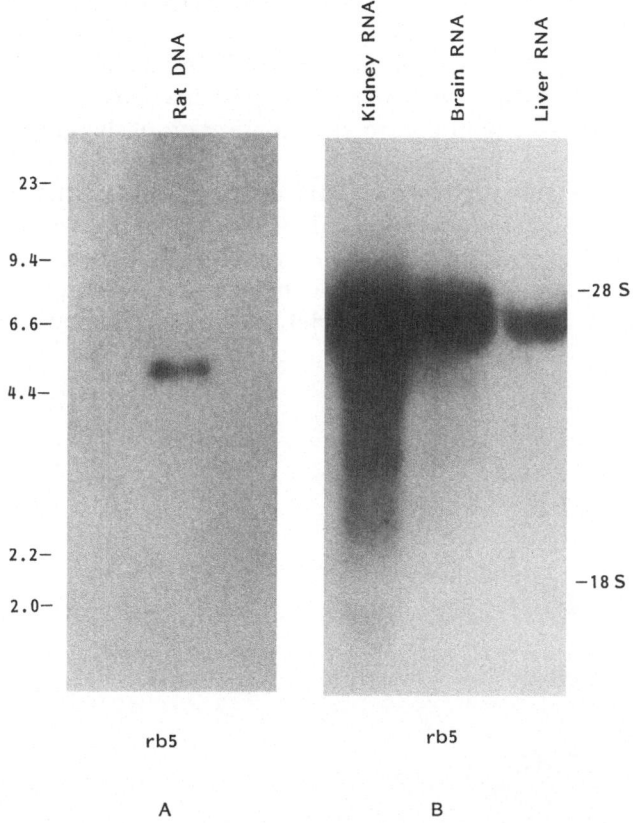

Figure 2. Organization and expression of the rat α-subunit gene. (A) Southern blot of rat genomic DNA sequences. Ten μg of total rat DNA was digested with EcoR1 and the DNA fragments resolved by electrophoresis through a 1% agarose gel. DNA was transferred to a nitrocellulose filter, and the filter was reacted with 10⁷ cpm of radiolabeled rb5 cDNA probe. Molecular-weight markers (in kb) are shown at the left. (B) Northern analysis of α-subunit mRNA in rat tissues. RNA was prepared from rat kidney, brain, and liver. Twenty-five μg of total cellular RNA was run in each lane. RNA was electrophoresed through a 1% formaldehyde-agarose gel, then transferred to a nitrocellulose filter. The filter was reacted with 2 × 10⁷ cpm of radiolabeled rb5 cDNA as probe. The positions of 28 and 18 S RNA markers are shown on the right.

were significantly amplified in C⁺ cells compared with the α-subunit
gene copy number in parental HeLa cells. As shown in Fig. 3, HeLa C⁺
DNA sequences coding for the sodium pump α-subunit gene are amplified at least 40-fold over the level in HeLa cells. Slot blot hybridization
analysis of mRNA sequences from HeLa C⁺ cells indicated that ATPase
mRNA was also amplified in C⁺ cells compared with ATPase mRNA
levels in parental HeLa cells. These results are shown in Fig. 4A. A
duplicate blot was screened with an actin cDNA clone. As shown in
Fig. 4B, the level of actin mRNA sequences in all samples was virtually
identical, indicating that equal amounts of RNA had been loaded in each
lane. Taken together, these results suggest that ouabain resistance in
HeLa C⁺ cells is due in part to an amplification of the gene coding for
the Na,K-ATPase.

Further analysis of HeLa C⁺ and HeLa cells has demonstrated several unexpected aspects of the Na,K-ATPase gene in these cells. Comparison of HeLa and HeLa C⁺ EcoR1 genomic DNA fragments reactive
with the ATPase cDNA probe has demonstrated a restriction site polymorphism in the ATPase gene of HeLa C⁺ cells. As shown in Fig. 5A,

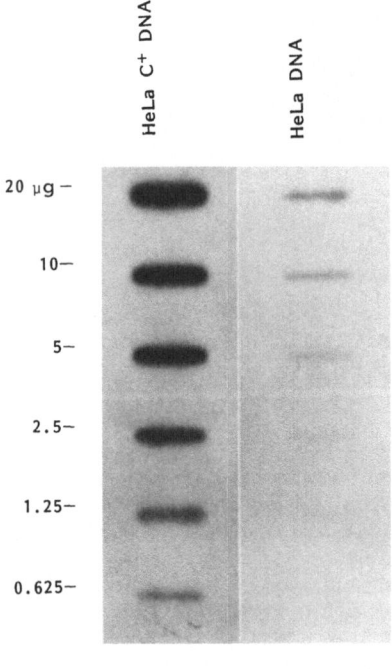

Figure 3. Slot blot analysis of HeLa and
ouabain-resistant HeLa C⁺ DNA sequences. DNA was prepared from HeLa and
HeLa C⁺ cells and serial dilutions of DNAs
were spotted onto a nitrocellulose filter. The
filter was reacted with 1.5 × 10⁷ cpm of radiolabeled rb5 cDNA as probe.

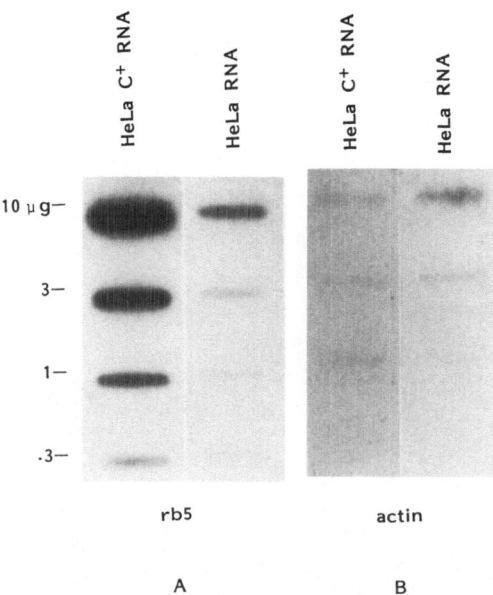

Figure 4. Slot blot analysis of ATPase mRNA levels in HeLa and HeLa C$^+$ cells. (A) Total cellular RNA was prepared from HeLa and HeLa C$^+$ cells. Serial dilutions of the RNA samples were spotted onto a nitrocellulose filter and the filter reacted with 10^7 cpm of radiolabeled rb5 cDNA as probe. (B) A duplicate blot was reacted with 1.5×10^7 cpm of a radiolabeled actin cDNA as probe.

HeLa cell DNA fragments of 4.8, 3, and 1.4 kb hybridize to the rb5 cDNA probe. In HeLa C$^+$ cells, an additional band of 6.6 kb reacts with the probe. The appearance of the new 6.6-kb band may reflect nucleotide sequence alterations in the ATPase gene or its regulatory regions. These changes could lead to increased levels of sodium pump expression and/or a decrease in the ouabain-binding properties of the ATPase during establishment of the ouabain-resistant phenotype.

Northern analysis of HeLa C$^+$ ATPase mRNA sequences is shown in Fig. 5B. Two mRNAs, one 5 and the other ~3 kb, hybridize to the ATPase probe. At this time, we do not know whether the 3-kb transcript represents a truncated α-subunit mRNA that arises from the α-subunit gene via alternative RNA splicing or whether this mRNA represents the product of another gene that shares sequence homology with the gene coding for the ATPase. One intriguing possibility is that the 3-kb mRNA codes for an α-subunit polypeptide that lacks all or part of the ouabain binding site. Preliminary studies we have carried out indicate that the

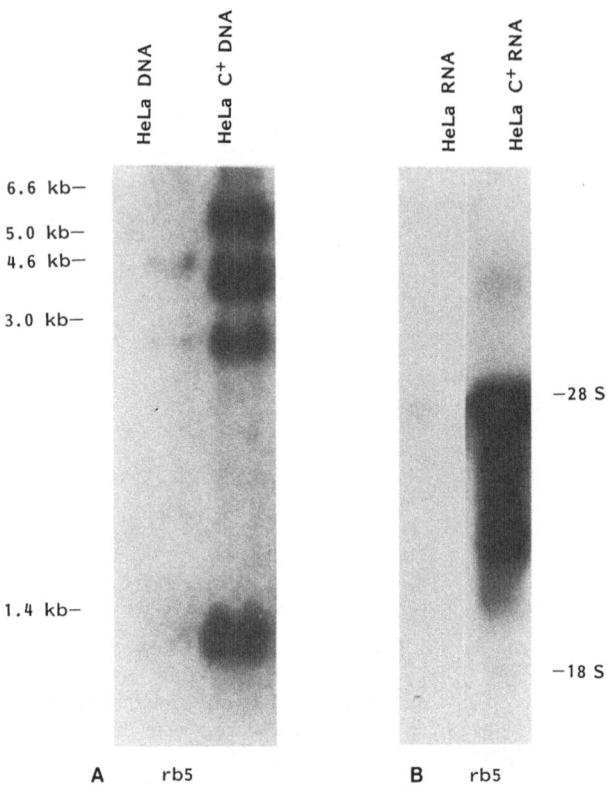

Figure 5. Organization of the ATPase gene and its expression in HeLa and HeLa C$^+$ cells. (A) Southern blot analysis. DNA was isolated from HeLa and HeLa C$^+$ cells and digested with EcoR1. Ten μg of DNA from each sample was electrophoresed through a 1% agarose gel and the DNA fragments were transferred to a nitrocellulose filter. The filter was reacted with 1.5 × 10^7 cpm rb5 cDNA probe. Molecular-weight markers (in kb) are shown at the left. (B) Northern analysis of ATPase mRNAs. Total RNA was prepared from HeLa and HeLa C$^+$ cells. Twenty-five μg of RNA was loaded in each lane of a 1% formaldehyde-agarose gel. Following electrophoresis, RNAs were transferred to a nitrocellulose filter and the filter reacted with 2 × 10^7 cpm of radiolabeled rb5 cDNA. The positions of 28 and 18 S RNA markers are shown at the right of the figure.

level of expression of the 3-kb mRNA closely parallels that of the 5-kb α-subunit mRNA during growth of HeLa C$^+$ cells in the presence or absence of ouabain. Thus the 3-kb mRNA appears to code for a form of the ATPase unique to C$^+$ cells. Whether the 3-kb mRNA plays a functional role in the establishment of ouabain resistance remains to be determined.

We have surveyed a variety of ouabain-resistant cell lines (Ash *et*

Table I. Summary of Ouabain-Resistant Cell Lines and Their Derivation

Cell line	Derivation	Reference
HeLa C[+]	UV mutagenesis of HeLa cells and step selection in ouabain	Ash et al. (1984)
oua[R]-6	Transfection of mouse 5-kb oua[R] gene into ouabain-sensitive CV-1 monkey cells	Levenson et al. (1984); English et al. (1985)
HEM-5	Microcell fusion of primary mouse fibroblasts and HeLa cells	Kozak et al. (1979)
MDCK[ouaR]	EMS mutagenesis of MDCK cells and step selection in ouabain	Soderberg et al. (1983)

al., 1984; Levenson et al., 1984; English et al., 1985; Soderberg et al., 1983; Kozak et al., 1979) (summarized in Table I) in order to determine whether these cell lines also contain amplified levels of sodium pump genes. The result of such an analysis is presented in Fig. 6. Ten μg of DNA from each drug-resistant cell line and its ouabain-sensitive parent was spotted onto a filter and hybridized with the rb5 cDNA probe. HeLa C[+] DNA, as expected, exhibited higher ATPase gene copy number than did HeLa DNA. No increase in ATPase gene copy number was observed in any

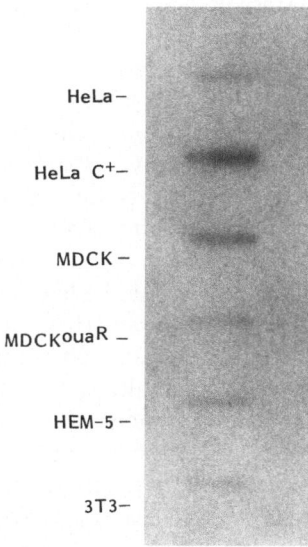

HeLa –

HeLa C[+]–

MDCK –

MDCK[ouaR] –

HEM-5 –

3T3–

rb5

Figure 6. ATPase α-subunit gene copy number in ouabain-resistant cell lines. DNA was prepared from the cell lines as described here and in Table I. Ten μg of DNA from each cell line was spotted onto a nitrocellulose filter. The filter was reacted with 2×10^7 cpm of radiolabeled rb5 cDNA.

other ouabain-resistant cell line. These findings suggest that ouabain resistance can arise from processes other than amplification of the gene coding for the sodium pump. One possible explanation for ouabain resistance in these cell lines is that mutations within the ATPase α-subunit lead to changes in the ouabain-binding properties of the pump. Direct DNA sequence analysis and expression of such mutated pump genes will be necessary to resolve this issue. Alternatively, ouabain resistance may result from the expression of other genes. This possibility is discussed below.

5.   ISOLATION AND CHARACTERIZATION OF A OUABAIN-RESISTANCE GENE

Experiments carried out by Baker and co-workers demonstrated that cell lines of various species differ considerably with respect to the level of ouabain that is cytotoxic to wild-type cells (Baker, 1976; Mankovitz *et al.*, 1974; Robbins and Baker, 1977). In particular, cells of human origin were shown to be quite sensitive to ouabain compared with rodent cells. We have found that monkey CV-1 cells exhibit sensitivity to ouabain at levels comparable to human cells (Levenson *et al.*, 1984). A dose of 5×10^{-8} M ouabain is sufficient to cause greater than 95% loss of viability when administered to CV-1 cells, whereas mouse cells are resistant to a dose of 2×10^{-4} M ouabain. This wide variation in ouabain sensitivity has made it possible to develop a gene transfer and selection system for identifying a mouse ouabain-resistance (oua^R) gene.

We have isolated a 5-kb murine genomic DNA sequence that can, in direct DNA-transfer experiments, confer ouabain resistance to ouabain sensitive CV-1 monkey cells (Levenson *et al.*, 1984). Since the α-subunit of the sodium pump is believed to be the cellular substrate for ouabain binding (Ruoho and Kyte, 1974), we expected that the DNA sequence responsible for conferring the ouabain-resistant phenotype to CV-1 cells would be likely to code for the α-subunit of the Na,K-ATPase. However, characterization of CV-1 transfectants with the ouabain-resistance gene we cloned suggests that ouabain resistance can be conferred to cells by an alternative mechanism.

Measurement of intracellular Na^+ and K^+ levels in CV-1 transfectants harboring the oua^R gene showed that these cells transported K^+ at a rate nearly equivalent to that of control cells despite a significant reduction in Na,K-ATPase activity (English *et al.*, 1985). K^+ transport activity declined in transfectants during the first 12 hr of exposure to ouabain. K^+ transport activity then increased so that by 36 hr after ouabain administration the rate of K$^+$ transport was nearly equivalent to that of untreated CV-1 cells. Analysis of mRNA sequences in transfected cells employing the oua^R gene as probe demonstrated that detectable levels of mRNA sequences homologous to the oua^R gene appeared in concert with the increase in K$^+$ transport activity (English *et al.*, 1985). These results suggest that the oua^R gene codes for a ouabain inducible K$^+$-transport system or a regulator of such a system. We have found that the oua^R is expressed at high levels in rat liver, rat Sertoli cells, and several ouabain-resistant mouse tumor cell lines (R. Kent, P. Hall, L. Cantley, R. Levinson, and D. Housman, unpublished data).

As shown in Fig. 7, the oua^R gene hybridizes to an mRNA of 1.2 kb. This mRNA is too small to code for a polypeptide the size of the ATPase α-subunit (100 kD) that requires ~3 kb of coding information. Although oua^R mRNA is too small to code for the α-subunit of the Na,K-ATPase, it is large enough to code for the β-subunit of the pump (40,000-M_r protein component). Alternatively, the oua^R gene could code for a polypeptide that serves to regulate the activity of the pump or interact with the ATPase in such a way as to alter its ion-transport properties. The nucleotide and corresponding amino acid sequence of a large portion

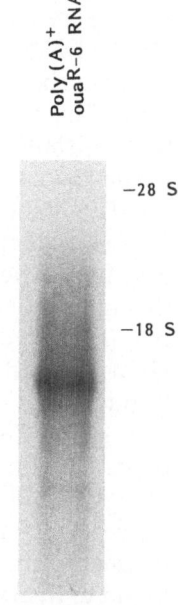

Figure 7. Northern analysis of the oua^R gene. Total cellular RNA was prepared from oua^R-6 cells. These cells are CV-1 cells harboring the mouse oua^R gene. RNA was fractionated on an oligo-dT cellulose column and 10 μg of poly (A)$^+$ RNA was electrophoresed on a 1% formaldehyde agarose gel. RNA was transferred to a nitrocellulose filter and the filter reacted with 1.5 × 10^7 cpm of a radiolabeled subfragment of the oua^R gene as probe. Molecular-weight markers are shown at the right.

of the oua^R gene has now been determined (V. R. Racaniello, R. Levenson, L. Cantley, unpublished data). The protein contains at least one hydrophobic region that may represent a transmembrane-spanning domain. We have compared the available amino acid sequence of the oua^R gene product with those proteins currently listed in the Dayhoff–NIH protein-sequence data base. This analysis revealed that the polypeptide product of the oua^R gene contains virtually no sequence homology to any protein in the data base. Identification of the oua^R gene product therefore remains a challenging biochemical problem.

6. TRANSFER OF THE SODIUM PUMP α-SUBUNIT GENE CONFERS OUABAIN RESISTANCE TO OUABAIN-SENSITIVE CELLS

Employing the technique of microcell fusion to transfer single mouse chromosomes to recipient HeLa cells, it was demonstrated (Kozak *et al.*, 1979) that the genetic locus responsible for ouabain resistance was located on mouse chromosome 3. In order to determine whether the oua^R gene we isolated was the same as the gene responsible for conferring ouabain resistance to HEM-5 cells, the oua^R gene was hybridized to HEM-5 DNA on Southern blots. No sequence homology was observed, however, between the oua^R gene and the mouse component of HEM-5 DNA. These results suggested that the mouse gene responsible for ouabain resistance in HEM-5 cells was different from the murine oua^R gene we had isolated by direct DNA transfer.

The ability to distinguish human versus mouse DNA restriction fragments that hybridize to the rat ATPase probe has enabled us to determine whether the mouse α-subunit gene confers ouabain resistance to human cells. Total cellular DNAs from HEM-5, HeLa, and mouse 3T3 cells (HeLa and mouse fibroblasts were the two parental lines from which HEM-5 was derived) were digested with EcoR1 and electrophoresed through a 0.8% agarose gel. Following immobilization of the DNA fragments on a Zetabind filter, the blot was hybridized with rb5 cDNA as probe. As shown in Fig. 8, HeLa cell DNA fragments of 4.8, 3, and 1.4 kb hybridized to the probe. In 3T3 cells, bands of 4.8, 3, 2.8, and 2 kb are reactive with rb5 cDNA. In the hybrid HEM-5 cell line, the 5 DNA fragments characteristic of both the mouse and human parents were found to hybridize to the ATPase probe. These results strongly suggest that transfer of the mouse sodium pump α-subunit gene is responsible for conferring ouabain resistance to ouabain-sensitive HeLa cells.

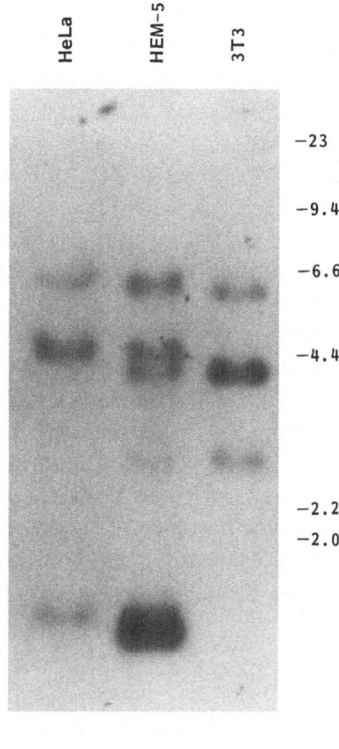

Figure 8. Ouabain-resistant somatic cell hybrids contain a ouabain-resistant ATPase α-subunit gene. DNA was prepared from HeLa, mouse 3T3, and HEM-5 cells. DNA was digested with EcoR1 and the fragments resolved by electrophoresis through a 0.8% agarose gel. DNA fragments were transferred to a Zetabind filter and the filter reacted with 1.5×10^7 cpm of radiolabeled rb5 cDNA. Molecular weight markers (in kb) are shown at the right.

7. ISOLATION OF GENES RELATED TO THE SODIUM PUMP AND THEIR EXPRESSION IN A OUABAIN-RESISTANT CELL LINE

Screening of the rat brain λgt11 library with sodium pump antibodies resulted in the isolation of an immunoreactive recombinant phage clone, λrb19, harboring a 950-bp cDNA insert. The fusion protein prepared from a lysogen of this clone was found to be immunoreactive with sodium pump antibodies on Western blots (Mercer *et al.*, 1986). Initial experiments that we carried out led us to believe that rb19 cDNA might contain additional nucleotide sequences coding for the α-subunit of the ATPase. Further characterization of the rb19 clone, however, demonstrated that this cDNA coded for a different gene. This conclusion is based on the following results: (1) rb19 and rb5 cDNAs do not hybridize to each other; (2) rb19 and rb5 cDNAs hybridize to different mRNAs; and (3) rb5 and rb19 cDNAs hybridize to different genomic DNA restriction fragments (Mercer *et al.*, 1986). Although the exact nature of the gene coded for by rb19 cDNA is now known, our data strongly suggests that the polypeptide product of rb19 cDNA is related to the sodium pump and may play a role in the establishment of ouabain resistance.

As shown in Fig. 9, two different mRNAs are transcribed from the rb19 gene. One mRNA of about 3 kb is the predominant species found in rat kidney, whereas the second mRNA of about 2.8 kb in size is the predominant species present in rat brain. Interestingly, no mRNA homologous to rb19 was detectable in the liver. Thus, there is a very specific pattern of expression of rb19 in different rat tissues that differs considerably from the pattern of expression of the ATPase α-subunit.

We also investigated the organization and expression of rb19 DNA sequences in human and ouabain-resistant human cells. As shown in Fig. 10A, rb19 hybridized to a single 9.4-kb DNA fragment in HeLa cells, whereas in HeLa C$^+$ cells, restriction fragments of 9.4, 5.5, 4.4, and 1.3 kb hybridized to the probe. The intensity of hybridization of these DNA fragments indicated that rb19 sequences were amplified in HeLa C$^+$ cells. Thus, as we observed in the case of rb5, rb19 is amplified and

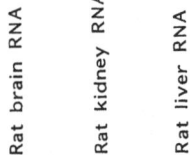

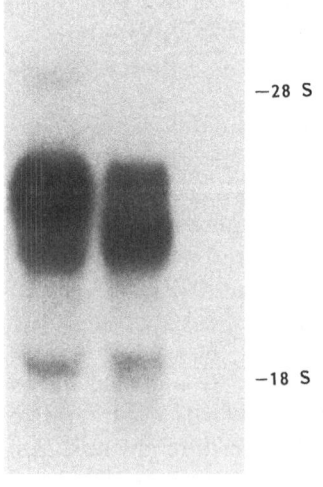

—28 S

—18 S

rb19

Figure 9. Expression of the rb19 gene in rat tissues. RNA was prepared from rat kidney, brain, and liver. Twenty-five μg of RNA from each tissue was loaded on separate lanes of a 1% formaldehyde-agarose gel. Following electrophoresis the RNA was transferred to a nitrocellulose filter and the filter was reacted with 2 × 10^7 cpm of radiolabeled rb19 cDNA as probe. The position of 28 and 18 S RNA markers are shown at the right.

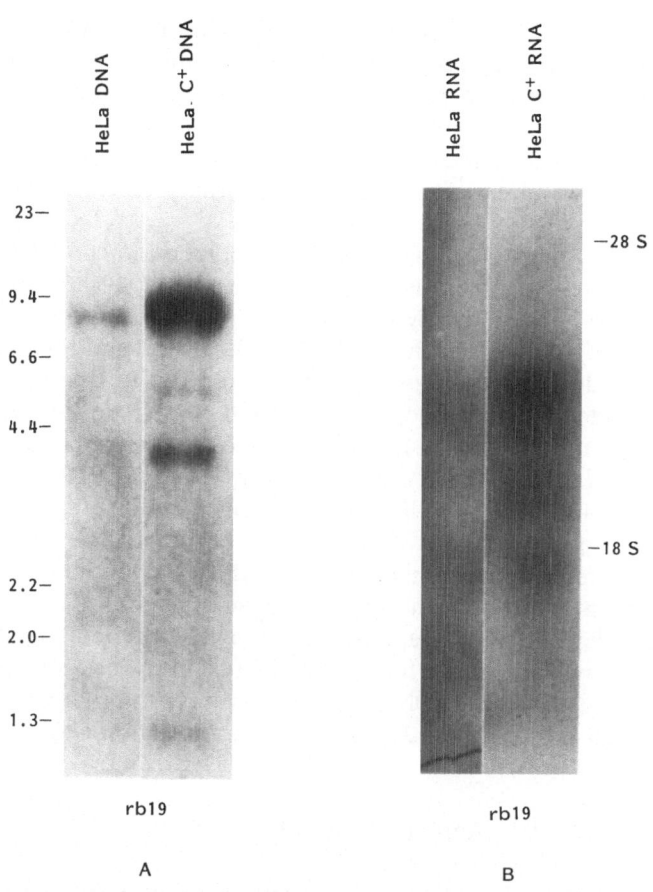

Figure 10. Organization and expression of the rb19 gene in HeLa and HeLa C⁺ cells. (A) Southern analysis. DNA was prepared from HeLa and HeLa C⁺ cells, digested with EcoR1, and the DNA fragments electrophoresed through a 1% agarose gel. After transfer to a nitrocellulose filter, the blot was hybridized with 1.5×10^7 cpm of radiolabeled rb19 cDNA. Molecular-weight markers (in kb) are shown on the left. (B) Northern analysis. Total cellular RNA was prepared from HeLa and HeLa C⁺ cells. Twenty-five μg of RNA from each sample was electrophoresed through a 1% formaldehyde-agarose gel, then transferred to a nitrocellulose filter. The filter was reacted with 2×10^7 cpm of radiolabeled rb19 cDNA as probe. The position of 28 and 18 S RNA markers is shown at the right.

exhibits restriction site polymorphisms in HeLa C⁺ cells. Slot blot analysis of HeLa C⁺ rb19 sequences showed that rb19 sequences were amplified to approximately the same extent (40-fold) as rb5 sequences (Mercer *et al.*, 1986).

Hybridization of rb19 cDNA to HeLa mRNA sequences on Northern blots demonstrated that two rb19 transcripts are present in HeLa cells. As shown in Fig. 10B, these mRNAs are ~3 and 2.8 kb in size. Both transcripts appear to be present at approximately equimolar ratios in these cells. In HeLa C$^+$ cells, at least four rb19 mRNAs ranging in size from ~2.4 kb to 3 kb are visible. Again, it is not clear whether these mRNAs arise via alternative splicing or are the products of related but different genes.

We do not know the nature of the polypeptide coded for by the rb19 gene. One possibility is that rb19 codes for a variant form of the α-subunit, i.e., one of the two brain-specific pump isoforms. However, rb19 cDNA does not appear to contain the FITC peptide coding sequence, nor does it hybridize to rb5 cDNA or mRNA. It is also possible that rb19 codes for a polypeptide unrelated to the sodium pump. We consider this unlikely, since rb19 was isolated using antibodies directed against the Na,K-ATPase. A third alternative is that rb19 cDNA codes for the β-subunit of the sodium pump. This view is supported by the fact that the β-subunit is amplified in HeLa C$^+$ cells (P. Powell and J. F. Ash, personal communication). In addition, it seems reasonable to assume that in C$^+$ cells both the α- and β-subunits of the ATPase might be amplified in order to generate an increased number of functional pumps. The availability of a full-length rb19 cDNA clone and direct DNA sequence analysis of the cDNA will be necessary in order to determine whether rb19 codes for the ATPase β-subunit or another polypeptide.

8.   CONCLUSIONS AND FUTURE PROSPECTS

The isolation of a cDNA clone encoding the catalytic subunit of the Na,K-ATPase provides a new avenue of approach for analyzing the structure, function, and regulation of this important enzyme. Recently, full-length cDNAs coding for the *Torpedo electroplax* and sheep kidney Na,K-ATPase α-subunit have been isolated and the complete amino acid sequence of the polypeptides deduced (Kawakami *et al.*, 1985; Shull *et al.*, 1985). It should now be possible for researchers to introduce the α-subunit gene or mutated versions thereof into a heterologous cell system and study the regulation of ATPase synthesis, assembly, and membrane insertion under controlled conditions.

We have used the cDNA clone coding for a portion of the rat sodium pump to isolate a nearly full-length rat ATPase cDNA molecule. We have also used the rat cDNA as a probe to isolate DNA sequences coding for the human Na,K-ATPase α-subunit gene. Comparison of the nucleotide and amino acid sequences of the rat and human forms of the α-subunit should reveal sites of structural differences between the two polypeptides. This approach may provide insight into the structural regions of the ATPase involved in ouabain binding and the basis for the wide variation in ouabain sensitivity exhibited by the ouabain resistant rodent and the drug-sensitive human forms of the ATPase. The availability of the full-length ouabain-resistant rat α-subunit gene will also enable us to transfect this gene into a ouabain-sensitive cell and assess whether expression of the rat gene is sufficient to confer ouabain resistance to a drug-sensitive host.

The ouabain-resistant human HeLa C$^+$ cell line provides a unique system for studying the relationship between the sodium pump and the underlying basis of ouabain resistance. We have demonstrated that ouabain resistance in this cell line is due in part to an amplification of the gene coding for the α-subunit of the Na,K-ATPase. Amplification of the ATPase gene leads to expression of increased levels of ATPase mRNA and α-subunit polypeptides. Because HeLa C$^+$ cells also exhibit decreased affinity for ouabain binding, it is possible that the amplified sodium pump genes may also contain mutations that result in altered ouabain affinity. This view is supported by the following observations: (1) HeLa C$^+$ DNA sequences reactive with the ATPase probe exhibit a restriction site polymorphism, and (2) a novel mRNA species of 3 kb has been identified in C$^+$ cells that hybridizes to the ATPase cDNA probe. It is tempting to speculate that the restriction site polymorphism we observe at the DNA level serves to generate a truncated 3-kb ATPase mRNA that contains an altered ouabain-binding site or that lacks this site altogether. Isolation and characterization of the ATPase gene(s) from HeLa C$^+$ cells will be necessary in order to resolve this issue.

The HeLa C$^+$ cell system may also provide a route toward identifying other genes that may be related to the sodium pump. We have defined a second amplified DNA locus in these cells employing a cDNA probe (rb19) that was isolated employing antibodies to the sodium pump. rb19 gene copy number and mRNA levels in C$^+$ cells closely parallel those found for the ATPase α-subunit. Taken together, these results suggest that this gene may code for a polypeptide associated with the activity of the sodium pump. Direct DNA sequence analysis may reveal whether this gene codes for the β-subunit of the pump, a modulator of the ATPase, or another transport system.

A central issue in the analysis of the sodium pump is the extent to which the structure of the pump is conserved in various cells and tissues of the same and unrelated organisms. The basic biochemical mechanism of the pump, the active transport of sodium and potassium across the plasma membrane, is the same in every system studied. This suggests that the structure of the catalytic subunit may be highly conserved. This view is supported by the following data: (1) antibodies raised against the catalytic subunit of the sodium pump are immunoreactive with the ATPase across both tissue and species lines (Cantley, 1981), and (2) the amino acid sequences of the α-subunit amino terminus and FITC binding peptide are virtually identical in the dog kidney, rat brain, sheep kidney, and duck salt gland (Cantley, 1981; Schneider *et al.*, 1985). We find, however, that the rat brain ATPase probe exhibits only weak sequence hybridization to the human form of the α-subunit in both Southern and Northern blotting experiments (see Figs. 4 and 5). This result suggests there may be significant nucleotide sequence divergence between the human and rat forms of the ATPase α-subunit. We have recently obtained human sodium pump α-subunit DNA clones. When these clones were used to probe several different rodent DNAs, no hybridization was observed under conditions of high stringency. Thus the human and rodent forms of the pump appear to have diverged markedly at the nucleotide sequence level.

The approaches to the study of the Na,K-ATPase described here have the potential to be applied to the analysis of other ion-transport systems. Once polypeptides that carry out ion-transport processes have been identified, antibodies generated against these polypeptides can be used to isolate corresponding cDNAs from a recombinant DNA library. Kopito and Lodish (1985) used this approach to isolate a cDNA coding for the band 3 anion-exchange protein from a λgt11 library. Once the cDNA coding for an ion-transport system is in hand, the cDNA can be used as a molecular probe to isolate other members of a family of genes coding for related transport systems. Lodish (personal communication) has used this method to identify several genes that appear to be related to band 3. Similar approaches may reveal whether the two brain sodium pump isoforms are derived from the same or different genes and what factors act to regulate the expression of these isoforms during development.

The coordinated activity of ion-transport systems play a key role in mediating electrical activity in the brain and other tissues. The isolation and characterization of the genes coding for these transport systems is an important first step in elucidating the mechanisms that regulate the function of these systems.

ACKNOWLEDGMENTS

Work in the authors' laboratory was supported by grants from the National Cancer Institute (CA-38992), American Heart Association, March of Dimes, and Swebilius Foundation (to R.L.) We thank our colleagues

(Ed Benz, Bob Mercer, Dave Housman, and Rick Ash) for their advice and encouragement and Marybeth Hicks for her excellent assistance in the preparation of this manuscript.

REFERENCES

Ash, J. F., Fineman, R. F., Kalka, T., Morgan, M., and Wire, B., 1984, Amplification of sodium- and potassium-activated adenosinetriphosphatase in HeLa cells by ouabain step selection, *J. Cell Biol.* **99**:971–983.

Baker, R. M., 1976, Genetic and cellular properties of ouabain resistant mutants, in: *Biogenesis and Turnover of Membrane Macromolecules* (J. S. Cook, ed.), pp. 93–103, Raven Press, New York.

Cantley, L. C., 1981, Structure and mechanism of the (Na,K)-ATPase, *Curr. Top. Bioenerget.* **11**:201–237.

English, L. H., Epstein, J., Cantley, L., Housman, D., and Levenson, R., 1985, Expression of a ouabain resistance gene in transfected cells: Ouabain treatment induces a K^+-transport system, *J. Biol. Chem.* **260**:1114–1119.

Farley, R. A., Tran, C. M., Carilli, C. T., Hawke, D., and Shively, J. E., 1984, The amino acid sequence of a fluorescein-labeled peptide from the active site of (Na, K)-ATPase, *J. Biol. Chem.* **259**: 9532–9535.

Hertz, L., 1977, Biochemistry of glial cells, in: *Cell, Tissue, and Organ Culture in Neurobiology* (S. Federoff and L. Hertz, eds.), pp. 37–71, Academic Press, New York.

Kawakami, A., Noguchi, S., Noda, M., Takahashi, H., Ohta, T., Kawamura, M., Nojima, H., Nagano, K., Hirose, T., Inayama, S., Hayashida, H., Miyata, T., and Numa, S., 1985, Primary structure of the α-subunit of Torpedo californica(Na^+, K^+)ATPase deduced from cDNA sequence. *Nature (Lond.)* **316**:733–736.

Iverson, L. L., and Kelly, J. S., 1975, Uptake and metabolism of gamma-aminobutyric by neurones and glial cells, *Biochem. Pharmacol.* **24**:933–938.

Kopito, R. R., and Lodish, H. F., 1985, Primary structure and transmembrane orientation of the murine anion exchange protein, *Nature (Lond.)* **316**:234–238.

Kozak, C. A., Fournier, R. E. K., Leinwand, L. A., and Ruddle, F. H., 1979, Assignment of the gene governing cellular ouabain resistance to Mus. musculus chromosome-3 using human/mouse microcell hybrids, *Biochem. Genet.* **17**:23–34.

Levenson, R., Racaniello, V., Albritton, L., and Housman, D., 1984, Molecular cloning of the mouse ouabain-resistance gene, *Proc. Natl. Acad. Sci. USA* **81**:1489–1493.

Mankovitz, R., Buchwald, M., and Baker, R. M., 1974, Isolation of ouabain-resistant human diploid fibroblasts, *Cell* **3**:221–226.

Mercer, R. W., Schneider, J. W., Savitz, A., Emanuel, J., Benz, E. J., Jr., and Levenson, R., 1986, Rat brain Na, K-ATPase β-chain gene: Primary structure, tissue specific expression and amplification on ouabain-resistant C^+ cells, *Mol. Cell. Biol.* **6**:3884–3890.

Robbins, A. R., and Baker, R. M., 1977, (Na,K)ATPase activity in membrane preparations of ouabain-resistant HeLa cells, *Biochemistry* **16**:5163–5168.

Ruoho, A., and Kyte, J., 1974, Photoaffinity labeling of the ouabain-binding site on (Na^+ + K^+) adenosine triphosphatase, *Proc. Natl. Acad. Sci. USA.* **71**:2352–2356.

Sanger, F., Nicklen, S., and Coulson, A. R., 1977, DNA sequencing with chain-terminating inhibitors, *Proc. Natl. Acad. Sci. USA* **74**:5463–5468.

Schneider, J. W., Mercer, R. W., Caplan, M., Emanuel, J. R., Sweadner, K. J., Benz, E. J., Jr., and Levenson, R., 1985, Molecular cloning of rabbit brain Na,K-ATPase α subunit cDNA, *Proc. Natl. Acad. Sci. (U.S.A.)* **82**:6357–6361.

Shull, G. E., Schwartz, A., and Lingrel, J. B., 1985, Amino acid sequence of the catalytic subunit of the (Na$^+$, K$^+$) ATPase deduced from complementary DNA, *Nature (Lond.)* **316:**691–695.

Soderberg, K., Rossi, B., Lazdunski, M., and Louvard, D., 1983, Characterization of ouabain-resistant mutants of a canine kidney cell line, MDCK, *J. Biol. Chem.* **258:**12300–12307.

Specht, S. C., and Sweadner, K. J., 1984, Two different Na,K-ATPase in the optic nerve: Cells of origin and axonal transport, *Proc. Natl. Acad. Sci. USA* **81:**1234–1238.

Sweadner, K. J., 1979, Two molecular forms of (Na$^+$ + K$^+$)-stimulated ATPase in brain: Separation and difference in affinity for strohanthidin, *J. Biol. Chem.* **254:**6060–6067.

Sweadner, K. S., and Goldin, S. M., 1980, Active transport of sodium and potassium ions, *N. Engl. J. Med.* **302:**777–783.

Thomas, R. C., 1972, Electrogenic sodium pump in nerve and muscle cells, *Physiol. Rev.* **52:**563–594.

Young, R. A., and Davis, R. W., 1983, Efficient isolation of genes by using antibody probes, *Proc. Natl. Acad. Sci. USA* **80:**1194–1198.

2

Molecular Biology of the Genes Encoding the Major Myelin Proteins

GREG LEMKE

1. INTRODUCTION

Nervous systems develop through a series of cell–cell interactions. Among the most striking of these interactions is that which occurs between neurons and glia and results in the elaboration of the myelin sheath, the electrical insulator surrounding rapidly conducting axons. Formation of this sheath entails a marked reorganization of both glial cell morphology and metabolism, most obvious in a massive increase in plasma membrane biosynthesis, and in the induction of a set of genes and proteins unique to myelinating cells. This chapter considers the recent preliminary efforts of several laboratories in which recombinant DNA methods have been employed to examine the structure and regulated expression of induced, myelin-specific genes; in particular, those coding for the major structural proteins of the sheath. Although only in its infancy, this work has already provided significant insights into myelin formation and structure. More importantly, it promises to eventually provide an understanding of the detailed molecular events that both trigger and accompany myelination.

GREG LEMKE • Molecular Neurobiology Laboratory, The Salk Institute for Biological Studies, La Jolla, California 92037.

2. FORMATION AND STRUCTURE OF THE MYELIN SHEATH

Myelin is formed in the central nervous system by oligodendrocytes (derived from the neural tube) (Wood and Bunge, 1984) and in the periphery by Schwann cells (derivatives of the neural crest) (Asbury, 1985). To a first approximation, the final structure elaborated by each of these cells is the same. It consists of a spirally wrapped, concentrically compacted stack of membrane bilayers, whose normal interperiod spacing is as tight as, or tighter than, any other known membrane assembly (e.g., chloroplast thylakoid membranes, rod outer segment discs) (Chabre, 1975; Caspar and Kirschner, 1971). This stacked ring of membranes is not acellular, but rather is contiguous with the plasma membrane of the myelinating cell (Geren, 1954). It is formed through repeated wrapping of an extended sheet of Schwann cell or oligodendrocyte plasma membrane about the nerve axon and subsequent compaction of this extended sheet at both apposed cytoplasmic and extracellular membrane surfaces. For some axons, this remarkable process may proceed until a sheath of more than 90 compacted bilayers is generated (Friede and Samorajski, 1967). The main feature of central myelination which distinguishes it from peripheral myelination is the capacity of the oligodendrocyte to myelinate many different axons independently, while a given Schwann cell myelinates only one (Raine, 1984).

Although all axons are associated with glia, not all axons are myelinated (Bray et al., 1981). A variety of elegant transplantation and cross-anastomosis experiments have demonstrated that the determinative factor with respect to whether a given axon will be myelinated is a signal intrinsic to that axon. Unmyelinated peripheral axons, for example, are not myelinated when confronted with Schwann cells from a myelinated nerve; also, all Schwann cells, those populating unmyelinated as well as myelinated fibers, seem equally capable of elaborating a myelin sheath (Weinberg and Spencer, 1976; Aguayo et al., 1976). The identity of the myelination signal of the axon and the mechanism by which that signal is conveyed to the glial cell are unknown.

3. MYELIN-SPECIFIC PROTEINS

A set of unique proteins are induced in Schwann cells and oligodendrocytes during the course of their differentiation into myelin-forming cells. The most prominent members of this set are the major myelin structural proteins: the proteolipid protein (PLP) (Lees et al., 1979) and the glycoprotein P_0 (Ishaque et al., 1980) (integral membrane proteins unique to oligodendrocytes and Schwann cells, respectively), and myelin basic protein [an extrinsic membrane protein found in both central nervous system (CNS) and peripheral nervous system (PNS) myelin] (Martenson et al., 1971; Greenfield et al., 1982). Also induced are less prevalent structural proteins, such as the myelin-associated glycoprotein (MAG) (Quarles et al., 1983), and enzymes involved in the de novo biosynthesis of myelin-associated lipids. The genes encoding the major myelin structural proteins are the subject of active investigation: Reports on the molecular cloning and analysis of myelin basic protein (MBP), P_0, and PLP cDNA and/or genomic clones have recently appeared. This chapter briefly summarizes the results of these efforts as they pertain to protein structure and function, and to the regulation of gene expression during myelination.

3.1. Myelin Basic Protein

3.1.1. Structure of the Gene and Generation of the MBP Family.   Myelin basic protein is a small, exceptionally basic, extrinsic membrane protein that accounts for 30–40% and 5–15% of CNS and PNS myelin proteins, respectively (Lees and Brostoff, 1984). It is localized at the cytoplasmic apposition of myelin lamellae (the major dense line) (Omlin et al., 1982) and has been implicated in myelin compaction at this surface (Braun, 1984). It has been very widely studied in the context of its ability to induce experimental allergic encephalomyelitis (a model for multiple sclerosis) when injected into laboratory animals (Alvord et al., 1984).

In some species, notably mice and rats, MBP is represented as a family of four proteins of molecular weights 14,000, 17,000, 18,500, and 21,500 (Yu and Campagnoni, 1982). The relative amounts of each of these proteins varies among species, although the 18,500-M_r form often predominates. MBP was first isolated in the early 1960s by delipidation and acid extraction of white matter, and the complete amino acid sequences of the bovine and human 18,500-M_r proteins have been known for more than a decade (Eylar et al., 1971; Carnegie, 1971). Comparison of these sequences with that subsequently determined for the 14,000-M_r molecule revealed that these proteins differ by virtue of a deletion of approximately 40 amino acids near the C-terminus of the 14,000-M_r form. Similarly, the rarer, 21,500- 17,000-M_r forms were later found identical to the 18,500- and 14,000-M_r molecules, respectively, with the exception of a ~30 amino acid insertion (Martenson et al., 1972; Carnegie and Dowse, 1984). The structural interrelationship of the four MBPs is diagrammed in Fig. 1.

Given the wide variability in the levels of each of these variant

Figure 1. Structural relationship among members of the myelin basic protein family. The 17,000- and 21,500-M_r forms carry a ~30-amino acid insertion in the amino-terminal half of the protein while the 18,500- and 21,500-M_r forms carry on unrelated 40-amino acid insertion near the carboxyl terminus. In the CNS myelin of adult mice, the ratio of the 21,500-, 18,500-, 17,000-, and 14,000-M_r forms is 1 : 10 : 3.5 : 35. (Barbarese et al., 1978.)

proteins, it is difficult to ascribe any functional significance to one form over another. The mechanism by which this family is generated, however, has been elucidated through the cloning of an MBP cDNA and the subsequent isolation of the MBP gene. Several laboratories have now reported on the isolation of an MBP cDNA from rat and mouse brain cDNA libraries via hybridization of a mixed synthetic oligonucleotide probe (Roach et al., 1983; Zeller et al., 1984; Kimura et al., 1985). Roach et al. (1983) used a set of 16 14-mers reverse translated from the sequence N-Gln-Asp-Glu-Asn-Pro-C (residues 78–82 of the 14,000-M_r protein, a region common to all four MBPs). As detected by low-resolution Northern blots, this set of oligonucleotides hybridized to a single 2.1-kb poly A$^+$ RNA species present in rat brain but absent from liver. A screen of a λgt10 cDNA library prepared from 18-day rat brain mRNA identified two clones, each encoding the 14,000-M_r MBP. Roach and colleagues sequenced a 1.45-kb fragment corresponding to the 5' end of one of these clones that contained the complete coding region of the MBP mRNA. In this sequence, the codon for the amino terminal alanine of MBP is immediately preceded by ATG methionine; it is therefore presumed that the mature MBP is generated directly (by removal of the initiator, followed by N-acetylation of the adjacent alanine). This presumption is consistent with the fact that MBP is not an integral membrane (or secreted) protein and with the observation that the mature MBPs can be synthesized on free ribosomes (Colman et al., 1982; Yu and Campagnoni, 1982). Subsequent sequence analysis of cosmid and λ phage clones of the mouse MBP gene have confirmed this presumption (see below in this section). The deduced sequence of the 14,000-M_r protein agrees perfectly with previous amino acid sequence data on the rat protein, with the exception of position 124, which is translated as isoleucine in the cDNA sequence, but which is methionine in the published protein sequence. Roach et al. (1983) suggest that this discrepancy may be accounted for by polymorphisms existing between the Sprague-Dawley (cDNA) and Buffalo (protein) strains of laboratory rat.

As to the origin of the MBP family, initial Southern blotting studies of Roach and colleagues and Kimura *et al.* (1985) could not distinguish conclusively between the possibility that (1) this family is generated via alternative RNA splicing of a single gene containing multiple (large) introns, or (2) the individual MBPs are encoded by RNAs transcribed from separate genes. [Earlier *in vitro* translation experiments had demonstrated that the individual MBPs are translated from separate mRNAs (Yu and Campagnoni, 1982).] Roach *et al.* observed a single broad band on Northern blots of rat brain mRNA. Zeller *et al.* (1984), examining mouse brain A^+ RNA, also report a rather broad band in the range 2.1–2.3 kb (suggestive of multiple species), although this group also detects somewhat fainter bands at 4.1, 1.9, 1.5, and 1.2 kb. In a set of recently published studies of mouse MBP cosmid and λ phage clones, Takahashi *et al.* (1985) and de Ferra *et al.* (1985) conclude that the former circumstance holds, that the MBP gene appears to be present only once per haploid genome and that the members of the MBP family are generated via exon choice during RNA processing. A map of the MBP gene, derived from these studies, is shown in Fig. 2. The banding pattern obtained upon Southern blot analysis of mouse genomic DNA is predicted nearly in its entirety from the restriction map of this cloned gene, strongly supporting the notion that there is a single MBP gene in the mouse. Southern blots with a cDNA probe (performed by Takahashi and co-workers) using 16 different restriction sites in several different mouse strains failed to detect any polymorphisms, making the possibility of additional, very closely related genes unlikely.

The MBP gene is large—greater than 30 kb—and is made up of

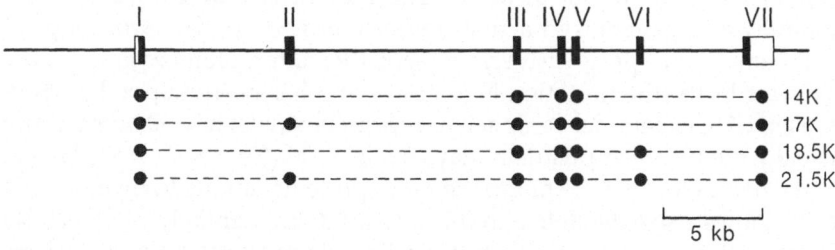

Figure 2. Exon/intron organization and proposed RNA splicing pattern of the mouse MBP gene. Filled boxes indicate exons corresponding to regions of translated mRNA, while open boxes indicate those portions of exons I and VII encoding 5′ and 3′ untranslated regions, respectively. A dot (●) under a given exon indicates its inclusion in the mRNA encoding specified members of the MBP family.

seven exons (de Ferra *et al.*, 1985), interrupted by introns varying in size from 0.5 to greater than 5 kb. The segregation of the MBP exons appears to be in good agreement with the organization of MBP structural domains predicted on the basis of Chou-Fasman and other secondary structure analyses of the MBP protein sequence (de Ferra *et al.*, 1985; Stoner, 1984).

The scheme of splicing events suggested to generate the four MBP RNAs is also illustrated in Fig. 2. cDNAs corresponding to each of these events (i.e., to mRNAs encoding each of the four MBPs in the mouse) have been isolated (de Ferra *et al.*, 1985). Alternative splicing events giving rise to a related set of proteins have been observed for several genes, including myosin light chains (Nabeshima *et al.*, 1984) and calcitonin/CGRP (Amara *et al.*, 1982). The pattern of splicing events observed for MBP is reminiscent of that which occurs in the formation of the individual members of the α-crystallin family of structural proteins (King and Piatigorsky, 1983) but is notable for the number of variant proteins produced.

3.1.2. MBP Gene Expression in the Shiverer Mouse.   A number of mutations have been identified in the mouse that primarily affect some aspect of central and/or peripheral myelination (Baumann, 1980). These include those expressed in trembler (Falconer, 1951), quaking (Samorajski *et al.*, 1970), and jimpy (Phillips, 1954) mice. Biochemical and ultrastructural analysis of these and other mutants have provided important insights into the mechanisms governing the elaboration and ultimate structural integrity of central and peripheral myelin sheaths, as well as those relating to several well-characterized neuropathies. The availability of a cloned probe for the MBP gene has directly implicated a lesion in this gene as the cause of the most widely studied of the mouse mutations affecting myelination, namely, shiverer. The shiverer mouse was initially identified by Biddle *et al.* (1973) as a mutant afflicted by a convulsive neurological disorder. The mutant is characterized phenotypically as a generalized action tremor that first appears around postnatal day 12 and progressively worsens thereafter. Homozygotes at the shiverer locus (the mutation is recessive and has been mapped to chromosome 18) experience tonic seizures from postnatal day 30 onward and usually die between postnatal days 50 and 100 (Chernoff, 1981, Hogan and Greenfield, 1984). Among the set of proteins unique to myelin, only MBP shows a specific deficit in this mutant: It is essentially undetectable in either the central or peripheral myelin of homozygous animals (Campagnoni *et al.*, 1984). Roach and colleagues and Kimura *et al.* (1985) found that this deficit extends to the level of mRNA expression. The 2.1-kb MBP mRNAs are absent from shiverer mice as detected by Northern blot. In comparing Southern blots of wild-type and shiverer genomic

DNA, it became clear that this RNA deficit in turns reflects the deletion of a large portion of the MBP gene from the Shiverer genome (Roach *et al.*, 1983; Kimura *et al.*, 1985). Roach *et al.* (1985) carefully mapped the extent of this deletion by probing a set of Southern blots of restriction digests of shiverer and wild-type genomic DNA with cloned restriction fragments spanning various portions of the MBP gene (as mapped in Fig. 2). They find that the shiverer deletion extends from approximately 12 kb 3' from exon I through the remainder of the MBP gene. Exons III through VII are thus eliminated. SI nuclease protection experiments employing a genomic fragment spanning exon I revealed that accurate initiation of MBP transcription does occur in shiverer homozygotes but that the steady state level of these transcribed RNAs (which are not correctly spliced or polyadenylated) is approximately 16-fold lower than that of wild type (Roach *et al.*, 1985). (This much-reduced level of what are likely to be RNAs of heterogeneous length would hybridize with only the exon I-encoded portion of the MBP cDNA, thereby accounting for their nondetection on Northern blots of brain RNA.)

By examining the hybridization pattern of cloned MBP probes to DNA from hamster–mouse hybrid cell lines, Roach *et al.* (1985) also found that the MBP gene maps to mouse chromosome 18. Sidman *et al.* (1985) mapped the shiverer mutation to the same chromosome by classic genetics. This coincidence of locations provides strong supporting evidence that the deletion of the MBP gene is a critical part of the shiverer lesion.

It is of interest to note that the absence of MBP in shiverer mice has markedly different consequences in the CNS as compared with the periphery. Ultrastructurally, CNS myelin, when present, appears as loose, cytoplasm-filled whorls of membrane, tightly associated at extracellular appositions (the intraperiod line), but not associated at all at cytoplasmic appositions (Ganser and Kirschner, 1980). By contrast, PNS myelin from the same animals is well elaborated, has a compact structure with nearly normal periodicity, and is functionally normal (Kirschner and Ganser, 1980). These observations have led to the suggestion that MBP plays a critical role in myelin compaction at apposed cytoplasmic membrane surfaces and that some functionally related component of peripheral myelin, not present in the CNS, is capable of substituting for MBP. The recent cloning and sequence analysis of a cDNA encoding the Schwann cell glycoprotein P_0 has suggested that the cytoplasmic portion of this transmembrane protein may be this component (see Section 3.2).

3.1.3. Control of MBP Gene Expression. Perhaps the most significant avenue of study which cDNA clones for the myelin genes open is that leading to an understanding of the cis- and trans-acting mechanisms that influence glial cell gene expression and glial cell differentiation. This sort of work is at a very early state, but for MBP we now have data on the time course of RNA expression during normal development as well as in developing oligodendrocytes cultured *in vitro*.

Zeller *et al.* (1984) have measured MBP RNA expression in developing mouse brain by dot blot using a short (92-bp) MBP cDNA fragment as a probe. These experiments demonstrate that the time course of induction of the MBP gene closely parallels the rate of myelin formation and slightly precedes the appearance of MBP itself. Low levels of RNA are first detected at postnatal day 4. Message levels rise approximately eight-fold to a peak around day 18 and then fall to an intermediate steady-state level in the adult. Carson *et al.* (1983) have also shown, using *in vitro* translation systems, that the distribution of the four MBPs in the mouse is not invariant during development. This variability may reflect a corresponding variation in the level of presumptive trans-acting factors regulating MBP splicing.

The role of neuronal axons in the induction and regulation of the myelin program is among the most intriguing features of myelination. The requirement for appropriate axons is well established in the periphery, but somewhat less so in the case of oligodendrocytes in the CNS. Full expression of major myelin proteins such as MBP and P_0 by Schwann cells, for example, is clearly dependent on contact with appropriate (i.e., normally myelinated) peripheral axons. When withdrawn from the influence of these axons (through denervation *in vivo*, or dissociation, purification, and culture *in vitro*), expression of MBP and P_0 is largely extinguished, as assayed biochemically or immunologically (Fryxell *et al.*, 1983; Politis *et al.*, 1982). This appears not to be the case with oligodendrocytes, at least as assayed *in vitro* with cells grown in the absence of neurons, and suggests that the mechanisms regulating the stable expression of myelin genes and proteins may differ in CNS and PNS glia. Zeller *et al.* (1985) examined this question by monitoring the expression of MBP RNA and protein in cultures of rat brain oligodendrocytes grown in the absence of identified neurons. They find that the MBP gene is expressed in these cultured cells apparently in accordance with a previously activated developmental clock. Cells dissociated at 5 days prenatal, at birth, and at 2 days postnatal, and cultured in the absence of neurons, all begin to express MBP RNA at postnatal-equivalent day 6 (i.e., after 11, 6, and 4 days in culture, respectively). In contrast to Schwann cells, however, these oligodendrocytes go on to express MBP protein, even in the absence of neurons (Wood and Bunge, 1984; Zeller *et al.*, 1985). These experiments suggest that, once activated in oligodendrocytes, the myelin program (apart from the formation of the sheath itself) unfolds in the absence of continuing neuronal cues.

3.2. P_0

3.2.1. Deduced Protein and Gene Structure: Topology in the Myelin Membrane. The glycoprotein P_0 is the major structural protein of peripheral myelin, accounting for over 50% of the protein present in the sheath (Greenfield *et al.*, 1973). It is restricted in its expression to myelin-forming Schwann cells and is not found in the CNS (Lees and Brostoff, 1984). It is a markedly hydrophobic, integral membrane glycoprotein of apparent 28,000–30,000 M_r, which is thought to serve as a bifunctional structural element linking adjacent lamellae and thereby stabilizing the myelin assembly (Braun, 1984). The molecular mechanism by which this stabilization is brought about is unknown. As with MBP, P_0 is often represented as proteins of varying (usually lower) molecular weight. However, the recent cloning and analysis of a P_0 cDNA has indicated that, unlike MBP, these variant forms appear to result from post-translational modifications.

Lemke and Axel (1985) identified a P_0 cDNA by the combined techniques of differential screening and hybrid selection. This approach was taken in light of the limited amino acid sequence data available for the P_0 protein. A peripheral nerve λgt10 cDNA library, prepared from 8–10 rat sciatic nerve mRNA, was first screened for clones representing mRNAs expressed in sciatic nerve but not in brain. (Preliminary *in vitro* translation experiments indicated that P_0 mRNA was particularly abundant in sciatic nerve and absent from brain.) The group of differentially expressed cDNAs thus identified was sorted in nonhomologous sets and the largest member of the major set, which accounted for approximately 70% of +/− clones, was subsequently screened for its ability to hybrid select P_0 mRNA. This hybrid selection was assayed by *in vitro* translation of the selected message followed by immunoprecipitation of the translation product using a monospecific rabbit anti-P_0 antiserum (Brockes *et al.*, 1980). This screen identified P_0 mRNA as the most abundant of the messages differentially expressed between peripheral nerve and brain.

The sequence of the protein-coding portion of the P_0 cDNA isolated by this procedure is presented in Fig. 3. Two portions of this deduced sequence had been previously identified by amino acid sequence analysis of the purified protein or fragments thereof. Ishaque *et al.* (1980) determined the 17 N-terminal amino acids of rabbit P_0. This sequence, with amino acid substitutions at four positions, is present in the above de-

```
                          o                        o                                    60
                                                                                        o
ATTCGGCTGGCCCTTGCCCCTACCCCAGCT ATG GCT CCT GGG GCT CCC TCA TCC AGC CCC AGC CCT ATC
                               Met Ala Pro Gly Ala Pro Ser Ser Ser Pro Ser Pro Ile


              o                               o                        120
                                                                       o
CTG GCT GCC CTG CTC TTC TCT TCT TTG GTG CTG TCC CCA ACC CTG GCC ATT GTG GTT TAC ACG
Leu Ala Ala Leu Leu Phe Ser Ser Leu Val Leu Ser Pro Thr Leu Ala Ile Val Val Tyr Thr (5)


              o                          o                     180
                                                              o
GAC AGG GAA GTC TAT GGT GCT GTG GGC TCC CAG GTG ACC CTG CAC TGC TCC TTC TGG TCC AGT
Asp Arg Glu Val Tyr Gly Ala Val Gly Ser Gln Val Thr Leu His Cys Ser Phe Trp Ser Ser (26)
+++


         o                              o                240
                                                        o
GAA TGG GTC TCA GAT GAC ATC TCT TTT ACC TGG CGC TAC CAG CCT GAA GGA GGC CGA GAT GCC
Glu Trp Val Ser Asp Asp Ile Ser Phe Thr Trp Arg Tyr Gln Pro Glu Gly Gly Arg Asp Ala (47)
                                                +++              +++          +++
    o                        o            300
                                         o
ATT TCA ATC TTC CAC TAT GCC AAG GGT CAA CCT TAC ATC GAT GAG GTG GGG ACC TTC AAG GAG
Ile Ser Ile Phe His Tyr Ala Lys Gly Gln Pro Tyr Ile Asp Glu Val Gly Thr Phe Lys Glu (68)
                              +++                                            +++
                           o                      360
                                                  o
CGC ATC CAG TGG GTA GGG GAC CCT AGC TGG AAG GAT GGC TCC ATT GTC ATA CAC AAC CTA GAC
Arg Ile Gln Trp Val Gly Asp Pro Ser Trp Lys Asp Gly Ser Ile Val Ile His Asn Leu Asp (89)
+++                                      +++
                        o                    420
                                             o
TAC AGT GAC AAC GGC ACT TTC ACA TGT GAT GTC AAA AAC CCA CCG GAC ATA GTG GGC AAG ACG
Tyr Ser Asp Asn Gly Thr Phe Thr Cys Asp Val Lys Asn Pro Pro Asp Ile Val Gly Lys Thr (110)
                                        +++                  +++            +++
                 o                      480
                                        o
TCT CAG GTC ACG CTC TAT GTC TTT GAA AAA GTG CCC ACT AGG TAT GGG GTG GTG TTG GGA GCC
Ser Gln Val Thr Leu Tyr Val Phe Glu Lys Val Pro Thr Arg Tyr Gly Val Val Leu Gly Ala (131)
                                        +++          +++
                             o          540
                                        o
GTG ATC GGT GGC ATC CTC GGG GTG GTG CTG TTG CTG CTG TTG CTC TTC TAC CTG ATC CGG TAC
Val Ile Gly Gly Ile Leu Gly Val Val Leu Leu Leu Leu Leu Phe Tyr Leu Ile Arg Tyr (152)
                                                                            +++
              o                     600
                                    o
TGC TGG CTG CGC AGG CAG GCT GCC CTG CAG AGG AGG CTC AGT GCC ATG GAG AAG GGG AAA TTT
Cys Trp Leu Arg Arg Gln Ala Ala Leu Gln Arg Arg Leu Ser Ala Met Glu Lys Gly Lys Phe (173)
              ++++++            ++++++            +++      +++
       o                      660
                              o
CAC AAG TCT TCT AAG GAC TCC TCG AAG CGC GGG CGG CAG ACG CCA GTG CTG TAT GCC ATG CTG
His Lys Ser Ser Lys Asp Ser Ser Lys Arg Gly Arg Gln Thr Pro Val Leu Tyr Ala Met Leu (194)
    +++      +++          ++++++      +++
                    720
                    o
GAC CAC AGC CGA AGC ACC AAA GCT GCC AGT GAG AAG AAA TCT AAA GGG CTG GGG GAG TCT CGC
Asp His Ser Arg Ser Thr Lys Ala Ala Ser Glu Lys Lys Ser Lys Gly Leu Gly Glu Ser Arg (215)
    +++      +++          ++++++      +++              +++
                 780                                                        840
                 o                    o                    o                o
AAG GAT AAG AAA TAG CGGTTAGCGGCCGGGCGGGGGGTCGGGGGTCTGCGATGGAGTCTTCCAAAGGCTCTCAGGTG
Lys Asp Lys Lys ---
+++      ++++++
```

duced sequence beginning with the isoleucine residue encoded at nucleotide 119. The variation between these sequences may reflect evolutionary divergence between rabbit and rat. Ishaque *et al.* also presented amino acid sequence data on an additional proteolytic fragment of the P_0 protein, the so-called P_0 glycopeptide. [Biochemical studies have indicated that this fragment contains a site for N-linked glycosylation of the protein (Kitamura *et al.*, 1979).] This sequence begins at Asp-80 in the deduced sequence. Its first 21 amino acids (determined by sequenator analysis) agree completely with the deduced sequence, except for the interposition for His-86 and Asp-87. The sequence of the seven remaining residues of this glycopeptide were determined by carboxypeptidase analysis and is not present anywhere in the deduced sequence. Since the order of these amino acids was ambiguous in the protein sequence, it seems likely that they are artifactual.

Several features of the primary structure of the P_0 protein presented earlier are relevant to its localization to the myelin membrane. The initiator ATG methionine is followed by 28 uncharged amino acids that precede the N-terminal isoleucine of the mature protein. This domain almost certainly comprises the signal sequence necessary for the translocation and appropriate insertion of nascent P_0 into the myelin membrane. Hydrophobicity profiles, obtained using the program of Kyte and Doolittle (1982), define a single, highly hydrophobic membrane-spanning region from Tyr-125 through Ile-150, bounded by charged anchors on either side. Assuming a conventional polarity of membrane insertion, this single traverse of the bilayer spatially divides the protein into an extracellular domain from Ile-I through Arg-124 and an intracellular domain from Arg-151 through Lys-219. The extracellular domain contains several uncharged stretches. It also contains (at Asn-93) the canonical acceptor sequence for N-linked glycosylation. This sequence occurs only once in the deduced P_0 structure. Its assignment is consistent with several experiments that indicate that P_0 is glycosylated at a single position via N-linked attachment and that the attached oligosaccharide is extracellularly positioned (Wood and McLaughlin, 1985).

←——

Figure 3. Nucleotide sequence of P_0 cDNA and deduced amino acid sequence of P_0 protein. The P_0 precursor is encoded in nucleotides 32–775. Uncharged amino acids are underlined. + + + denotes basic amino acids. The putative membrane-spanning domain extends from Tyr-125 through Ile-150 (nucleotides 491–568). The site for N-linked glycosylation is Asn-93. Numbers in parentheses at the end of each line are amino acids of the mature protein. (From Lemke and Axel, 1985.)

Lemke and Axel (1985) note that the neutral ionic properties and glycosylation of the P_0 extracellular domain are characteristic of "self-adhesive" proteins and suggest that this domain brings about the association and subsequent compaction of apposed extracellular membrane surfaces through homophilic interactions. They suggest that such interactions may both compact the mature myelin sheath and also provide an adhesive pathway which might guide the growing sheet of membrane during the elaboration of myelin. Support for this hypothesis comes from the observation that (1) the purified protein readily aggregates and becomes insoluble in aqueous solutions and neutral organic solvents (Brostoff et al., 1975) and (2) the protein is expressed at very early stages of myelin formation (Trapp et al., 1981).

Of the 69 residues comprising the P_0 cytoplasmic (C-terminal) domain, 21 are basic and only 6 are acidic. This portion of the molecule thus carries a very strong net positive charge (which is evenly distributed through to the C-terminus). This highly basic domain occupies the same myelin space (at the major dense line) as does MBP. It therefore seems likely, given the observations with MBP-deficient mice noted above, that MBP and the P_0 cytoplasmic domain perform the same function, namely, the compaction of apposed cytoplasmic membrane surfaces. In the CNS, this function would be performed by MBP exclusively, whereas in the PNS the task would be shared by MBP and P_0. According to this model, MBP is essential to the structure of the CNS sheath but is largely superfluous in the periphery. This is the conclusion drawn from ultrastructural observations of the shiverer mouse (Kirschner and Ganser, 1980). The relative levels of MBP in the CNS (abundant) and the periphery (less so) may reflect these differences. A computer-assisted comparison of the sequence of the P_0 cytoplasmic domain to MBP reveals only short patches of incomplete homology. The mouse MBP sequence from residue 1 through 7, for example, shares homology with a P_0 sequence from residue 160 through 166 (MBP: Ala-Ser-Gln-Lys-Arg-Pro-Ser; P_0: Ala-Leu-Gln-Arg-Arg-Leu-Ser). The extent of these homologies, however, is inadequate to suggest a common origin for these two genes. The mechanism by which MBP and the P_0 cytoplasmic tail form the major dense line is not established. One possibility is that these very basic polypeptides interact electrostatically with acidic lipids present in the cytoplasmic face of the apposed myelin bilayer. Acidic lipids are present in myelin (Norton and Cammer, 1984), and there is good evidence in other systems (e.g., the erythrocyte membrane) that these negatively charged lipids may be preferentially localized to the cytoplasmic leaflet of the bilayer (Verkleij et al., 1973). Furthermore, when P_0 is purified from peripheral myelin, it is complexed with a set of very tightly bound acidic lipids (Ishaque *et al.*, 1980). The most abundant of these associated lipids is phosphotidylserine, which in the red cell is confined almost exclusively to the cytoplasmic leaflet of the bilayer. An alternative hypothesis is that P_0 compacts apposed cytoplasmic membrane surfaces through direct protein–protein interactions of apposed cytoplasmic domains. P_0 is thus suggested to be a bifunctional molecule that promotes the compaction of both apposed extracellular and cytoplasmic membrane surfaces in the peripheral myelin sheath.

Preliminary characterization of P_0 genomic clones isolated from a rat λ phage library indicate that the P_0 gene is considerably smaller than that encoding MBP (Lemke, unpublished data). It is divided into multiple exons: one or more encoding 5' untranslated sequences along with most of the signal sequence, two encoding the extracellular domain, one the transmembrane domain, one a small stretch of the cytoplasmic domain, and the final exon the remainder of this domain along with all the 3' untranslated sequences. As with MBP, this exon/intron organization of the gene is well correlated with the spatial segregation of functional domains of the P_0 protein.

3.2.2. P_0 Gene Expression During Development.   Lemke and Axel (1985) measured P_0 mRNA levels during postnatal development by Northern blots of total RNA prepared from the sciatic nerves of rats at various times after birth. They find that the appearance of P_0 RNA roughly parallels the extent of myelination in the nerve. P_0 RNA is present at low but detectable levels at birth, rises seven- to eightfold to a peak around day 14–21, and then falls to a lower steady-state level in the adult. There are no readily detectable changes in P_0 message size as a function of developmental time. This message appears as a single band on Northern blots. Together with the observation that sciatic nerve A^+ RNA yields a single protein product upon *in vitro* translation, this result suggests that the lower-molecular-weight forms of P_0 commonly observed in purified preparations of myelin (Lees and Brostoff, 1984) do not derive from multiple mRNAs. This same sort of temporal coordination of gene expression with the appearance of myelin is observed for mRNA encoding HMG-CoA reductase, an enzyme that catalyzes the rate-limiting step in *de novo* cholesterol biosynthesis (Lemke and Axel, 1985) (cholesterol is a major myelin lipid). These results are illustrated in the Northern blots of Fig. 4.

These data are thus entirely consistent with the appearance of MBP RNA in developing mouse brain observed by Zeller *et al.* (1984). Although they suggest that myelin gene expression and myelination are

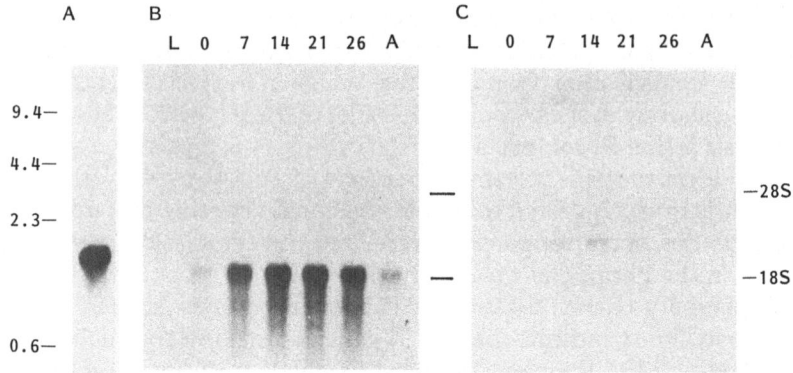

Figure 4. Expression P_0 mRNA. (A) Northern blot of the P_0 message; 0.5 μg of poly A⁺
RNA from the sciatic nerves of 8–10 rats was eletrophoresed through a 1% aga-
rose/formaldehyde gel, blotted, and probed with ³²P-labeled P_0 cDNA. (B) Northern blot
of P_0 RNA levels during development; 3 μg of total RNA, isolated from the sciatic nerves
of rats at various times after birth, was probed with radiolabeled P_0 cDNA. Numbers above
each lane represent the time, in postnatal days, at which sciatic nerves were dissected for
RNA isolation. Lane L corresponds to 3 μg of total RNA prepared from adult rat liver.
(Exposure time 5 hr with an intensifying screen.) (C) Northern blot of HMG-CoA reductase
RNA levels during development. The RNA samples in Fig. 3B were electrophoresed in
parallel lanes of the same gel and independently probed with the radiolabeled cDNA insert
from the plasmid pRed-10, which corresponds to the 3' region of the HMG-CoA reductase
mRNA as described by Liscum *et al.* (1983). (Exposure time 92 hr with an intensifying
screen.) (From Lemke and Axel, 1985.)

mutually dependent events *in vivo*, this is an important point that will
have to be addressed directly.

3.3. Proteolipid Protein

The proteolipid protein—also referred to as the Folch-Lees proteolipid, the myelin proteolipid, or lipophilin—is the major integral membrane protein of central myelin (Lees *et al.*, 1979; Lees and Brostoff,
1984). [The term *proteolipid* was originally coined by Folch and Lees (1951)
to describe a set of lipoproteins distinguished by their solubility in organic solvents such as chloroform/methanol and insolubility in aqueous
solutions. This set of lipoproteins is notably diverse; PLP is physically
and chemically distinct from other lipoproteins.] PLP expression is restricted to oligodendrocytes and to CNS myelin (Lees and Brostoff, 1984).
The protein is an exceptionally hydrophobic molecule of apparent
26,000–30,000 M_r, portions of which seem to be presented at both the
extracellular and cytoplasmic membrane surfaces of myelin lamellae (Golds

and Braun, 1951). Biochemical analyses indicate that the protein contains covalently attached fatty acids (2–4% by weight) (Stoffyn and Folch-Pi, 1970) and that it is not glycosylated. PLP has been suggested to play a structural role in the association of apposed extracellular membrane surfaces. It has also been suggested to function as an ionophore in myelin, in that both DCCD-inhibitable proton transport and voltage-dependent conductance changes are observed when purified PLP is reconstituted into liposomes or planar bilayers (Lin and Lees, 1982; Ting-Beal et al., 1979).

Milner and colleagues have isolated cDNA clones encoding rat PLP in the course of a large-scale differential screen of a rat brain cDNA library carried in the plasmid vector pBR322 (Milner et al., 1985). This screen was undertaken with the objective of identifying brain-specific transcripts, which might then, through a combination of biochemical and anatomical techniques, be analyzed as to function (Milner and Sutcliffe, 1983). The most prominent of the brain-specific cDNAs identified by this screen was a clone that recognized two mRNAs of 3.2 and 1.6 kb. These transcripts accounted for 2% and 1% of brain messenger RNAs, respectively (Milner et al., 1985). Full-length clones corresponding to each of these mRNAs were isolated, the nucleotide sequence of each clone determined, and a long open reading frame present in both clones identified. The amino acid sequence deduced from this reading frame was then compared with protein sequences entered into both the EMBO and Dayhoff amino acid sequence databases. A computer-assisted search performed in early 1983 revealed no proteins with sequences homologous to that deduced for the brain-specific cDNAs. A subsequent human-assisted inspection, however, revealed that the open reading frame of these cDNAs corresponded quite closely to bovine PLP, the amino acid sequence of which, while largely determined (Lees and Brostoff, 1984), had not been entered into the aforementioned databases. [This protein sequence was established through the efforts of several laboratories (Stoffel et al., 1982a–c; Lees et al., 1983; Jolles et al., 1983).]

Several features of the primary and secondary structure of PLP as deduced from the cDNA and amino acid sequences are of interest. First, in spite of a similar molecular weight and biochemical profile, the protein seems remarkably dissimilar to P_0, its presumed analogue in the PNS. Instead of a single membrane-spanning domain, there are several (three in the model of Laursen et al., 1984). There is no highly basic carboxy-terminal domain inside the cell, as appears to be the case for P_0 (although a moderately basic internal domain confined to the cytoplasm is proposed by Laursen et al., 1984). Together with the observation that the major dense line is not formed in the CNS myelin of shiverer mice, this finding tends to reinforce the notion that PLP does not by itself play a direct role in the compaction of apposed cytoplasmic membrane surfaces. Second, Milner et al. (1985) find no evidence of a cleaved, amino-terminal signal sequence for PLP. This is somewhat unusual for an integral membrane protein [but is characteristic of membrane proteins most of whose mass is contained within the lipid bilayer (Anderson et al., 1983)] and suggests that the manner in which this protein is assembled into the membrane must involve signals present in internal domains of the molecule. The hydrophobic domains of PLP are extensive: 44% of the amino acids in the rat PLP are hydrophobic (Leu, IIe, Val, Phe, Tyr, Met, Cys, or Trp). Third, in accordance with biochemical studies, the PLP primary sequence reveals no candidate acceptor sites for N-linked glycosylation. Finally, PLP appears to have been remarkably conserved during evolution. The sequence of the rat protein as deduced by Milner et al. (1985) is 97–99% homologous to the sequence of the bovine protein (there are a few discrepancies between laboratories for the latter). Such an extreme conservation suggests a strong structure–function relationship for PLP.

Through restriction endonuclease and sequence analysis, Milner and colleagues demonstrated that the two species of PLP mRNA arise not from alternative splicing, but rather from alternative recognition of two polyadenylation signals, one (AAUAUA) 1388 bp from the 5' end of the message, and a second (AAUAAA) 3005 bp into the mRNA. Thus, unlike MBP, both messages encode the same protein, the only difference between them being in the extent of 3' untranslated sequences. This finding is consistent with Southern blots, which indicate that the PLP gene is present only once per haploid rat genome (Milner et al., 1985). It is of interest to note that mRNAs for each of the major myelin proteins are characterized by rather long 3' untranslated regions.

Milner et al. also find that PLP mRNA is expressed in the rat C6 glioma cell line. This observation is consistant with earlier demonstrations that C6 cells, often referred to as astrocytoma cells in light of their production of GFAP, express many phenotypic characters unique to oligodendrocytes, including 2'3' cyclic nucleotide phosphohydrolase, proteolipid protein, and myelin basic protein (Volpe et al., 1975; Maltese and Volpe, 1979). It would therefore appear that the mutagenesis procedure through which this clonal tumor cell line was produced targeted a glioblast precursor. The fact that these cells express a repertoire of genes and proteins unique to terminally differentiated astrocytes and oligodendrocytes is not without interest.

Recent work from several laboratories has demonstrated that the PLP gene maps to the same position on the mouse X chromosome as does jimpy, a CNS-specific mutation which results in pronounced dysmyelination (Hogan and Greenfield, 1984; Willary and Riordan, 1985; Dautigny *et al.*, 1986). Dautigny and colleagues have shown that this mutation results in an altered structure in the PLP mRNA. Nave and co-workers (1986) have gone on to identify this alteration directly: the PLP mRNA produced by *jimpy* mice contains a small (74 base pair) deletion which removes 24 amino acids and also alters the reading frame in the carboxy terminal portion of the PLP protein. These studies have resolved a longstanding debate as to the primary biochemical defect in *jimpy* and have again illustrated the power of molecular genetics to directly address problems in neural development.

3.4. Other Myelin Proteins

3.4.1. The Myelin-Associated Glycoprotein. The myelin-associated glycoprotein (MAG) described by Quarles and colleagues is a 100,000-M_r integral membrane protein that is a quantitatively minor component of CNS and PNS myelin (approximately 1% in the CNS) (Quarles *et al.*, 1983; Quarles, 1980). It appears to be localized specifically at the innermost loops of the myelin sheath (Sternberger *et al.*, 1979), the periaxonal region, and to be excluded from compact myelin, although this is somewhat controversial (Webster *et al.*, 1983). Largely due to this unusual distribution and to its early expression, MAG has been suggested to play a role in the glial–axon recognition events that precede myelination. The protein is heavily glycosylated, with approximately one-third of the MAG molecule carbohydrate by weight (Lees and Brostoff, 1984), is particularly rich in sialic acid, and shares a carbohydrate determinant with N-CAM and at least two other cell adhesion molecules (Kruse *et al.*, 1985). Several laboratories are attempting to isolate cDNA clones for MAG by screening λgt11 expression libraries with monoclonal and polyclonal anti-MAG antibodies.

3.4.2. P2. P2 is a 131-amino acid extrinsic membrane protein found predominantly at the major dense line of PNS myelin (Lees and Brostoff, 1984; Weise *et al.*, 1980). Like MBP, to which it is not structurally related, P2 is exceptionally basic. Its abundance varies significantly between species. The complete amino acid sequences of the bovine, rabbit, and human P2 proteins have been determined. Analysis of these proteins has revealed biochemical and sequence homologies between portions of the P2 protein and certain lipid and fatty acid binding proteins such as the rat liver Z protein and cellular retinoid proteins. Berlohr *et al.* (1984) also found that one of the major RNAs induced during the *in vitro* differentiation of 3T3-L1 cells into adipocytes is a message encoding a 131-amino acid protein that shares 69% homology with rabbit myelin

P2. These observations have been taken to suggest that P2 is a member of a rather diverse family of lipid-, sterol-, and retinoid-binding proteins, and that it may fulfill an important transport role in myelin assembly and maintenance. The molecular biology of this very interesting family of proteins is being actively pursued by several laboratories, although reports on the cloning of myelin P2 have yet to appear.

4. CONCLUSION

The experiments summarized above have yielded important information as to the structure and function of the genes encoding the major myelin proteins, of the likely roles of these proteins in the organization and maintenance of the remarkable structure of the sheath, and of the genetic basis of a mutation directly affecting the formation and function of the sheath. These important results notwithstanding, clones for the myelin-specific genes may ultimately prove to be most significant not from the standpoint of what they reveal of the structure and function of sheath proteins, but rather of what they provide in the way of tools necessary for a molecular dissection of the development program itself. Together with *in vitro* cell culture and *in vivo* anatomical techniques, they permit us to ask: How is expression of the myelin genes restricted to myelinating glia? How and when is expression of these genes and the remainder of the myelin program triggered? What role do nerve axons play in this process and at what level of glial cell differentiation do they operate? What are the glial and neuronal molecules mediating cell recognition and signaling? and What are the intracellular mechanisms by which the signal to differentiate is transduced to the glial genome? Given clones for the P_0, MBP, and PLP genes, several aspects of each of these questions can now be addressed directly.

REFERENCES

Aguayo, A. J., Epps, J., Charron, L., and Bray, G. M., 1976, Multipotentiality of Schwann cells in cross-anastomosed and grafted myelinated and unmyelinated nerves, *Brain Res.* **104**:1–20.

Alvord, E. C., Kies, M. W., and Suckling, A. J., 1984, *Experimental Allergic Encephalomyelitis*, Alan R. Liss, New York.

Amara, S. G., Jonas, V., Rosenfeld, M. G., Ong, E. S., and Evans, R. M., 1982, Alternative RNA splicing in calcitonin gene expression generates different polypeptide products, *Nature (Lond.)* **298**:240–244.

Anderson, D. J., Mostov, K. E., and Blobel, G., 1983, Mechanisms of integration of *de novo*-synthesized polypeptides into membranes, *Proc. Natl. Acad. Sci. USA* **80:**7249–7253.

Asbury, A. K., 1975, The biology of Schwann cells, in: *Peripheral Neuropathy*, Vol. I (P. J. Dyck, P. K. Thomas, and E. H. Lambert, eds.), pp. 201–212, W. B. Saunders, Philadelphia.

Barbarese, E., Carson, J. H., and Braun, P. E., 1978, Accumulation of the four myelin basic proteins in mouse brain during development, *J. Neurochem.* **31:**779–782.

Baumann, N., 1980, *Neurological Mutations Affecting Myelination*, Elsevier/North-Holland Biomedical Press, Amsterdam.

Berlohr, D. A., Angus, C. W., Lane, M. D., Bolanowski, M. A., and Kelly, T. J., 1984, Expression of specific mRNAs during adipose differentiation: Identification of an mRNA encoding a homologue of myelin P2 protein, *Proc. Natl. Acad. Sci. USA* **81:**5468–5472.

Biddle, F., March, E., and Miller, J. R., 1973, Research news, *Mouse Newslett.* **48:**24–25.

Braun, P. E., 1984, Molecular organization of myelin, in: *Myelin*, 2nd ed. (P. Morell, ed.), pp. 97–116, Plenum Press, New York.

Bray, G. M., Rasminsky, M., and Aguayo, A. J., 1981, Interactions between axons and their sheath cells, *Ann. Rev. Neurosci.* **4:**127–162.

Brockes, J. P., Raff, M. C., Nishiguchi, D. J., and Winter, J., 1980, Studies on cultured rat schwann cells. III. Assays for peripheral myelin proteins. *J. Neurocytol.* **9:**67–77.

Brostoff, S. W., Karkhanis, Y. D., Carlo, D. J., Reuter, W., and Eylar, E., 1975, Isolation and partial characterization of the major proteins of rabbit sciatic nerve myelin, *Brain Res.* **86:**449–458.

Campagnoni, A. T., Campagnoni, C. W., Boune, J. M., Jacque, C., and Bauman, N., 1984, Cell-free synthesis of myelin basis protein in normal and dysmyelinating mutant mice, *J. Neurochem.* **42:**733–739.

Carnegie, P. R., 1971, Amino acid sequence of the encephalitogenic protein of human myelin, *Biochem. J.* **123:**157–162.

Carnegie, P. R., and Dowse, C. A., 1984, Partial characterization of the 21.5K myelin basic protein from sheep brain, *Science* **223:**936–938.

Carson, J. H., Nielson, M. L., and Barbarese, E., 1983, Developmental regulation of myelin basic protein expression in mouse brain, *Dev. Biol.* **96:**485–492.

Chabre, M., 1975, X-ray diffraction of retinal rods, *Biochim. Biophys. Acta* **382:**322–326.

Caspar, D. L. D., and Kirschner, D. A., 1971, Myelin membrane structure at 10A resolution, *Nature (New Biol.)* **231:**46–52.

Chernoff, G. F., 1981, Shiverer: An autosomal recessive mutant mouse with myelin deficiency, *J. Hered.* **72:**128–132.

Colman, D. R., Kreibich, G., Frey, A. B., and Sabatini, D. D., 1982, Synthesis and incorporation of myelin polypeptides into CNS myelin, *J. Cell Biol.* **95:**958.

De Ferra, F., Engh, H., Hundson, L., Kamholz, J., Puckett, C., Molineaux, S., and Lazzarini, R. A., 1985, Alternative splicing accounts for the four forms of myelin basic protein, *Cell* **43:**721–727.

Dautigny, A., Mattei, M.-G., Morello, D., Alliel, P. M., Pham-Dinh, D., Amar, L., Arnaud, D., Simon, D., Mattei, J.-F., Guenet, J.-L., Jolles, P., and Avner, P., 1986. The structural gene coding for myelin-associated proteolipid protein is mutated in *jimpy* mice. *Nature (Lond.)* **321:**867–869.

deF. Webster, H., Paklovits, C. G., Stoner, G. L., Favilla, J. T., Frail, D. E., and Braun, P. E., 1983, Myelin-associated glycoprotein: Electron microscopic immunocytochemical localization in compact developing and adult central nervous system myelin, *J. Neurochem.* **41:**1469–1479.

Eylar, E. H., Brostoff, S. W., Hashim, G., Caccam, J., and Burnett, P., 1971, Basic A1 protein of the myelin membrane: The complete amino acid sequence, *J. Biol. Chem.* **246:**5770–5774.

Falconer, D. S., 1951, Two new mutants, "trembler" and "reeler" with neurological action in the house mouse, *J. Genet.* **50:**192–196.

Folch, J., and Lees, M., 1951, Proteolipids, a new type of tissue lipoproteins, *J. Biol. Chem.* **191:**807–813.

Fryxell, K., Balzer, D. R., and Brockes, J. P., 1983, Development and applications of a solid phase radioimmunoassay for the P_0 protein of peripheral myelin, *J. Neurochem.* **40:**538–546.

Ganser, A. L., and Kirschner, D. A., 1980, Myelin structure in the absence of basic protein in the shiverer mouse, in: *Neurological Mutations Affecing Myelination* (N. Baumann, ed.), pp. 171–176, Elsevier/North-Holland Biomedical Press, Amsterdam.

Geren, B. B., 1954, The formation of the Schwann cell surface of myelin in peripheral nerves of chick embryos, *Exp. Cell Res.* **7:**558–562.

Friede, R. L., and Samorajski, T., 1967, Relation between the number of myelin lamellae and axon circumference in fibers of vagus and sciatic nerves of mice, *J. Comp. Neurol.* **130:**223–232.

Golds, E. E., and Braun, P. E., 1978, Protein associations and basic protein conformation in the myelin membrane, *J. Biol. Chem.* **253:**8162–8170.

Greenfield, S., Brostoff, S., Eylar, E. H., and Morell, P., 1973, Protein composition of myelin of the peripheral nervous system. *J. Neurochem.* **20:**1207–1216.

Greenfield, S., Weise, M. J., Gantt, G., Hogan, E. L., and Brostoff, S. W., 1982, Basic proteins of rodent peripheral nerve myelin, *J. Neurochem.* **39:**1278–1282.

Hogan, E. L., and Greenfield, S., 1984, Animal models of genetic disorders of myelin, in: *Myelin*, 2nd ed. (P. Morell, ed.), pp. 510–515, Plenum Press, New York.

Ishaque, A., Roomi, M. W., Szymanska, I., Kowalski, S., and Eylar, E. H., 1980, The P_0 glycoprotein of peripheral nerve myelin, *Can. J. Biochem.* **58:**913–921.

Jolles, J., Nussbaum, J.-L., and Jolles, P., 1983, Enzymic and chemical fragmentation of the apoprotein of the major rat brain myelin proteolipid, Sequence data, *Biochim. Biophys. Acta* **742:**33.

Kimura, M., Inoko, H., Katsuki, M., Ando, A., Sato, T., Hirose, T., Takashima, H., Inayama, S., Okano, H., Takamatsu, K., Mikoshiba, K., Tsukada, Y., and Wanatabe, I., 1985, Molecular genetic analysis of myelin-deficient mice: Shiverer mutant mice show deletion of gene(s) coding for myelin basic protein, *J. Neurochem.* **44:**692–696.

King, C. R., and Piatigorsky, J., 1983, Alternative RNA splicing of the murine alpha A-crystallin gene: Protein-coding information within an intron, *Cell* **32:**707–712.

Kirschner, D. A., and Ganser, A. L., 1980, Compact myelin exists in the absence of myelin basic protein in the shiverer mutant mouse, *Nature (Lond.)* **283:**207–210.

Kitamura, K., Suzuki, A., Suzuki, M., and Uyemura, K., 1979, Amino acid sequence of the glycopeptide derived from a major glycoprotein in bovine peripheral nerve myelin. *FEBS Lett.* **100:**67–70.

Kyte, J., and Doolittle, R. F., 1982, A simple method for displaying the hydropathic character of a protein, *J. Mol. Biol.* **157:**105–132.

Kruse, J., Keilhauer, G., Faissner, A., Timpl, R., and Schachner, M., 1985, The J1 glycoprotein—a novel nervous system cell adhesion molecule of the L2/HNK-1 family, *Nature (Lond.)* **316:**146–148.

Laursen, R. L., Samiullah, M., and Lees, M., 1984, Structure of bovine white matter proteolipid and its organization in myelin, *Proc. Natl. Acad. Sci. USA* **81:**2912–2916.

Lees, M. B., and Brostoff, S. W., 1984, Proteins of myelin, in: *Myelin*, 2nd ed. (P. Morell, ed.), pp. 197–224, Plenum Press, New York.

Lees, M. B., Sakura, J. D., Sapirstein, V. S., and Curatolo, W., 1979, Structure and function of proteolipids in myelin and non-myelin membranes, *Biochem. Biophys. Acta*. **559**:209–230.

Lees, M. B., Chao, B., Lin, L.-H., Samiullah, M., and Laursen, R., 1983, Amino acid sequence of bovine white matter proteolipid, *Arch. Biochem. Biophys.* **226**:643–656.

Lemke, G., and Axel, R., 1985, Isolation and sequence of a cDNA encoding the major structural protein of peripheral myelin, *Cell* **40**:501–508.

Lin, L.-F. H., and Lees, M. B., 1982, Interactions of dicyclohexylcarbodiimide with myelin proteolipid, *Proc. Natl. Acad. Sci. USA* **79**:941–945.

Liscum, L., Luskey, K. L., Chin, D. J., Ho, Y. K., Goldstein, J. L., and Brown, M. S., 1983, Regulation of 3-hydroxy-3-methyl-glutaryl coenzyme A reductase and its mRNA in rat liver studied with a monoclonal antibody and a cDNA probe, *J. Biol. Chem.* **258**:8450–8455.

Maltese, W. A., and Volpe, J. J., 1979, Induction of an oligodendroglial enzyme in C-6 glioma cells maintained at high density or in serum-free medium, *J. Cell. Physiol.* **101**:459–470.

Martenson, R. E., Deibler, G. E., and Kies, M. W., 1971, The occurrence of two myelin basic proteins in the central nervous system of rodents of the suborders *myomorpha* and *sciuromorpha*, *J. Neurochem.* **18**:2427–2433.

Martenson, R. E., Diebler, G. E., Kies, M. W., McNeally, S. S., Shapira, R., and Kibler, R. F., 1972, Differences between the two myelin basic proteins of the rat central nervous system, *Biochim. Biophys. Acta* **263**:193–203.

Milner, R. J., and Sutcliffe, J. G., 1983, Gene expression in rat brain, *Nucl. Acid Res.* **11**:5497–5520.

Milner, R. J., Lai, C., Nave, K.-A., Lenoir, D., Ogata, J., and Sutcliffe, J. G., 1985, Nucleotide sequences for two mRNAs for rat brain proteolipid protein, *Cell* **42**:931–939.

Nabeshima, Y., Fujii-Kuriyama, Y., Muramatsu, M., and Ogata, K., 1984, Alternative transcription and two modes of splicing result in two myosin light chains from one gene, *Nature (Lond.)* **308**:333–338.

Nave, K.-A., Lai, C., Bloom, F. E., Milner, R. J., 1986, Jimpy mutant mouse: a 74 base deletion in the mRNA for myelin proteolipid protein and evidence for a primary defect in RNA splicing. *Proc. Natl. Acad. Sci. (U.S.A.)*, in press.

Norton, W. T., and Cammer, W., 1984, Isolation and characterization of myelin, in: *Myelin*, 2nd ed. (P. Morell, ed.), pp. 147–195, Plenum Press, New York.

Omlin, F. X., deF. Webster, H., Palkovitz, G. G., and Cohen, S. R., 1982, Immunocytochemical localization of basic protein in the major dense line regions of central and peripheral myelin, *J. Cell Biol.* **95**:242–248.

Phillips, J. S. R., 1954, Jimpy: A new totally sex-linked gene in the house mouse, *Arch. Indukt. Abstammungs-Vererbungsl* **86**:322–325.

Politis, M. J., Sternberger, N., Ederle, K., and Spencer, P. S., 1982, Studies on the control of myelogenesis, *J. Neurosci.* **2**:1252–1266.

Quarles, R. H., 1980, Glycoproteins from central and peripheral myelin, in: *Myelin: Chemistry and Biology* (G. Hashim, ed.), pp. 55–77, Alan R. Liss, New York.

Quarles, R. H., Barbarash, G. R., Figlewicz, D. A., and McIntyre, L. J., 1983, Purification and partial characterization of the myelin-associated glycoprotein from adult rat brain, *Biochim. Biophys. Acta* **757**:140–149.

Raine, C. S., 1984, Morphology of myelin and myelination, in: *Myelin*, 2nd ed. (P. Morell, ed.), pp. 1–50, Plenum Press, New York.

Roach, A., Boylan, K., Horvath, S., Prusiner, S. B., and Hood, L. E., 1983, Characterization of a cloned cDNA representing rat myelin basic protein: Absence of expression in shiverer mutant mice, *Cell* **34**:799–806.

Roach, A., Takahashi, N., Pravtcheva, D., Ruddle, F., and Hood, L., 1985, Chromosomal mapping of the mouse myelin basic protein gene and structure and transcription of the partially deleted gene in shiverer mutant mice, *Cell* **42**:149–157.

Samorajski, T., Freide, R. L., and Reimer, P. R., 1970, Hypomyelination of the quaking mouse, *J. Neuropathol. Exp. Neurol.* **29**:507–512.

Sidman, R. L., Conover, C. S., and Carson, J. H., 1985, Shiverer gene maps near the distal end of chromosome 18 in the house mouse, *Cytogenet. Cell. Genet.* **39**:241–245.

Sternberger, N. H., Quarles, R. H., Itoyama, Y., and deF. Webster, H., 1979, Myelin-associated glycoprotein demonstrated immunocytochemically in myelin and myelin-forming cells of developing rat, *Proc. Natl. Acad. Sci. USA* **76**:1510–1514.

Stoffel, W., Hillen, H., Schroeder, W., and Deutzmann, R., 1982a, Primary structure of the C-terminal cyanogen bromide fragments II, III, and IV from bovine brain proteo-lipid apoprotein, *Hoppe-Seylers Z. Physiol. Chem.* **363**:855–864.

Stoffel, W., Schroeder, W., Hillen, H., and Deutzmann, R., 1982b, Analysis of the primary structure of the strongly hydrophobic brain myelin proteolipid protein (lipophilin), *Hoppe-Seylers Z. Physiol. Chem.* **363**:1117–1122.

Stoffel, W., Hillen, H., Schroeder, W., and Deutzmann, R. 1982c, Lipophilin of brain white matter. Purification and amino acid sequence studies of the four tryptophan fragments, *Hoppe-Seylers Z. Physiol. Chem.* **363**:1397–1407.

Stoffyn, P., and Folch-Pi, J., 1971, On the type of linkage binding fatty acids present in the brain white matter proteolipid apoprotein, *Biochem. Biophys. Res. Commun.* **44**:157–160.

Stoner, G. L., 1984, Predicted folding of beta-structure in myelin basic protein, *J. Neurochem.* **43**:433–447.

Takahashi, N., Roach, A., Teplow, D. B., Prusiner, S. B., and Hood, L. E., 1985, Cloning and characterization of the myelin basic protein gene from mouse: One gene can encode both 14kd and 18.5kd myelin basic proteins by alternate use of exons, *Cell* **43**:139–148.

Ting-Beal, H. P., Lees, M. B., and Robertson, J. D., 1979, Interactions of Folch-Lees proteolipid apoprotein with planar lipid bilayers, *J. Membr. Biol.* **51**:33–46.

Trapp, B. D., Itoyama, Y., Sternberger, N. H., Quarles, R. H., and deF. Webster, H., 1981, Immunocytochemical localization of P_0 protein in Golgi complex membranes and myelin of developing rat Schwann cells, *J. Cell Biol.* **90**:1–6.

Verkleij, A. J., Zwaal, R. F. A., Roelofsen, B., Corfurfius, P., Kastelijn, D., and Van Deenen, L. L. M., 1973, The asymmetric distribution of phospholipids in the human red cell membrane, *Biochim. Biophys. Acta* **323**:178–193.

Volpe, J. J., Fijimoto, K., Maras, J. C., and Agrawal, H. C., 1975, Relation of C-6 glial cells in culture to myelin, *Biochem. J.* **152**:701–703.

Weinberg, H. J., and Spencer, P. S., 1976, Studies on the control of myelogenesis. II. Evidence for neuronal regulation of myelin production, *Brain Res.* **113**:363–378.

Weise, M. J., Hsieh, D. L., Levit, S., and Brostoff, S. W., 1980, Bovine P2 protein: Sequence of NH_2-terminal of the protein, *J. Neurochem.* **35**:388.

Willard, H. F., and Riordan, J. R., 1985, Assignment of the gene for myelin proteolipid protein to the X chromosome: implications for X-linked myelin disorders. *Science* **230**:940–942.

Wood, P., and Bunge, R. P., 1984, The biology of the oligodendrocyte, in: *Advances in Neurochemistry*, Vol. 5: *Oligodendroglia* (W. T. Norton, ed.), pp. 1–46, Plenum Press, New York.

Wood, J. G., and McLaughlin, J., 1975, The visualization of concanavalin A binding sites in the intraperiod line of rat sciatic nerve myelin, *J. Neurochem.* **24:**233–235.

Yu, Y.-T., and Campagnoni, A. T., 1982, *In vitro* synthesis of the four mouse myelin basic proteins: Evidence for a lack of a metabolic relationship, *J. Neurochem.* **39:**1559–1564.

Zeller, N. K., Hunkeler, M. J., Campagnoni, A. T., Sprague, J., and Lazzarini, R. A., 1984, Characterization of mouse myelin basic protein messenger RNAs with a myelin basic protein cDNA clone, *Proc. Natl. Acad. Sci. USA* **81:**18–22.

Zeller, N. K., Behar, T. N., Dubois-Dalcq, M. E., and Lazzarini, R. A., 1985, Timely expression of myelin basic protein gene in rat brain oligodendrocytes cultured in the absence of neurons, *J. Neurosci.* **5:**2955–2962.

3

Molecular Biology of the Neural and Muscle Nicotinic Acetylcholine Receptors

STEVE HEINEMANN,
GIGI ASOULINE, MARC BALLIVET, JIM BOULTER,
JOHN CONNOLLY, EVAN DENERIS, KAREN EVANS,
SYLVIA EVANS, JOHN FORREST, PAUL GARDNER,
DAN GOLDMAN, ABHA KOCHHAR,
WALTER LUYTEN, PAM MASON, DOUG TRECO,
KEIJI WADA, and JIM PATRICK

1. INTRODUCTION

Most theories of nervous system function depend heavily on the existence and properties of the synapse. For this reason, this structure has been a focal point for neuroscience research for many decades. The best understood synapse is the neuromuscular junction because of its accessibility to biochemical and electrophysiological techniques and because of its elegant, well-defined structure. The nicotinic acetylcholine (ACh) receptor found in the postsynaptic membrane binds ACh released from

STEVE HEINEMANN, GIGI ASOULINE, MARC BALLIVET, JIM BOULTER, JOHN CON-
NOLLY, EVAN DENERIS, KAREN EVANS, SYLVIA EVANS, JOHN FORREST, PAUL
GARDNER, DAN GOLDMAN, ABHA KOCHHAR, WALTER LUYTEN, PAM MASON,
DOUG TRECO, KEIJI WADA, AND JIM PATRICK • Molecular Neurobiology Labo-
ratory, The Salk Institute, La Jolla, California 92037.

the nerve. The binding of ACh results in a conformational change in the receptor that opens a channel permeable to cations. The resulting ion flux depolarizes the muscle and ultimately leads to muscle contraction. Thus the ACh receptor contains both a ligand-binding domain as well as a channel domain. Biological and structural studies have shown that the muscle nicotinic ACh receptor is a glycoprotein made up of five subunits with the stoichiometry $\alpha_2\beta\gamma\delta$; each of these subunits has a molecular weight between 40,000 and 60,000, and is encoded by a separate gene. This complex has been shown in reconstitution experiments to be a functional receptor containing both a ligand-binding site and a ligand-gated channel (for recent reviews, see Conti-Tronconi and Raftery, 1982; Popot and Changeux, 1984; Stroud and Finer-Moore, 1985; Karlin *et al.*, 1986; McCarthy *et al.*, 1986).

Developmental studies have demonstrated that the ACh receptor found in skeletal muscle is regulated in an interesting way that may involve mechanisms unique to the nervous system. Muscle fibers are formed by the fusion of many single cells called myoblasts. When the myoblasts fuse, the rate of receptor incorporation into the plasma membrane increases dramatically (Patrick *et al.*, 1972). Initially the receptors are incorporated throughout the extent of the membrane and reach a density of about 200 per square micron (μm^2). Within a day or so after fusion, motor neurons make contact with the muscle and begin to form functional synapses. As the synapse forms, the ACh receptors begin to cluster at the synapse, eventually reaching a density of about $10,000/\mu m^2$. As the innervated muscle matures, the receptors located away from the synapse slowly disappear, so that by 3 weeks after birth, in the rat, they have declined to a density of less than a few per square micron (Bevan and Steinbach, 1977; Steinbach, 1981). Thus, during the time that the synapse forms, the distribution of the receptors undergoes this dramatic shift from a diffuse distribution to one highly localized to the synapse. While these events are taking place, the properties of the individual receptor molecules undergo a gradual transformation. Initially, the receptors exhibit a high metabolic turnover rate with a half-life ($t_{\frac{1}{2}}$) of about 1 day. As the synapse matures, the receptor becomes very stable ($t_{\frac{1}{2}} > 10$ days) (Berg and Hall, 1974, 1975; Chang and Huang, 1975; Steinbach *et al.*, 1979). A number of other properties of the receptors change as well. Specifically, the time that the channel stays open, after binding ACh, shortens from a few milliseconds to around 1 msec (Sakmann, 1978). In addition, the isoelectric focusing point changes (Brockes and Hall, 1975a) and an antigenic determinant is lost (Hall *et al.*, 1973). Taken together, these changes in receptor properties, which take place as the neuromuscular junction forms and matures, suggest that in some way the structure of the receptor is altered during synaptogenesis. To some extent, these developmental changes are reversible. If synaptic activity is blocked either by surgical denervation or by pharmacological block of the ACh receptor, the synthesis of receptor is dramatically increased (Miledi, 1960*a,b*; Cohen and Fischbach, 1973; Brockes and Hall, 1975*b*). Newly made receptors are incorporated over the whole extent of the muscle membrane and are responsible for making denervated muscle supersensitive to ACh. Extrasynaptic receptors are metabolically unstable and have the biochemical, biophysical, and antigenic properties of the receptors found before innervation. One of the most intriguing aspects of muscle physiology is the observation that direct stimulation of denervated muscle by electrodes placed near the muscle suppresses the increase in receptor synthesis (Cohen and Fischbach, 1973; Lomo and Westgaard, 1975; Hall and Reiness, 1977). This demonstrates that the level of electrical activity in the muscle can regulate the synthesis of macromolecules important for synaptic function. Studying how these mechanisms work in muscle may provide insight into how similar mechanisms might work in the brain where it is known that electrical activity is important for the normal development of specific nerve networks.

The development of the neuromuscular junction and the changes in receptor distribution and properties have in the past been studied by electrophysiological and biochemical methods. This approach has provided some important insights but has not led to an understanding of a number of central issues:

1. How does electrical activity regulate the synthesis of ACh receptor?
2. What is the relationship between the receptors found at the mature neuromuscular junction and those present early in development or after denervation?
3. Are these two classes of receptors found in muscle the product of one set of genes, i.e., one for each subunit, or are different genes expressed during development and after denervation?
4. Are the receptor genes regulated by the level of electrical activity in muscle?
5. What is the primary sequence of the receptor, and how does the structure relate to the functions carried out by that the receptor, i.e., ligand binding and channel formation?

It is clear that one way to obtain answers to these problems is to clone the mRNAs and genes coding for the receptor and to use the clones in experiments designed to answer specific questions. For this reason, our group first isolated cDNA and genomic clones coding for the ACh receptor expressed in the *Torpedo* electric organ, a highly enriched source of receptor (Ballivet *et al.*, 1982; Claudio *et al.*, 1983; Patrick *et al.*, 1983). The cDNA clones isolated from the fish electric organ were then used to obtain cDNA and genomic clones for the four subunits of the ACh receptor expressed in mouse skeletal muscle.

The availability of cDNA clones coding for the muscle nicotinic ACh receptor has made it possible to extend these studies in an important direction. Both the peripheral and central nervous systems are known to contain nicotinic ACh receptors at synapses between nerve cells. These receptors have been identified by functional and binding studies using cholinergic ligands such as ACh, nicotine, and α-bungarotoxin (e.g., see Clarke *et al.*, 1985). Neuronal receptors have been difficult to study because of their low abundance and the lack of good probes to aid in their purification. Molecular cloning makes possible another approach to study these receptors. We have used the cDNA clones coding for the muscle nicotinic receptor to screen rat and mouse cDNA and genomic libraries for related mRNAs or genes. The assumption behind this approach is that the neural nicotinic receptors are evolutionarily related to the muscle receptors, as reflected at the genetic level by a significant level of nucleotide sequence homology. This method has resulted in the isolation of a cDNA clone coding for a peptide that has about 50% amino acid sequence homology with the muscle α-subunit. In addition, this receptorlike protein has a number of structural features seen in the muscle α-subunits of the skeletal muscle ACh receptors. RNA homologous to these cDNA clones is expressed in specific areas of the peripheral and central nervous systems. The working hypothesis is that they represent mRNAs coding for neuronal nicotinic ACh receptors.

2. ISOLATION OF cDNA CLONES CODING FOR THE ACETYLCHOLINE RECEPTOR EXPRESSED IN THE *TORPEDO* ELECTRIC ORGAN

The richest known source of nicotinic ACh receptor is the *Torpedo* electric organ. This receptor, expressed in the electric organ, is pharmacologically and structurally closely related to the receptor at the mammalian neuromuscular junction. The fact that the fish electric organ receptor and its counterpart at the neuromuscular junction are related has been confirmed by recent molecular cloning experiments (see Section 3) demonstrating a high degree of amino acid sequence homology between species as far apart as cartilaginous fish, chick, and man.

We decided to first clone cDNAs prepared from poly(A)$^+$ RNA isolated from the *Torpedo* electric organ because this organ was known to synthesize large amounts of ACh receptor. Furthermore, large quantities (milligrams) of this receptor had been available for years, so that highly specific antibodies that recognized each subunit were available. Most importantly, the amino-terminal sequence of each peptide was known for the *Torpedo* receptor (Raftery *et al.*, 1980). This made it possible to identify clones initially by hybridization selection and finally by sequence analysis. Poly(A)$^+$ RNA was purified from *Torpedo* electric organ and size-fractionated on a sucrose gradient. Each fraction was added to a reticulocyte *in vitro* translation system and assayed for its ability to direct the synthesis of receptor subunits in the presence of [^{35}S]methionine. The products of *in vitro* translation were incubated with anti-ACh receptor antibody, the resulting immune precipitate was solubilized and run on a sodium dodecyl sulfate–polyacrylamide gel (SDS-PAGE). The ^{35}S-labeled receptor subunits migrate with apparent molecular weights of 40,000–60,000 M_r in this system (Ballivet *et al.*, 1982). Figure 1 shows the results of such an *in vitro* translation assay. This experiment dem-

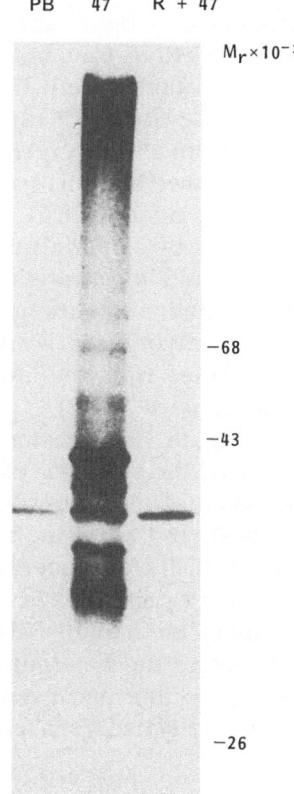

Figure 1. *In vitro* translation of sized electric organ mRNA. Poly(A)$^+$ RNA purified from electric organ RNA was layered on a 15–30% sucrose gradient and sedimented at 36,000 rpm in an SW-41 rotor for 17 hr. Fractions were collected and assayed for their ability to direct the synthesis of protein that was immunoprecipitable by antibody prepared against SDS-denatured *Torpedo* acetylcholine receptor (AChR). Also shown are translation products that are immunoprecipitable by preimmune serum, antireceptor antiserum, and antireceptor antiserum plus excess purified AChR. Immune serum precipitates four major bands between 35 kD and 50 kD plus the higher-molecular-weight material seen at the top of the gel. The band of about 43 kD present in each lane represents a major band in the total *in vitro* translation product of this fraction.

onstrates that a number of different-sized polypeptides are made and precipitated by the antireceptor antibody and not by preimmune sera. When poly(A)$^+$ RNA extracted from brain was assayed in this system, no bands were observed. From this experiment we estimated that the receptor message constituted about 1–2% of the translatable mRNA extracted from electric organ and that the sucrose gradient size-fractionation resulted in about threefold enrichment. Thus we could expect that several percent of the cDNA clones made from this RNA would code for receptor, making a hybridization-selection approach feasible.

2.1. Torpedo γ-Subunit

To increase the chance of obtaining full-length cDNA clones, the cDNA library was prepared by the method developed by Okayama and Berg (1982). This cloning method uses the vector to prime first- and second-strand synthesis and does not use S1 nuclease digestion, which is thought to shorten clones at the 5′ end. Initially, an ordered library of 960 bacteria was screened. To enrich for receptor clones, the library was first screened with ^{32}P-labeled cDNA made from poly(A)$^+$ RNA from electric organ and separately with ^{32}P-labeled cDNA made from poly(A)$^+$ RNA from brain. Figure 2 shows the results of this differential hybridization screen. Clones labeled with the arrows were picked as candidate clones for further testing, since they hybridized strongly to RNA from electric organ and weakly to brain mRNA. Subsequent analysis showed that this screening procedure gave a 15-fold enrichment for receptor clones. Of 31 clones picked as candidate clones and further screened by a hybridization-selection procedure, the DNA was purified and bound to nitrocellulose filters. The filters were incubated under hybridization conditions with poly(A)$^+$ RNA from *Torpedo* electric organ. The nonhybridized RNA was washed away and the hybridized RNA was eluted from the filter and added to an *in vitro* translation system (Ballivet *et al.*, 1982).

Figure 3 demonstrates that clone 12A1 hybridized to an RNA that directed the synthesis of a peptide immunoprecipitated by antireceptor antibody. This clone (12A1) was used to screen the complete library (960 clones), and six more hybridizing clones were identified. The longest clone (4D8) was sequenced and found to code for a 506-residue protein, including a typical 17-residue hydrophobic leader peptide with an N-terminal methionine (Claudio *et al.*, 1983). The deduced amino acid sequence of the N-terminal of the mature protein corresponded to the sequence that had been determined by protein sequencing of the γ-subunit isolated from *Torpedo* electric organ. mRNA made from this clone

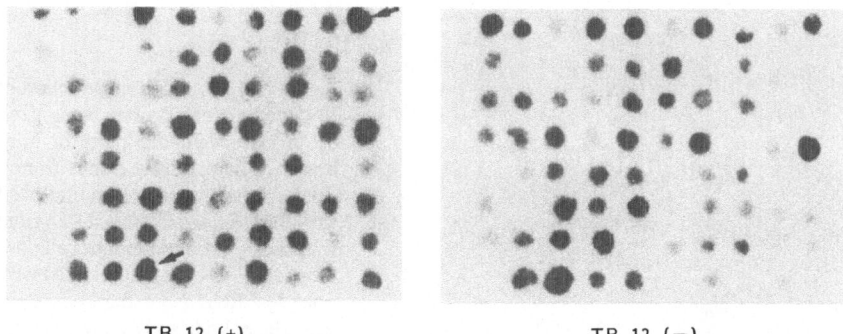

TB 12 (+)　　　　　　　TB 12 (−)

Figure 2. Differential–hybridization screening of cDNA clones. Two copies of the library were prepared by spotting duplicate ordered sets of 80 clones onto Luria broth plates containing 40 μg ampicillin per ml. The colonies were transferred to Whatman 540 filter paper, amplified with chloramphenicol, and lysed *in situ*. The filters were prehybridized with salmon sperm DNA (200 μg/ml) in 50% formamide/0.75 M sodium chloride/0.075 M sodium citrate, pH 7/0.1% polyvinyl pyrrolidone/0.1% Ficoll/1% bovine serum albumin (BSA) at 42°C for 2–6 hr. Hybridization was done in 50% formamide/0.75 M sodium chloride/0.075 M sodium citrate, pH 7/0.02% polyvinyl pyrrolidone/0.02%, Ficoll/0.2% BSA at 42°C for 18–24 hr. One copy of the library was hybridized with cDNA (5×10^4 cpm/ml) made from electric organ 18–20S poly(A)$^+$ RNA, and the other copy was hybridized with cDNA of identical specific activity synthesized from brain poly(A)$^+$ RNA. TB12 (+) is an array of 80 clones hybridized to electric organ probe, and TB12 (−) is the identical set hybridized with the brain probe. Clones identified by an arrow were picked as candidates. Twelve such paired arrays were processed in the same way to yield a collection of 31 candidates.

in vitro codes for a functional γ-subunit when injected into *Xenopus* oocytes (J. Boulter, J. Connolly, S. Heinemann, and J. Patrick, unpublished results).

2.2. Torpedo α-Subunit

A *Torpedo* α-subunit cDNA clone was initially cloned from a cDNA library containing short inserts—200–300 base pairs (bp)—made by conventional methods employing S1 nuclease digestion. An ordered library of bacteria containing cDNA inserts was screened by the following technique of hybridization selection. Batches of seven clones were grown up and their DNA fixed to nitrocellulose and hybridized to poly(A)$^+$ RNA from electric organ. The RNA was eluted from the filter and tested for its ability to direct the synthesis of receptor polypeptides in the *in vitro* translation system. One batch that proved positive in this assay contained a clone 1C9, that hybridized to RNA, which directed the synthesis of a receptor polypeptide (Patrick *et al.*, 1983). Figure 3 shows

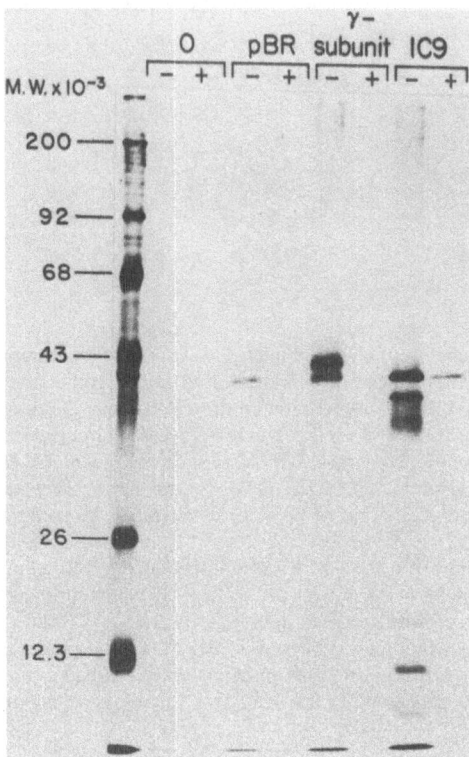

Figure 3. Hybridization–selection of RNA coding for a protein recognized by anti-acetylcholine receptor antibody. Pools of seven clones from a cDNA library were grown and used to prepare a cleared lysate. DNA from the pools was bound to nitrocellulose filters that were then hybridized with poly(A)$^+$ RNA from the electric organ. The hybridized RNA was eluted, translated *in vitro*, and the translation products tested for their ability to bind to anti-acetylcholine receptor antibody. Once immunoprecipitable products were detected, the individual clones in a pool were grown and tested separately. Also shown are immunoprecipitable proteins produced by mRNA that hybridized to DNA from various clones. Lanes marked with a plus sign (+) are those in which the immunoprecipitation was done in the presence of unlabeled competing purified AChR. The 43,000-M_r (43-kD) band present in all lanes is a contaminating protein.

the results of the *in vitro* translation experiment using clone 1C9 to select the RNA. Vector DNA (pBR322) was used as a control DNA. It is clear that 1C9 hybridizes to an RNA that directs the synthesis of a receptor peptide with a lower apparent molecular weight than the peptide made from RNA hybridized to the γ-subunit clone 12A1 (see Fig. 3). The insert from clone 1C9 was used to screen the cDNA library containing long inserts made by the Okayama and Berg (1982) method. Surprisingly, clone 1C9 hybridized to two distinct classes of cDNA clones as determined by their restriction maps. Clone 1C9 and the longest clone in each class were sequenced. Analysis of the sequence data demonstrated that clone 1C9 is actually a hybrid cDNA containing sequences from two mRNAs. Figure 5 presents a restriction map of the clones that depicts the structure of each cDNA. Clone 1C9 hybridized to clones 2D8 and 3C1 because its nucleotide sequence is 100% homologous at the 5′ end with clone 2D8 and at the 3′ end with clone 3C1. Clone 1C9 has a short internal sequence of 10 bp, which is found in both 2D8 and 3C1 (see

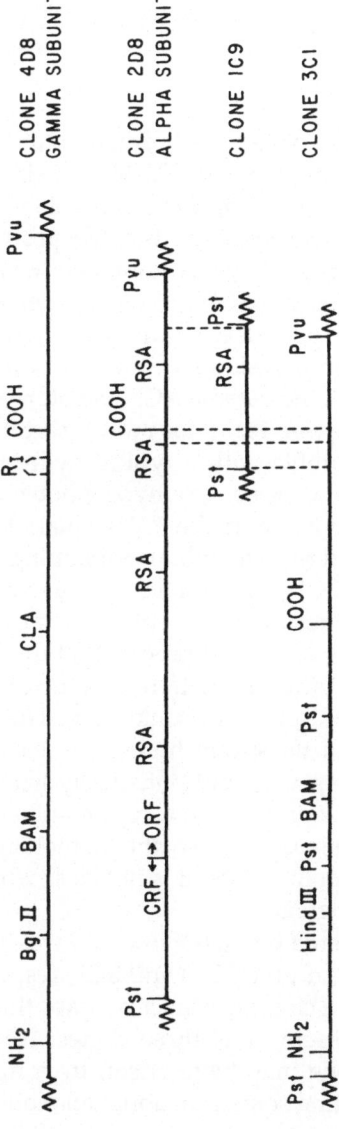

Figure 4. General structures of four clones isolated from a cDNA library prepared using *Torpedo* electric organ mRNA. Clone 4D8 was identified by hybridization to a short clone previously identified by hybridization selection. Clone 1C9 was identified by hybridization selection and clones 2D8 and 3C1 by hybridization with clone 1C9. The dotted lines indicate the portions of the 1C9 sequence homologous with 3C1 and 2D8.

Fig. 4: dotted lines joining clones 1C9, 2D8, and 3C1. Analysis of the sequence from clone 2D8 demonstrated that it contains an open reading frame coding for all but the amino-terminal 84 amino acids of the α-subunit. Recently we isolated a full-length α-subunit clone from a λgt10 cDNA library. mRNA made from this clone produces a functional α-subunit when injected into *Xenopus* oocytes (unpublished results).

2.3.   *Torpedo* Clone 3C1

Analysis of the sequence of clone 3C1 demonstrated that it contains an open reading frame coding for a 38,000-M_r protein of 337 residues. Except for the short stretch of 10 bp, there is no obvious nucleotide or amino acid sequence homology between the 3C1 protein and the ACh receptor. There is, however, some weak homology with the α-subunit. Analysis of the structure of the 3C1 protein suggests that it is very different from the receptor peptides. Figure 5 shows a plot of the hydropathy index as a function of residue for the 3C1 protein as compared with the α and γ-subunits of the *Torpedo* ACh receptor. All ACh receptor subunits sequenced to date (see Fig. 8 for mouse skeletal muscle receptor) have a characteristic profile with a typical hydrophobic signal sequence at the N-terminal and three very hydrophobic regions near the middle of the molecule and one at the C-terminal. The hydrophobic regions have been proposed to be membrane-spanning regions (see Section 4.2). By contrast, the 3C1 protein does not have a signal sequence and has one rather long (28-residue) hydrophobic region near the amino-terminal end. On this basis we would predict that the 3C1 product is a membrane protein but with a very different structure from the ACh receptor. The role of the 3C1 protein is unknown. The Dayhoff protein database was screened and no strong homology was found with any sequenced protein. However, weak but statistically significant homology was seen with the α-subunit of the ACh receptor and a number of other proteins including some kinases. Since the homology involves short sequences of five to eight amino acids, it is not clear what the functional significance of this homology is.

Clone 3C1 was discovered because clone 1C9 was a hybrid between two mRNAs coding for the α unit (2D8) and 3C1, respectively. The 1C9 cDNA may have arisen by a cloning artifact perhaps through the chance homology of the 10 bp present in all three clones: 1C9, 2D8, and 3C1. Alternatively, the 1C9 clone may have arisen by a mistake in mRNA processing, perhaps reflecting some functional relationship between the genes coding for the 3C1 protein and the α-subunit. If the latter is correct, one might expect the expression of the genes to be co-regulated. To test

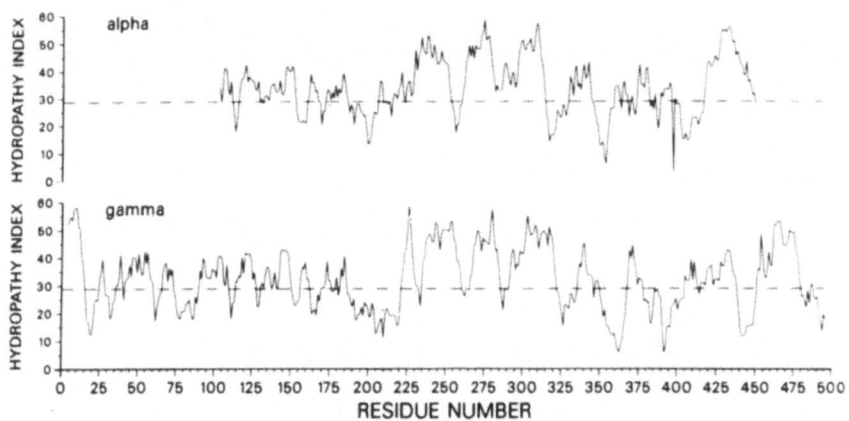

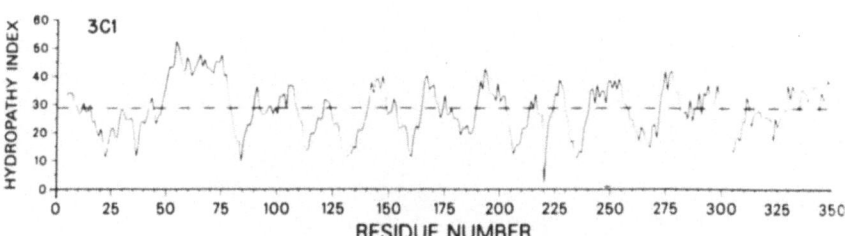

Figure 5. Relative hydropathy of the proteins encoded by clones 4D8, 2D8, and 3Cl as a function of their sequence number. The data were generated by the SOAP program devised by Kyte and Doolittle (1982). Hydrophobic regions of the protein lie above the midline and hydrophilic regions lie below the midline. The midline corresponds to the average hydropathy of the proteins in the Dayhoff database.

this idea, RNA was isolated from *Torpedo* tissue, electric organ, muscle, and liver. The RNA was size-fractionated on an agarose gel and transferred to nitrocellulose. The nitrocellulose blots were incubated with ^{32}P-labeled cDNA made by nick translation of cDNA clones encoding the 3Cl protein and the α and γ-subunit of the ACh receptor. The results of this experiment are shown in Fig. 6, which demonstrates that RNA specific for each clone is present in similar amounts in electric organ. The electric organ is known to make much more receptor than does muscle. This is also true for the 3C1 RNA. Figure 7 demonstrates that RNA hybridizing to the 3C1 clone is present in large amounts in electric organ as compared with muscle and cannot be detected in liver. An interesting possibility is that the 3C1 protein is a postsynaptic protein associated with the receptor in some way much like the 43,000-M_r protein present in the postsynaptic membrane, which has been identified by immunological and biochemical methods (Sobel *et al.*, 1978; Porter and Froehner, 1983, 1985).

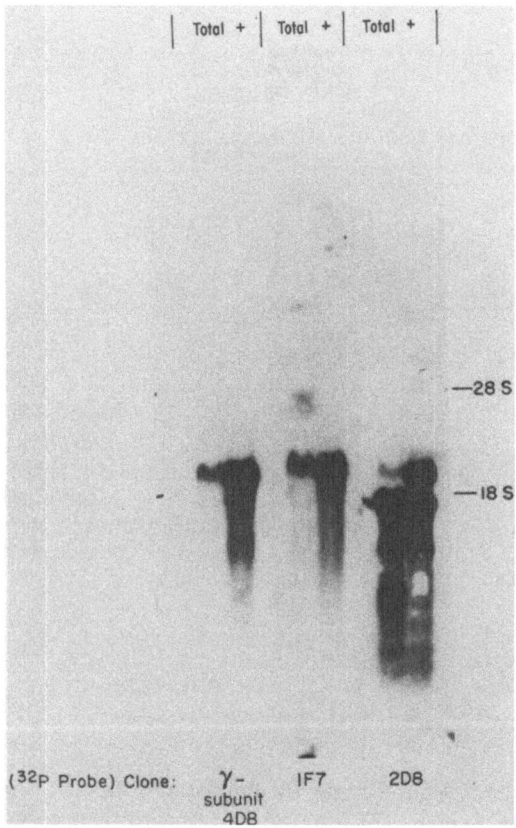

Figure 6. Hybridization of clones 4D8, 3Cl (or 1F7), and 2D8 to total and poly(A)$^+$ RNA isolated from *Torpedo* electric organ. The RNA was fractionated in denaturing agarose gels and transferred to nitrocellulose. The transferred RNA was hybridized to probes prepared from either clone 4D8, 3Cl, or 2D8. The lower band hybridizing with clone 2D8 in the total RNA lane is probably not mRNA coding for receptor, since it does not appear when the probe is derived from regions of clone 2D8 that contain only protein-coding sequences.

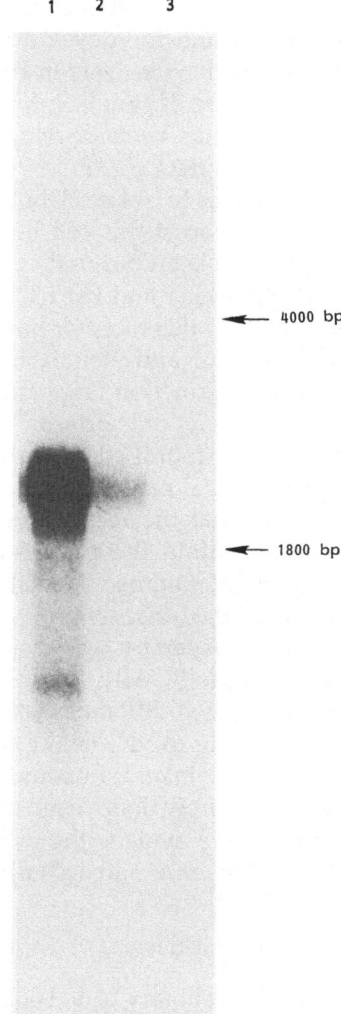

Figure 7. Hybridization of clone 3Cl to size-fractionated poly(A)$^+$ RNA isolated from *Torpedo* electric organ, muscle, and liver. Two of poly(A)$^+$ RNA from each source was fractionated in denaturing agarose gels, transferred to nitrocellulose, and hybridized with a probe made by nick translation of the entire 3Cl clone. Lane 1 is RNA extracted from *Torpedo* electric organ, lane 2 from *Torpedo* muscle, and lane 3 from *Torpedo* liver.

3. ISOLATION OF cDNA CLONES CODING FOR MOUSE SKELETAL MUSCLE ACETYLCHOLINE RECEPTOR

The questions concerning receptor expression during development and synaptogenesis are difficult to study in *Torpedo*. We therefore decided to isolate cDNA and genomic clones from mouse and rat. Receptor physiology and the development of the neuromuscular junction have been studied extensively in these species. Furthermore, the availability of a large base of genetic information and the technology to manipulate genes that has been developed in the mouse makes this species an appealing system.

We chose to isolate cDNA clones from mouse muscle cell line BC_3H-1. We developed this cell line more than a dozen years ago; its ACh receptor has been purified and characterized biochemically (Schubert *et al.*, 1974; Boulter and Patrick, 1977). In addition, the pharmacology and biophysics of this receptor have been studied extensively (Sine and Taylor, 1980; Sine and Steinbach, 1984*a,b*, 1985*a,b*, 1986). The BC_3H-1 cell line can be grown in large amounts and incorporates a high density of receptors on its cell surface. The properties of this receptor are those of the receptor found early in development or after denervation of adult mouse skeletal muscle. Sequence and S1 nuclease analysis have demonstrated that the receptor mRNAs made in the BC_3H-1 cell line are identical, within the protein-coding region, to the RNAs made after denervation of mouse skeletal muscle (Goldman *et al.*, 1985) (see Section 5.2 for detailed discussion).

A cDNA library was made from poly(A)$^+$ RNA purified from stationary-phase BC_3H-1 cells. Double-stranded cDNA was made and inserted into two different vector systems. The first library was put into pBR322 using dC:dG tailing. Subsequent libraries were made in λgt10 using EcoR1 linkers. In some cases, S1 nuclease digestion was used to trim the hairpin loop used to prime second-strand synthesis. Other libraries were made without S1 nuclease digestion using RNase H and DNA polymerase 1 to accomplish second-strand synthesis.

3.1. Mouse α-Subunit

A cDNA library of 50,000 clones in pBR322 was screened with a ^{32}P-labeled probe made by nick translation of a portion of a genomic clone isolated from chicken. The chick genomic fragment encoded α-subunit amino acid residues 161–239, which include the highly conserved putative membrane-spanning regions (see Section 4 for discussion of receptor structure). The chicken gene was isolated by Ballivet and collaborators using a probe made from the cDNA encoding the *Torpedo* α-subunit (Ballivet *et al.*, 1983).

Screening of the cDNA library resulted in the isolation of a clone (pMARα15) with a 1700-base insert, encoding a protein with high amino acid sequence homology to the *Torpedo*, chick, calf, and human α-subunits (Boulter *et al.*, 1985) (see Fig. 8 and Table I). On the basis of this homology and the distribution of homologous RNA in various tissues, we propose that this clone encodes the α-subunit of the ACh receptor expressed in BC_3H-1 muscle cells and skeletal muscle.

3.2. Mouse β-Subunit

A cDNA library inserted into λgt10 was screened with a fragment of a cDNA clone coding for the *Torpedo* β-subunit. This *Torpedo* clone was provided to us by Dr. Kathy Mixter and Dr. Norman Davidson, of the California Institute of Technology, and contains sequences encoding the putative membrane-spanning regions.

Screening the cDNA library at low stringency (5 × SSPE at 65°C) resulted in the identification of a clone encoding.a protein with amino acid sequence homology to the β-subunits of *Torpedo* and calf (see Figure 9). mRNA homologous to this cDNA is expressed in BC_3H-1 cells and skeletal muscle (Boulter *et al.*, 1986a).

3.3. Mouse γ-Subunit

A mouse γ-subunit was identified by screening the λgt10 cDNA library described in Section 3 with a probe made from the coding region of a mouse genomic clone coding for the γ-subunit. The mouse genomic clone was obtained by screening a genomic library with a probe made from a coding region from a human genomic clone given to us by Dr. Ballivet (see Boulter *et al.*, 1986a). A number of overlapping cDNA clones were isolated and sequenced. The deduced amino acid sequence is presented in Figure 10, and it is clear that there is a high degree of homology with *Torpedo*, chick, calf, and human γ-subunit sequences (Boulter *et al.*, 1986b).

3.4. Mouse δ-Subunit

The isolation of a cDNA clone coding for the muscle δ-subunit began by first isolating genomic clones from a mouse genomic library. The library was screened with a ^{32}P-labeled probe made by nick translation of a human genomic clone fragment containing sequences encoding the membrane-spanning regions of the human γ-subunit. The human genomic clone was isolated by Dr. Ballivet's group using a probe made from the cDNA clone encoding the *Torpedo* γ-subunit. Two classes of mouse genomic clones were isolated from the screening using a human γ-subunit probe. One member of each class was partially sequenced and a clear result was obtained. One class codes for the mouse γ-subunit of the muscle receptor. The second class codes for the δ-subunit. At this

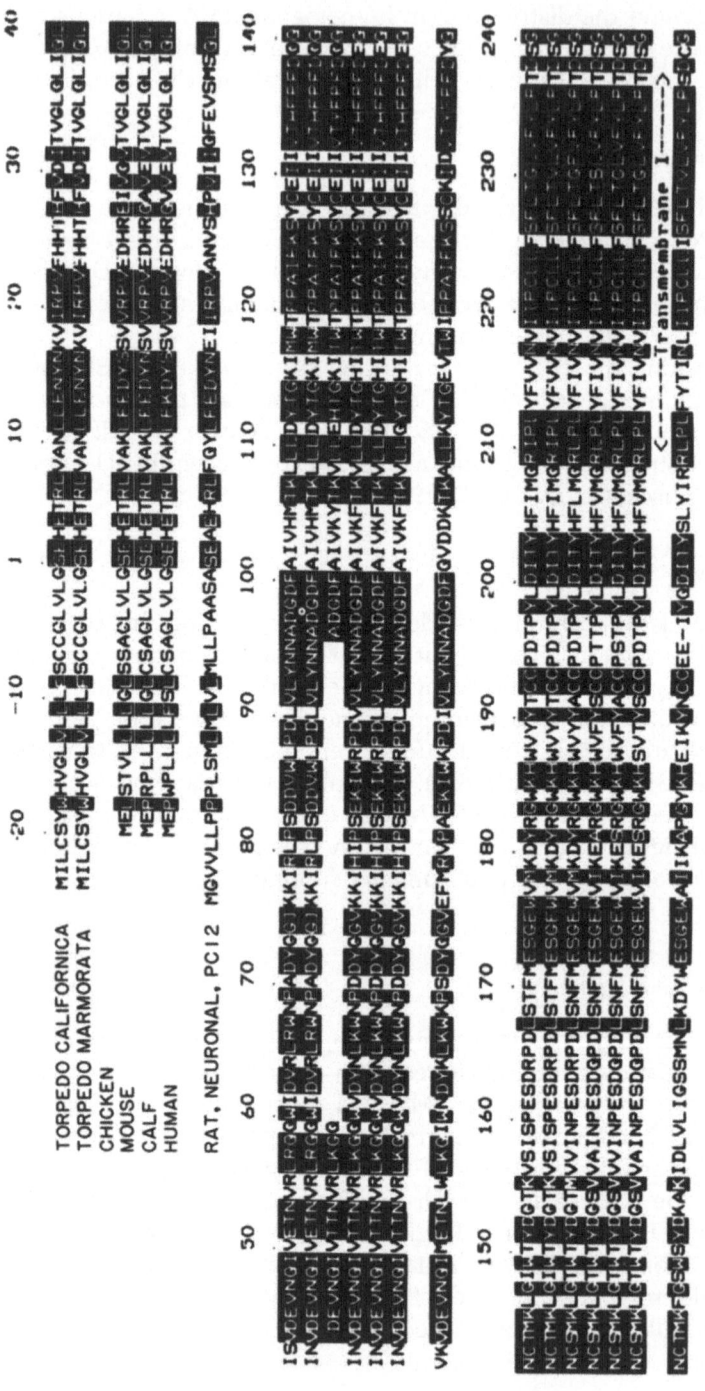

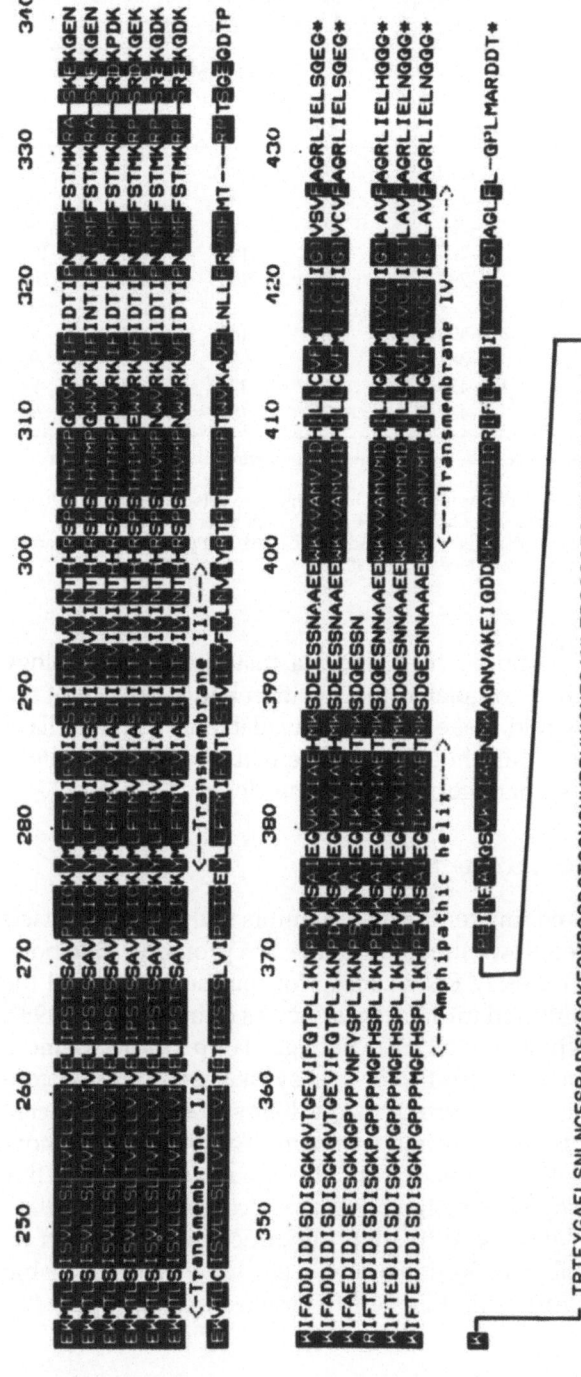

Figure 8. Amino acid sequence alignment of the α-subunits of the nicotinic ACh receptor (AChR) from muscle and electric organ of various species (1–6) with a putative neural nicotinic AChR α-subunit from rat (7). The sequences are represented in the one-letter code (*indicates the presence of a stop codon in the cloned sequence). Residues are numbered starting with the first amino acid of the mature polypeptide; the signal peptide residues are assigned negative numbers. Residues are outlined in black if at that alignment position at least one of the non-neural sequences (1–6) has a residue identical to that found in the putative neural receptor sequence (7). On the line between sequences 6 and 7, the predicted transmembrane domains are indicated by horizontal arrows. Sequences were obtained from the literature (except for 3, which was a personal communication from Dr. M. Ballivet): (1) Noda et al., 1982; (2) Devillers-Thiery et al., 1983; (3) Devillers-Thiery et al., 1983; (4) Boulter et al., 1985; (5) Noda et al., 1983; (6) Noda et al., 1983; (7) Boulter et al., 1986c.

Table I. Amino Acid Sequence Homology for Various Domains of the BC₃H–1 ACh Receptor α-Subunit and Calf, Human, Chick, and Torpedo Californica[a]

Protein domain	Amino acid residues	Homology (%)			
		Torpedo	Chick	Calf	Human
Amino terminal	1–210	77	(98)	94	95
Acetylcholine-binding site	128–142	93	87	100	100
MSR I[b]	211–236	92	92	96	96
MSR II	244–265	100	100	100	100
MSR III	279–297	79	100	100	100
Region between MSR III–IV	298–400	76	(82)	94	93
Amphipathic segment[c]	362–387	80	84	96	96
MSR IV	401–433	81	NA[d]	94	97
Mature protein	1–437	80	(86)	95	96

[a] Numbers in parentheses are for segments of the chick α-subunit for which the entire sequence of the region in question is unknown.
[b] Membrane-spanning regions (MSR I–IV) for the BC₃H-1 α-subunit were determined by inspection of hydrophobicity profiles and comparisons of homologous regions with other α-subunits.
[c] The amphipathic region is as described by Finer-Moore and Stroud (1984). For purposes of comparison, only the *T. californica* sequences are used (Noda *et al.*, 1982).
[d] NA, not available.

point, Dr. Davidson's group at the California Institute of Technology informed us that they had completed the isolation and sequence of the mouse δ-subunit and would make this clone available to us (La Polla *et al.*, 1984). Comparison of our limited sequence with that of the Caltech group confirmed that we had isolated the same clone.

3.5. Expression of Mouse Receptor in Oocytes

Each of the cDNAs coding for the four subunits of the mouse muscle ACh receptor was placed downstream from the SP6 promoter. SP6 polymerase was used to synthesize RNA coding for four subunits, and the RNA was capped and injected into *Xenopus* oocytes (Barnard *et al.*, 1982; Mishina *et al.*, 1984; White *et al.*, 1985). Voltage-clamp studies demonstrated that oocytes injected with the *in vitro*-synthesized mRNAs contained functional nicotinic ACh receptors on their surface. The injected oocytes also bound ^{125}I-labeled α-bungarotoxin. These results demonstrate that the four cDNAs cloned from the mouse muscle cell line BC₃H-1 encode the complete functional mouse receptor (Boulter *et al.*, 1986a). This system is being used to study the function of the brain receptorlike protein described in Section 6, and to elucidate the relationship between structure and function utilizing site-directed mutagenesis.

4. STRUCTURE OF THE ACETYLCHOLINE RECEPTOR

4.1. *Primary Structure*

The isolation of cDNA clones encoding the subunits of the ACh receptor made it possible to deduce the primary structure in a number of important species. Figure 11 shows the complete sequence of the four subunits of the mouse skeletal muscle ACh receptor. The sequence of the δ-subunit was obtained by Dr. Davidson's group (La Polla *et al.*, 1984). The α-, β- and γ-sequences are from our work. The most striking result from comparing the sequences of the four subunits is that they are highly homologous over their whole length and clearly are encoded by genes that are evolutionarily related. Furthermore, the degree of conservation between subunits is not uniform over their length. Some regions are extremely conserved between subunits (i.e., the putative membrane spanning regions), while other regions have diverged from one another. It is not clear what the significance of the presence of highly conserved regions is, but it is possible to speculate that they represent domains that must interact closely with other subunits or alternatively with regions that carry out a function common to all the subunits. A similar pattern of sequence conservation is seen when an individual subunit is compared between species (see Figs. 8–10). Analysis of the primary structure has led to models of the distribution of the subunit polypeptides across the membrane. These ideas have stimulated a number of experiments that attempt to test the models in order to determine the relationship between structure and function. This work is just beginning; as some of the early experiments conflict with one another, only a brief discussion of the model that we have proposed is presented.

4.2. *Folding through the Membrane*

In order to predict how the protein partitions in the membrane, it is necessary to identify regions that span the lipid bilayer. No direct means of identifying membrane-spanning regions has yet been found, so we chose to look for long stretches of hydrophobic amino acids using a program developed by Kyte and Doolitte (1982). Each amino acid is assigned a hydrophobicity value related to some measure of the hydrophobic properties of its side chain. The program calculates an average hydrophobicity of a segment of seven amino acids as it moves along the polypeptide from the amino-terminal to the carboxy-terminal. The result of this analysis is presented in Fig. 12 for all four mouse subunits. Hydrophobic regions lie above the midline. It is clear that all four sub-

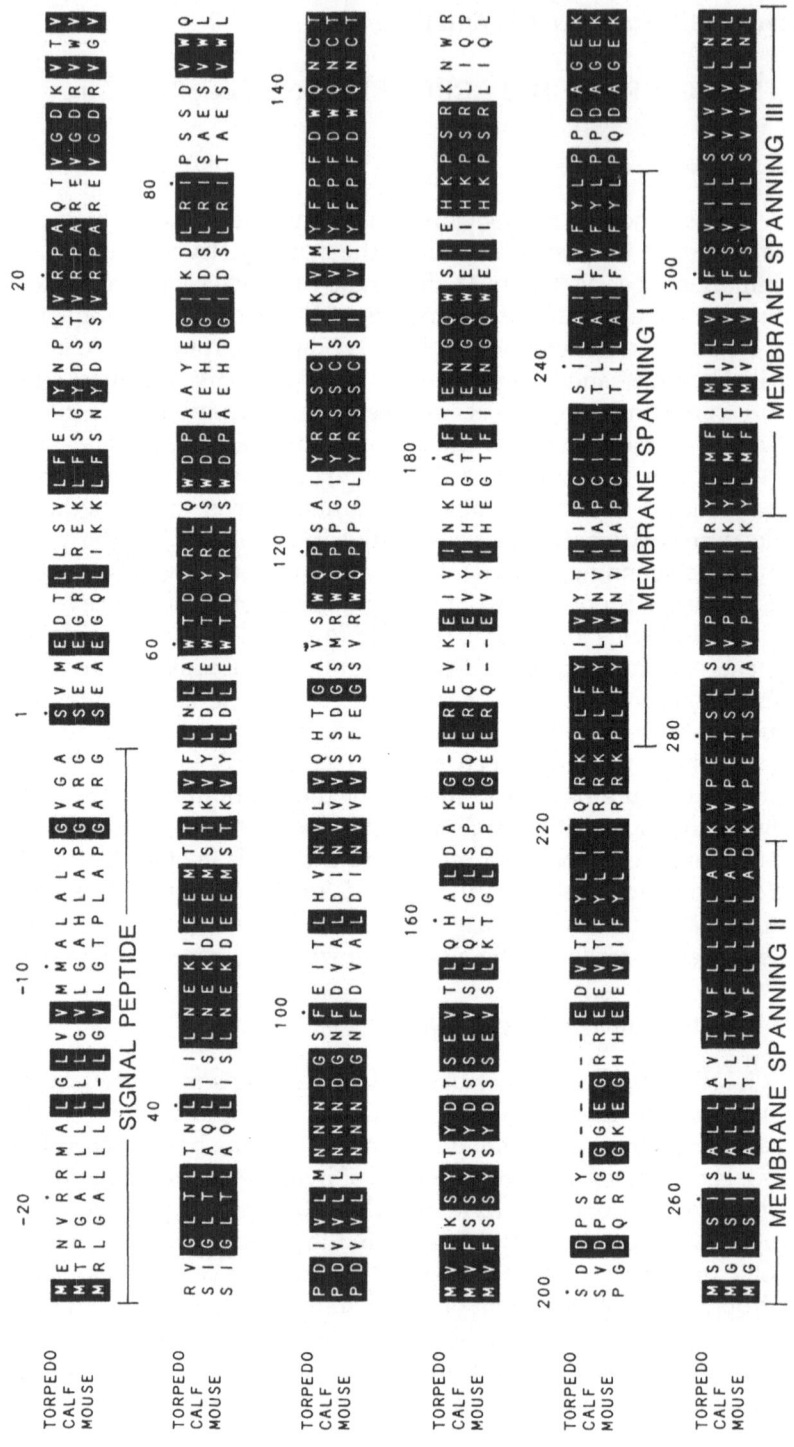

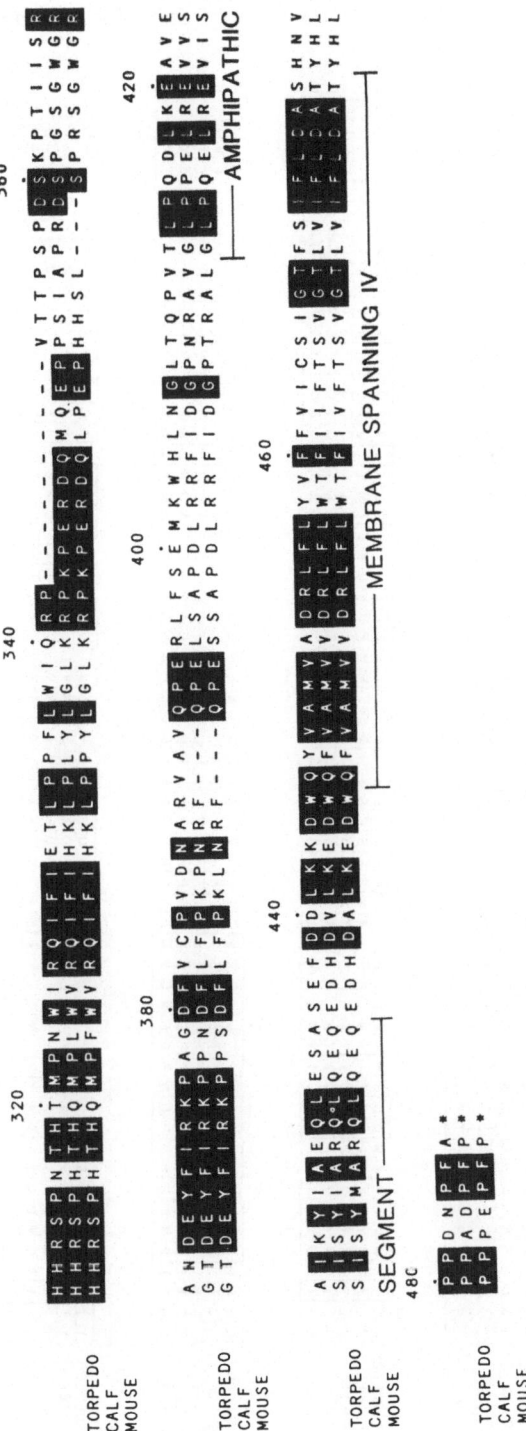

Figure 9. Interspecies comparison of aligned amino acid sequences for the nicotinic acetylcholine receptor (AChR) β-subunit. Positions with perfect homology for all species compared are shown with black background. Short gaps have been introduced to optimize alignment between species. Amino acid positions have been numbered starting with the amino-terminal amino acid for the mature β-subunit for each species. Amino acids in the signal peptide are preceded by negative numbers. The boundaries of putative structural domains are as indicated. Termination codons are designated by asterisks (*). *Torpedo californica* data are from Noda *et al.* (1983); calf data are from Tanabe *et al.* (1984).

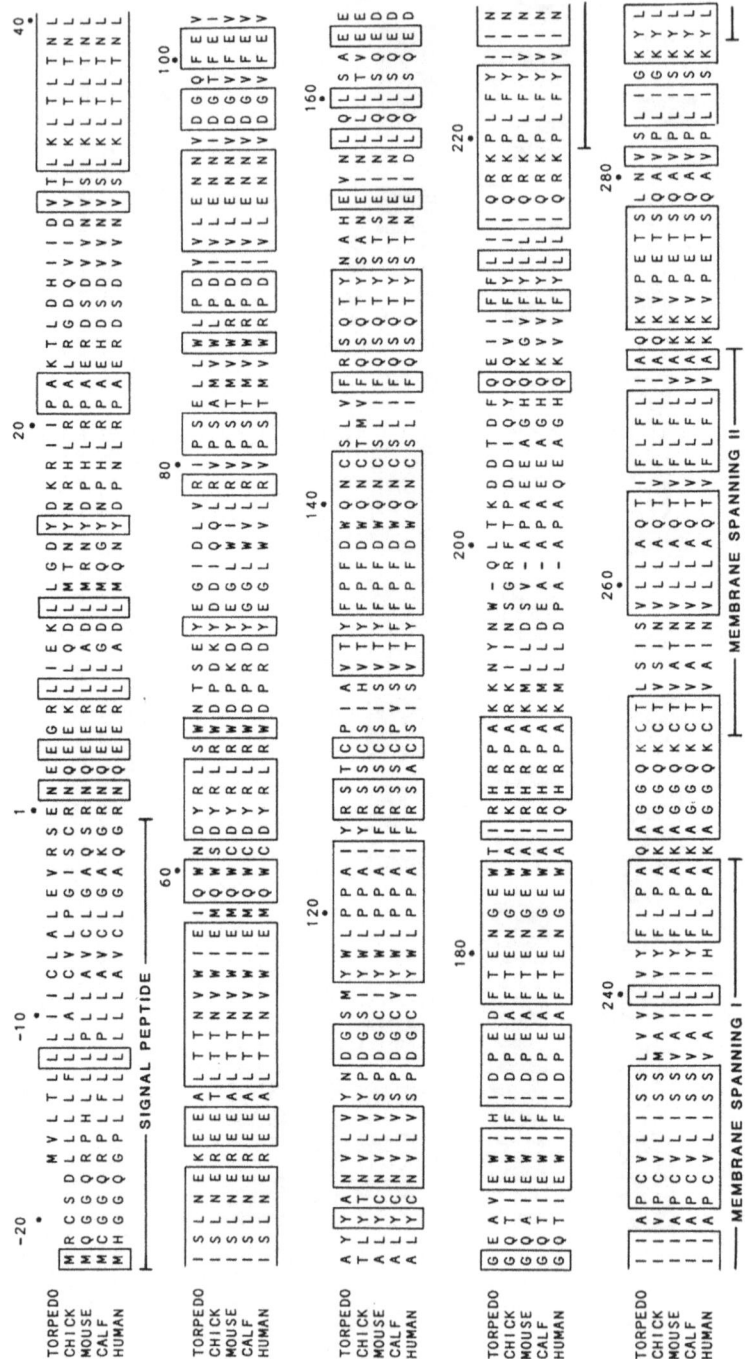

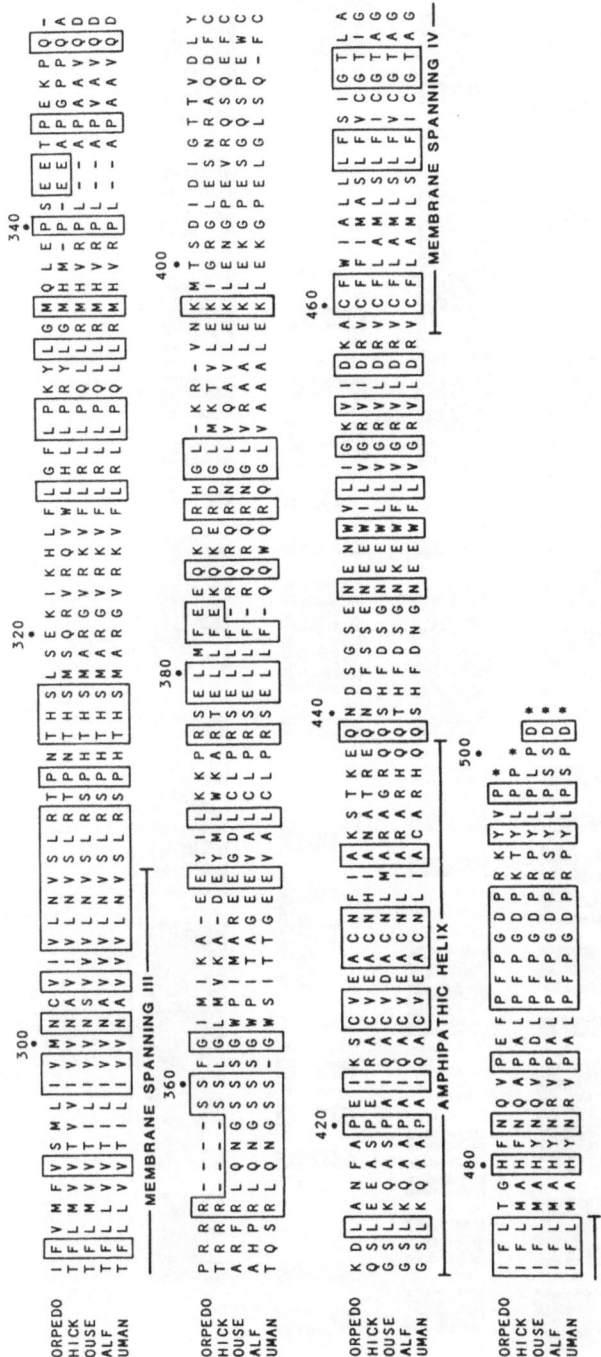

Figure 10. Interspecies comparison of aligned amino acid sequences for the nicotinic acetylcholine receptor γ-subunit. Positions with perfect homology for all species compared have been boxed. Short gaps have been introduced to optimize alignment between species. Amino acid positions have been numbered starting with the amino-terminal amino acid for the mature α-subunit for each species. Amino acids in the signal peptide are indicated by negative numbers. Termination codons are designated by asterisks (*). *Torpedo californica* data are from Claudio *et al.* (1983); chick data are from Nef *et al.* (1984); calf data are from Takai *et al.* (1985); human data are from Shibahara *et al.* (1985).

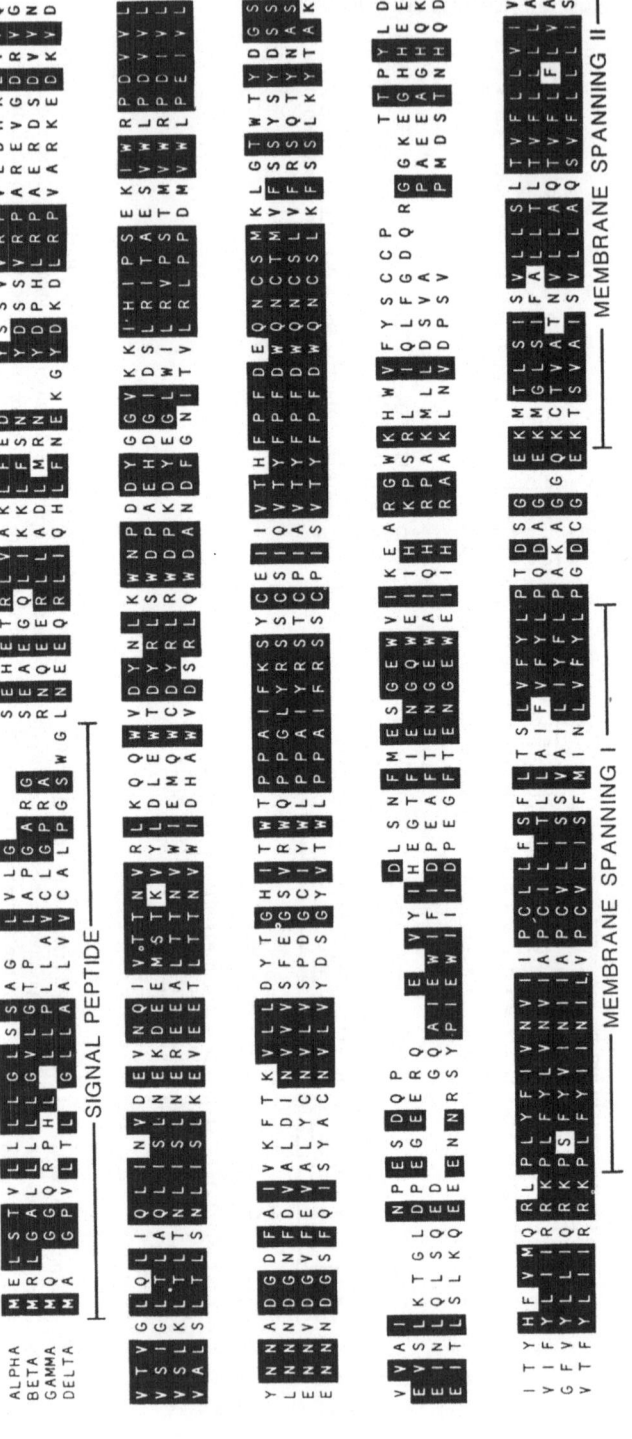

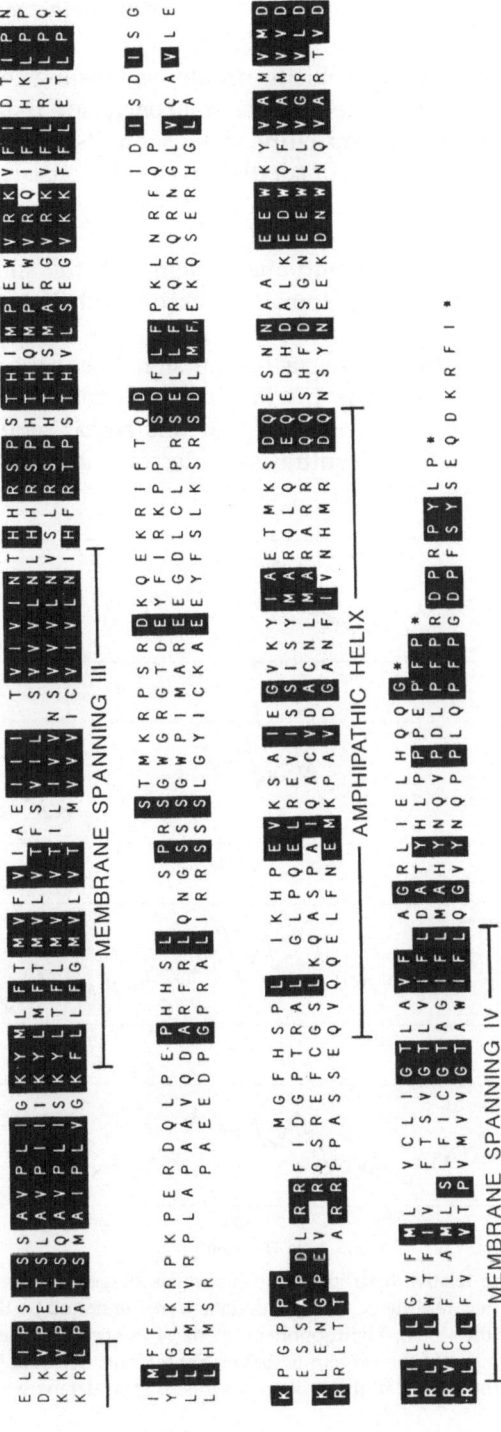

Figure 11. Comparison of the sequences of the mouse acetylcholine receptor subunits. α-Sequence data are from Boulter et al. (1985); β- and γ-sequence data are from Boulter et al. (1986 a,b); δ-sequence data are from La Polla et al. (1984). Putative transmembrane regions and the proposed amphipathic helix are indicated. Identical residues and conservative substitutions are shown against a black background.

units have a very similar hydrophobicity profile, suggesting that they fold in a similar fashion. This is the first general conclusion that comes out of an analysis of the primary structure of the subunits. Recent analysis of electron microscopic images of tubular crystals is consistent with this conclusion that each subunit folds in a similar way (Brisson and Unwin, 1985).

As expected for an integral membrane protein with the amino-terminal located in the extracellular domain (Jackson and Blobel, 1977; Wennogle *et al.*, 1980, 1981; Anderson *et al.*, 1982), there is a short stretch of hydrophobic amino acids forming a typical signal sequence at the amino-terminal of each subunit (Fig. 12). *In vitro* translation experiments confirm this prediction and have shown that this signal sequence is cleaved from each subunit during synthesis (Anderson *et al.*, 1981, 1982).

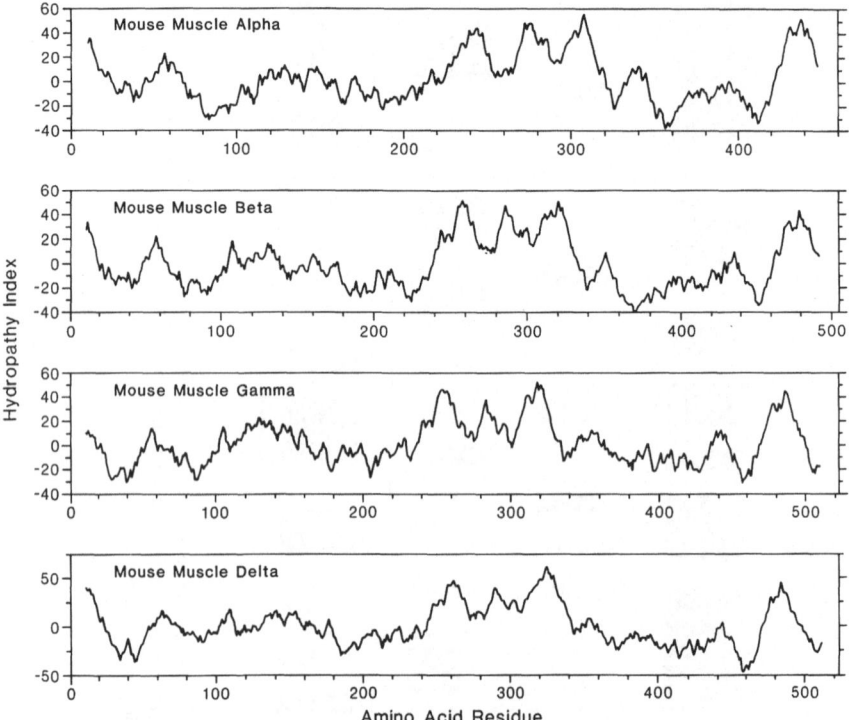

Figure 12. Figure shows the relative hydropathy of the mouse skeletal muscle receptor subunits as a function of their residue number. The data were generated by the SOAP program of Kyte and Doolittle (1982). Hydrophobic regions of the protein lie above the midline and hydrophilic regions of the protein lie below the midline. The midline corresponds to the average hydropathy of all the proteins in the Dayhoff database.

The hydrophobicity plot identifies three long hydrophobic stretches, about 20 amino acids long, near the middle of each subunit, and a fourth hydrophobic region near the carboxy-terminal (Fig. 12). Each hydrophobic region is flanked by charged residues thought to stop the transfer of polypeptides through the membrane and to serve to anchor the hydrophobic region within the membrane. In addition, a proline residue near the borders of each hydrophobic region would break the α-helix and allow the peptide to dip back into the membrane. On the basis of this sequence analysis, we proposed that each hydrophobic region is a membrane-spanning region, labeled membrane-spanning region I, II, III, and IV in Fig. 11, and that each subunit spans the membrane four times with the topology shown in Fig. 13 (Claudio *et al.*, 1983). However, there is an inherent problem with the Kyte and Doolittle (1982) analysis for predicting membrane-spanning regions in that it misses amphipathic helices that might be involved in channel formation. Finer-Moore and

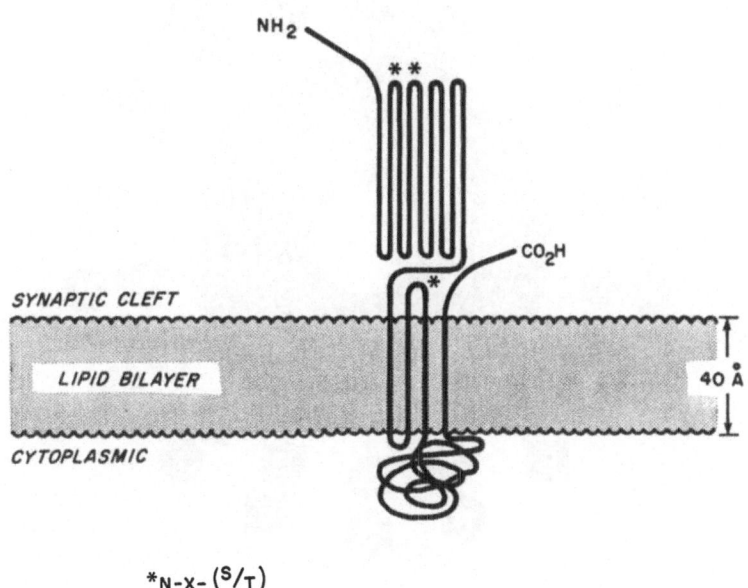

Figure 13. Partitioning of the γ-subunit polypeptide chain of an ACh receptor across the lipid bilayer based on the identification of membrane-spanning sequences shown in Fig. 12. Asterisks (*) indicate the location of canonical glycosylation sites located on the extracellular side of the membrane.

Stroud (1984) and Guy (1984) have proposed that the ACh receptor uses an amphipathic helix to form the ion channel. These investigators propose that the charged or hydrophilic surface of such an amphipathic helix lines the channel and that its hydrophobic surface interacts with other membrane-spanning segments. They have identified a region between the proposed membrane-spanning regions. Membrane-spanning region III and membrane-spanning region IV (see Fig. 11), as a candidate channel-forming amphipathic helix. See Fig. 14 for a helical wheel diagram (Schiffer and Edmundson, 1967) of the amphipathic helix from the homologous region of the mouse muscle α-subunit as well as a recently identified neuronal α-subunit (see Section 6 for discussion of this subunit). It is clear from this diagram that it is possible to line up the charged residues on one face and the hydrophobic residues on the opposite face of an α-helix. If this amphipathic helix spans the membrane, the model

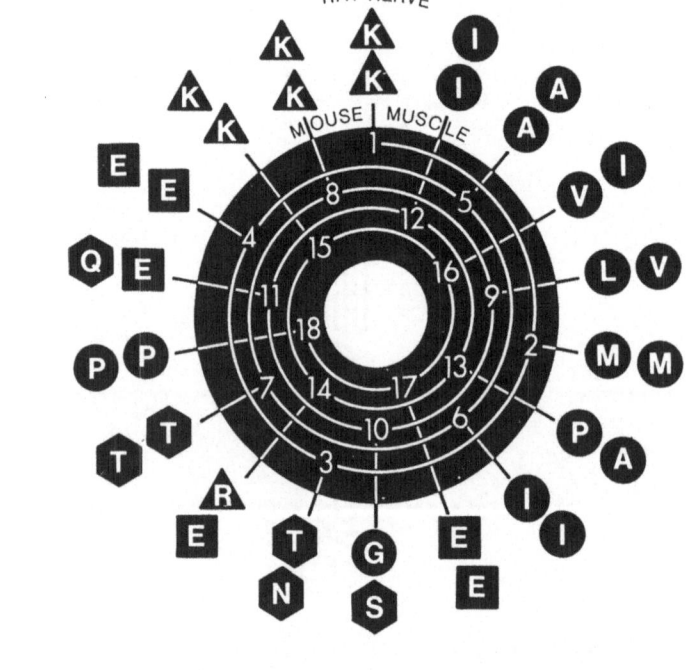

Figure 14. Helical wheel diagram of the amphipathic helix identified in the mouse skeletal muscle α-subunit and the rat neuronal α-subunit (see Fig. 8).

predicts that the carboxy-terminal of each subunit is intracellular, rather than extracellular, as we proposed (Fig. 13). Experiments using antibodies to localize the carboxy-terminal of the δ-subunit indicate that it is on the cytoplasmic face of the membrane consistent with five or any odd number of membrane-spanning regions (Ratnam and Lindstrom, 1984; Young *et al.*, 1984). However, site-directed mutagenesis experiments demonstrate that the receptor can still function even with a deletion of the amphipathic helix (Mishina *et al.*, 1984). This casts doubt on the idea that the amphipathic helix forms the ion channel. It is clear that an important problem is to identify by direct methods the membrane-spanning regions before any model of the channel can be constructed with confidence.

Our present model for how the subunits partition across the membrane is derived from analysis of the primary structure, and it is consistent with a large body of biochemical and structural data. The model predicts that each subunit spans the membrane five times, assuming that the amphipathic helix spans the membrane; it also predicts the following mass distribution: 50% of the peptide mass is located on the extracellular side of the membrane, 30% is cytoplasmic, and 20% is buried in the membrane. This is consistent with X-ray diffraction (Ross *et al.*, 1977; Brisson and Unwin, 1985), neutron scattering, and electron microscopy data (Wise *et al.*, 1981), which demonstrate that the bulk of the ACh receptor lies in the synaptic cleft. A reasonable working hypothesis is that all five subunits are arranged like staves in a barrel around a central pore, forming the ion channel (see Fig. 15).

Further tests of the model come from examining the post-translational modifications that occur during receptor synthesis. All four subunits of the *Torpedo* receptor are glycosylated. Asparagine-linked glycosylation sites are known to have the sequence asparagine-X-Ser/Thr. Analysis of the *Torpedo* γ-subunit sequence reveals five such potential glycosylation sites. However, only three are extracellular, according to the model shown in Fig. 14. This prediction is consistent with data suggesting that the γ-subunit is glycosylated in three positions (Anderson and Blobel, 1981). The other subunits all have at least one potential glycosylation site.

The *Torpedo* ACh receptor, like many proteins, can be phosphorylated (Gordon *et al.*, 1977; Teichberg *et al.*, 1977; Huganir *et al.*, 1984). Three different systems have been shown to phosphorylate the receptor: (1) a tyrosine-specific kinase phosphorylates the γ-, β-, and δ-subunits; (2) a cAMP-dependent system phosphorylates the γ- and δ-subunits; and (3) a protein kinase C phosphorylates the α- and δ-subunits. In each case, there is a characteristic potential phosphorylation site within the

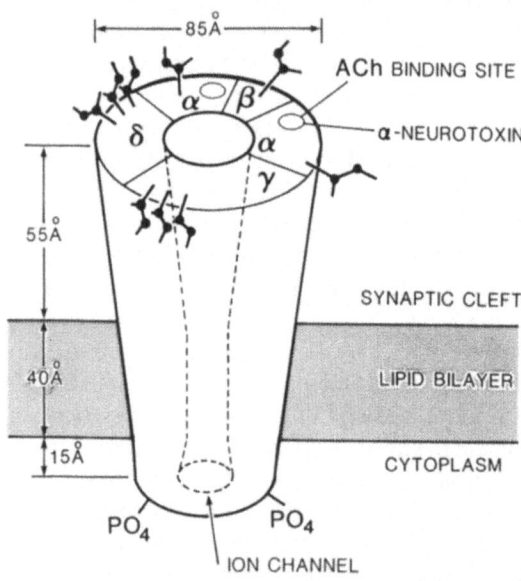

Figure 15. Diagrammatic representation of the nicotinic acetylcholine (ACh) receptor based on the work of Karlin and collaborators (Wise *et al.*, 1981) and Stroud and collaborators (Klymkowsky and Stroud, 1979; Ross *et al.*, 1977).

proposed cytoplasmic region of the subunit. Recently antibodies have been made against a peptide corresponding to the *Torpedo* δ-residues 354–366, including a potential phosphorylation site; arginine-arginine-serine-serine (Huganir *et al.*, 1984). This antibody blocks the cAMP-dependent phosphorylation of the δ- and γ-subunits, suggesting that this is the site that is phosphorylated in this system. Thus, the glycosylation and phosphorylation modifications are consistent with the model presented in Fig. 13.

Direct tests of the model have been performed using antibodies directed against specific sequences of the ACh receptor. Froehner and co-workers made an antibody that binds to a synthetic peptide corresponding to the *Torpedo* γ-subunit residues 360–377 (La Rochelle *et al.*, 1985). This antibody bound to the cytoplasmic face of membranes rich in *Torpedo* receptor consistent with the model presented in Fig. 13. On the other hand, Lindstrom and colleagues (Criado *et al.*, 1985) used a similar antibody approach and, on the basis of their results, have proposed that residues 143 to 191 form two additional membrane-spanning regions, an extended chain followed by an amphipathic helix. This is not consistent with the model in Fig. 13 and, if true, would force a complete reevaluation of the model. However, these antibody experiments were indirect and were not based on direct electron microscopic localization of the antibody. Furthermore, two observations argue against this interpretation of the antibody results. A peptide corresponding to the sequence 173–204 has been synthesized and shown to bind α-bungarotoxin (Wilson *et al.*, 1985), suggesting that this region of the protein contributes to an α-bungarotoxin binding site and cannot be a membrane-spanning region. This interpretation is strengthened by the finding of Karlin and co-workers (Kao *et al.*, 1984) that cysteines 192 and 193 are labeled by an affinity analogue of ACh, suggesting that the ACh-binding site is close to this region.

4.3. Acetylcholine-Binding Site

It was demonstrated a decade ago that ACh analogues bind near a disulfide bond on the α-subunit (Karlin *et al.*, 1976). The availability of the amino acid sequence and a model for the partitioning of the protein across the membrane made it possible to make an educated guess as to the location of the ACh-binding site. Numa and co-workers (Noda *et al.*, 1982) suggested that the binding site is located either near cysteines 128 and 142 or alternatively near cysteines 192 and 193. The model for the partitioning of the receptor across the membrane predicts that these cysteines are the only cysteines in the synaptic cleft available to form a disulfide bond. These four cysteines are conserved in all seven α-subunits that have been sequenced, including the neural α-subunit expressed in the central nervous system (Fig. 8). Free sulfhydryl groups are rare in extracellular proteins (Fayhey *et al.*, 1977) and, if present, are not conserved between species unless they form part of an active site (Schulz and Schirmer, 1979; Thornton, 1981). For this reason, we have proposed that all four cysteines participate to form two disulfide bonds (Boulter *et al.*, 1985). It is unlikely that a disulfide bond forms between the adjacent cysteines 192 and 193, as this has not been seen in other proteins (Schulz and Schirmer, 1979; Thornton, 1981). However, many proteins contain double cysteine bridges (Brown, 1976) in which two adjacent cysteines (i.e., Cys 192 and 193) each form a disulfide bond with another cysteine (i.e., Cys 128 and 142). Thus, we have proposed that cysteines 128 and 193 form one disulfide bond and cysteines 142 and 192 form the second disulfide bond. Recently, Karlin and co-workers presented evidence that the ACh-binding site is near Cys 192 and 193. Furthermore, the results of their experiments are most consistent with the absence of a double disulfide bond and suggests that our proposal is not correct. The data support a model in which disulfide bonds are formed between Cys 128 and 142 and between the adjacent cysteines 192 and 193 (Kao *et al.*, 1984; Kao and Karlin, 1986; Karlin *et al.*, 1986). These results imply that the ACh receptor contains a disulfide bond between adjacent cysteines that is rarely if ever found in large proteins.

5. EXPRESSION OF THE ACETYLCHOLINE RECEPTOR GENES IN SKELETAL MUSCLE

The expression of the ACh-receptor genes was studied in mouse and rat by measuring the level of receptor-specific RNA in various tissues (Goldman *et al.*, 1985). RNA was purified, size-fractionated on agarose gels, and transferred to nitrocellulose. The nitrocellulose blot was then hybridized to a ^{32}P-labeled probe made by nick-translating the cDNA (pMARα15) encoding the muscle α-subunit. The autoradiograph of this Northern blot is presented in Fig. 16. As expected, the α-subunit cDNA hybridized strongly to RNA of about 2 kb in length from the BC$_3$H-1 cell line, from which the cDNA was derived, and a cell known to make large amounts of ACh receptor. A band of lower intensity but similar size was seen in innervated mouse diaphragm and leg muscle. Innervated muscle is known to make a small amount of receptor when compared with denervated or cultured muscle. Surprisingly, a band of similar size was also seen in brain tissue (see Section 6). No detectable hybridizing RNA was found in liver.

5.1. Skeletal Muscle Denervation

Denervation of skeletal muscle is known to lead to a large (5- to 30-fold) increase in the amount of ACh receptor in the plasma membrane. This large increase in receptor incorporation can be prevented by stimulating the muscle directly with electrodes placed near the muscle. This demonstrates that the rate of ACh receptor incorporation into the membrane is regulated by the level of electrical activity in muscle. To test the possibility that this regulation might be at the level of gene expression, RNA was purified from innervated and denervated skeletal muscle. The RNA was size-fractionated, transferred to nitrocellulose, and probed with labeled cDNA encoding the muscle α-subunit. Figure 17 shows that denervation results in a large increase in the level of RNA coding for the α-subunit (Goldman *et al.*, 1985). A similar increase was seen for the other subunit RNAs in mouse (data not shown). In rat the level of β-subunit RNA did not change much after denervation. This may be ex-

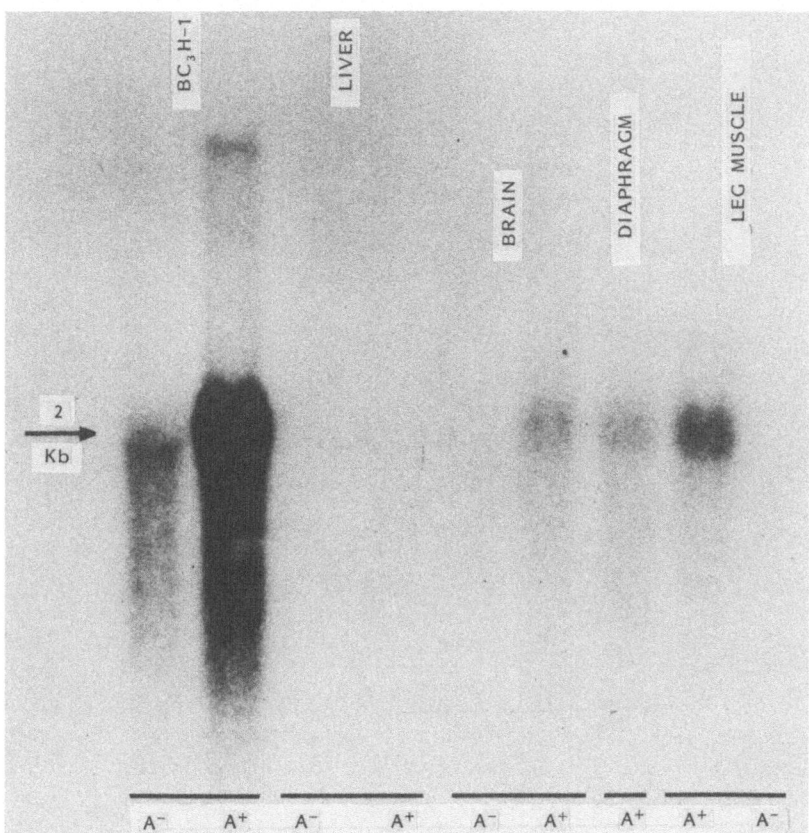

Figure 16. Autoradiograph of Northern blot hybridization using radiolabeled pMARα15 and mRNA isolated from the BC_3H-1 cells, mouse liver, brain, diaphragm, and leg muscle. RNA samples were prepared, electrophoresed, blotted, and hybridized to radiolabeled pMARα15. The amount of mRNA per lane is as follows: BC_3H-l, 1 µg; liver, 20 µg; brain, 20 µg; diaphragm and leg muscle, 10 µg each. A^+ refers to material eluted from an oligo-(dT) cellulose column; A^- denotes material that did not adhere to the column. Length of hybridizing species was estimated from the positions in the gel of 18 S and 28 S ribosomal subunits.

plained by the fact that β-subunit RNA is present in higher levels than the other subunits in innervated rat muscle. These results are consistent with the idea that the level of receptor gene expression depends on the level of electrical activity in muscle. If this if the case, the promoter regions of the receptor genes must have sequences sensitive to some signal that is a function of electrical activity. Similar results have been obtained in rodents by Merlie *et al.* (1984) and in chick cultures (Klarsfeld and Changeux, 1985).

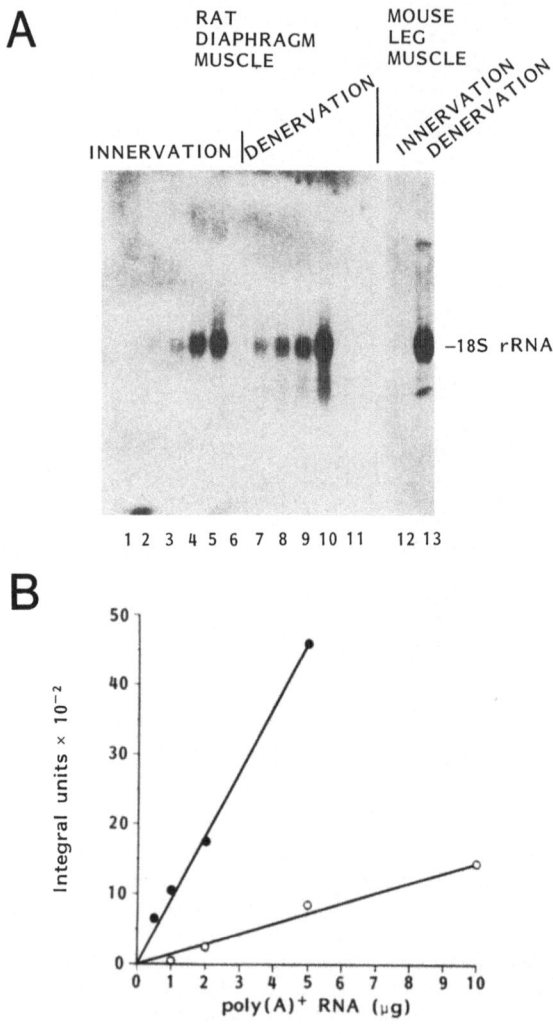

Figure 17. Effect of denervation on levels of mRNA coding for the α-subunit of the ace-tylcholine receptor (AChR). (A) Lanes 1–6 are RNA isolated from innervated rat hemi-diaphragms. Lanes 1–5 contain 0.5, 1, 2, 5, and 10 μg poly(A)$^+$ RNA, respectively. Lane 6 contains 10 μg poly(A)$^-$ RNA. Lanes 7–11 are RNA isolated from denervated rat hem-idiaphragms. Lanes 7–10 contain 0.5, 1, 2, and 5 μg poly(A)$^+$ RNA, respectively. Lane 11 contains 5 μg poly(A)$^-$ RNA. Lanes 12 and 13 contain 4.6 and 3.3 μg poly(A)$^+$ RNA isolated from innervated and denervated mouse leg muscle, respectively. (B) Quantitation of relative levels of rat diaphragm poly(A)$^+$ RNA coding for the subunit of the AChR. The autoradiogram in (A) was scanned with a densitometer and the integral of the peaks was determined. The integral of the scan is plotted against micrograms of poly(A)$^+$ RNA applied to the gel. RNA was isolated from 5-day denervated hemidiaphragms and inner-vated hemidiaphragms.

5.2. Synaptic versus Extrasynaptic Acetylcholine Receptor

After denervation, there is a large increase in receptor synthesis. These newly made extrasynaptic receptors are incorporated into the full extent of the muscle membrane and have the properties of the receptors found early in the development of muscle before the synapse matures. The fact that the biochemical, biophysical, and antigenic properties of these extrasynaptic receptors are different from those found at the mature synapse suggests that they may be the product of different genes. To address this question, RNA from innervated and denervated muscle was hybridized to the cDNA clone encoding the α-subunit made from RNA expressed in the muscle cell line BC_3H-1. Single-stranded cDNA corresponding to the antisense strand was used so that it would hybridize to the mRNA. The hybrid was subjected to single-strand specific S1 nuclease digestion and then run on a denaturing gel. The fragments protected from S1 nuclease digestion were visualized by hybridizing them to ^{32}P-labeled cDNA encoding the α-subunit. Figure 18 demonstrates that the 450-bp 5' fragment from the cDNA is protected by the RNA from both innervated and denervated skeletal muscle. A similar result is seen when the 1270-pb 3' fragment is used. This demonstrates that the cDNA cloned from the BC_3H-1 muscle cell line is the product of the same gene expressed in both innervated and denervated muscle (Goldman *et al.*, 1985). This is consistent with the idea that only one gene codes for the α-subunits, found at the mature synapse, and the extrasynaptic receptor incorporated into the membrane after denervation. The difference in the properties of these receptors must then be due either to post-transcriptional modification, local environmental differences, or a change in one of the other subunits: β, γ, or δ. It is still possible, however, that there exists a second gene or set of genes coding for the junctional receptor that we may have missed because of low homology or low mRNA abundance.

The S1 analysis revealed another fact. The 3' end of the mRNA coding for the α-subunit is heterogeneous. Two bands differing by about 130 bases were observed when the 3' 1270-bp fragment was used in the S1 nuclease protection experiment (Fig. 18). This difference can be explained by the finding that the 3' untranslated region of the mRNA contains three polyadenylation signals. Two of these signals are separated by 124 and 128 bp from the third, which presumably accounts for the heterogeneity seen at the 3' end of the mRNA. Both the long and short mRNAs increase upon denervation (Fig. 18), so the significance of the presence of two different RNAs is unclear. A similar heterogeneity is also observed in the 3' untranslated region of the γ-subunit. There

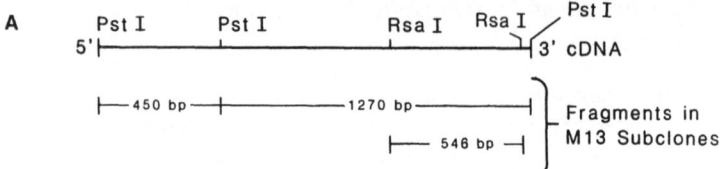

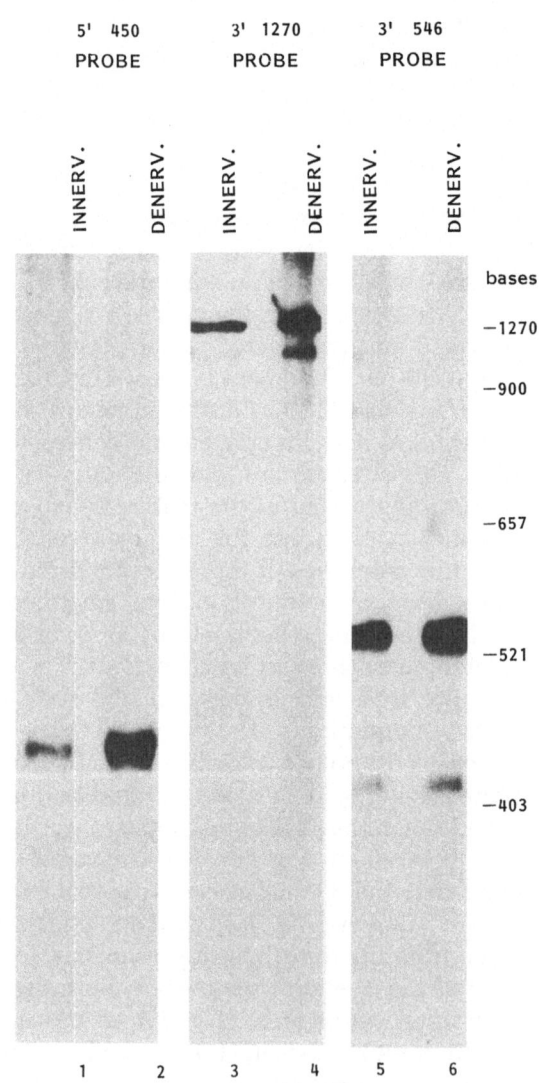

are a number of examples in which a single-copy gene is known to produce different RNAs by making use of multiple polyadenylations signals (Zehner and Paterson, 1983; Legace et al., 1983; Parnes et al., 1983). The functional significance of multiple polyadenylation signals is unknown.

6. BRAIN RECEPTORS

Nicotinic ACh receptors are known to be present in the vertebrate CNS. They are also found on sympathetic ganglion cells and chromaffin cells in the peripheral nervous system. These neuronal nicotinic receptors are pharmacologically different from the receptors found at the neuromuscular junction. For example, α-bungarotoxin, which blocks the nicotinic receptor at the neuromuscular junction, does not block the nicotinic receptor on a number of neural cells that have been examined. One of the best-studied systems, because of its accessibility, is the chromaffin cell and the PC12 cell line, which was derived from an adrenal tumor. These cells have a nicotinic ACh receptor that can be assayed physiologically. These cells also bind α-bungarotoxin in a specific and saturable fashion. However, the binding of toxin does not inactivate the receptor. Analysis of this system has demonstrated that the ACh receptor responsible for the ACh-induced ion fluxes in these cells is not the same molecule that binds toxin (Patrick and Stallcup, 1977a,b). Thus, these cells contain two different molecules that have ligand-binding sites with nicotinic-binding characteristics. One site gates an ion channel; the function of the toxin-binding site is unknown. A similar situation also occurs in chick sympathetic ganglia cells and Renshaw cells. Little is known about these receptors because they are present in small quantities and there is no specific high-affinity ligand that has proved useful for their purification. Furthermore, the finding that chromaffin cells have two different molecules that contain nicotinic binding sites indicates that more than one type of nicotinic receptor can coexist on one cell type.

←——————————————————————————

Figure 18. Sl nuclease analysis of mouse mRNAs. (A) Fragments of cDNA subcloned into M13 for hybridization with mRNA. (B) Gel profile of Sl nuclease-protected fragments generated by Sl nuclease digestion of heteroduplexes formed between innervated or denervated mouse leg muscle mRNA and the various M13 subclones in (A). Lanes 1, 3, and 5 represent Sl nuclease-resistant fragments generated by hybridizing innervated mouse leg muscle mRNA (40 μg) with the 5' 450-, 3' 1270-, and 3' 546-nucleotide-long subclones, respectively. Lanes 2, 4, and 6 represent Sl nuclease-resistant fragments generated by hybridizing denervated mouse mRNA (1 μg) with the 5' 450-, 3' 1270-, and 3' 546-nucleotide-long subclones, respectively.

This suggests that the CNS, with all its complexity, might contain a large set of nicotinic receptors, each having different properties. In an attempt to learn more about these receptors, we decided to apply the molecular genetic approach that has worked for the muscle nicotinic receptor. We reasoned that the neuronal nicotinic receptors could be related evolutionarily to the muscle nicotinic receptor and that this would be reflected in nucleotide sequence homology at the genetic level. If this is the case, cDNA clones encoding the muscle nicotinic receptor might hybridize to the mRNA or genes encoding neuronal receptors. In fact, the cDNA coding for the muscle α-subunit does hybridize to RNA extracted from mouse brain (see Fig. 16).

6.1. Neuronal α-Subunit Clone

In an effort to isolate a cDNA clone coding for a neuronal nicotinic receptor, we screened a cDNA library made from RNA purified from the PC12 cell line. This cell line is known to express a functional nicotinic receptor as well as an α-bungarotoxin-binding site. The cDNA library was screened with a ^{32}P-labeled probe made by nick translation of the cDNA clone pMARα15 encoding the mouse muscle α-subunit of the nicotinic receptor. Approximately one million λgt10 clones were screened at low stringency (5 × SSPE, 50°C); the T_m in this concentration of salt is about 100°C. Three clones were isolated and one, λPCA48, was sequenced. This clone contains an open reading frame coding for a protein of 499 amino acids with considerable (47%) amino acid sequence homology to the α-subunit of muscle (see Fig. 8). On the basis of sequence and structural homology and the tissue distribution of homologous RNA (discussed in Section 6.2), we have proposed that this clone, λPCA48, encodes a neural nicotinic receptor (Boulter *et al.*, 1986c).

6.2. Structure of the Neuronal α-Subunit

The protein encoded by the λPCA48 cDNA isolated from the PC12 cell line is 499 residues long; a hydrophobicity analysis shows that it has a hydrophobicity profile very similar to that of the muscle ACh receptor subunits (Fig. 19). This makes it very likely that it partitions across the membrane in a similar fashion and folds in the same way. The most striking difference between the neural α-subunit and the muscle α-subunit is its length. The neural subunit is longer than the six muscle α-subunits that have been sequenced. Alignment of the neural and muscle α-subunits demonstrates that there is an insertion of 41 amino acids in the region predicted by our model to be cytoplasmic (Fig. 8). This makes the neural α-subunit approximately the same length as the other (β, γ,

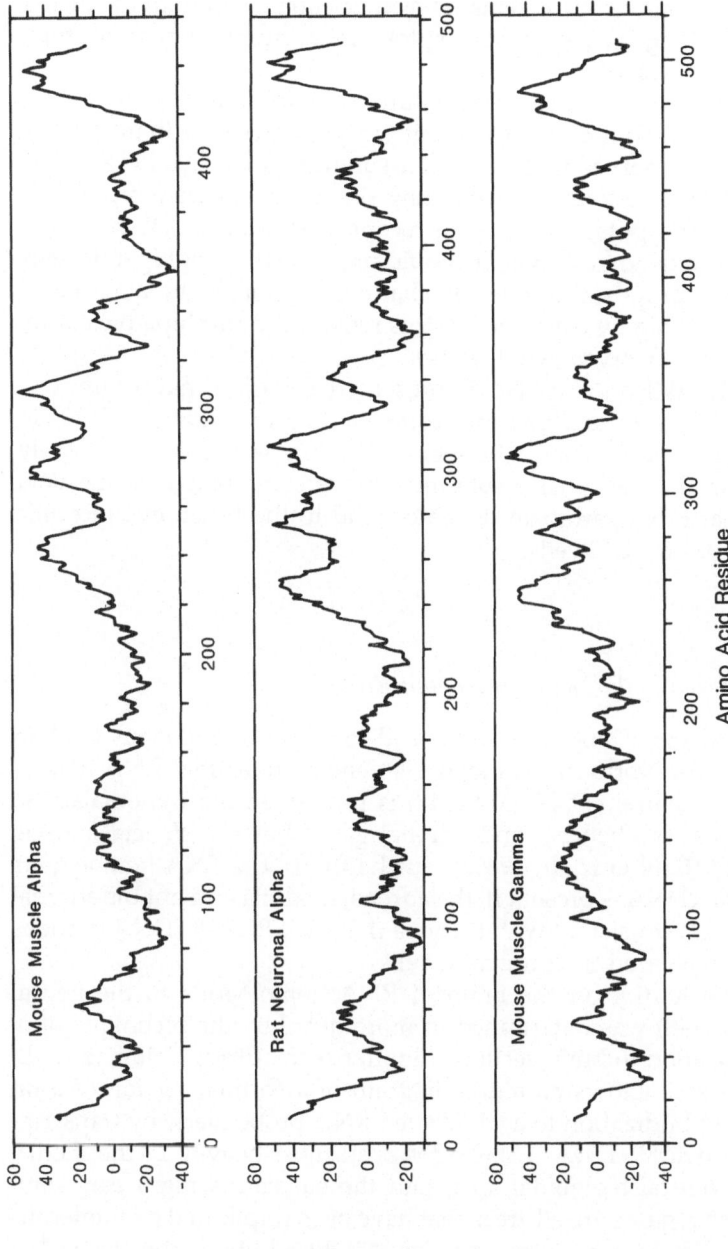

Figure 19. Relative hydropathy of the mouse skeletal muscle α- and γ-subunits and the rat neuronal α-subunit as a function of their sequence number. The data were generated by the SOAP program of Kyte and Doolittle (1982). Hydrophobic regions of the protein lie above the midline and hydrophilic regions lie below the midline. The midline corresponds to the average hydropathy of all the proteins in the Dayhoff database.

δ) subunits of the muscle receptor. This could be explained if the muscle α-subunit were derived evolutionarily by a deletion event from the neural α-subunit, making the neuronal receptor a more primitive receptor type compared with muscle.

Another important structural feature is the location of potential disulfide bonds. All four cysteines thought to be extracellular in the muscle α-subunit are conserved in the neural α-subunit. Thus, Cys 192 and Cys 193, which are close to the ACh-binding site in the muscle receptor, are conserved in the putative neuronal receptor. If the λPCA48 clone encodes a functional ion channel, it should preserve the amphipathic helix that has been proposed to line the channel. Although the amino acid sequence is not highly conserved in this region, the amphipathic nature of the helix is conserved (see Fig. 14).

Recently, Wilson *et al.* (1985) presented evidence that α-bungarotoxin binds to a synthetic peptide corresponding to residues 173–204 in the *Torpedo* α-subunit. As can be seen in Fig. 8, this region is not highly conserved in the neuronal α-subunit. This is consistent with the data showing that α-bungarotoxin does not bind to the functional nicotinic ACh receptor on PC12 cells.

6.3. *Expression of the Neural α-Subunit RNA*

The λPCA38 cDNA clone was used to detect homologous RNA in various tissues. Northern blot analysis demonstrates that RNA homologous to the neural cDNA (λPCA48) is present in the hypothalamus, hippocampus, cerebellum, and adrenal gland but not skeletal muscle (see Fig. 20) (Boulter *et al.*, 1986c). Furthermore, the RNA homologous to the neural cDNA is present in the adrenal medulla and not the adrenal cortex (Fig. 21) consistent with the idea that the λPCA48 cDNA encodes a receptor expressed in chromaffin cells.

The distribution in the brain of RNA homologous to the neural cDNA (λPCA48) was determined in more detail by the technique of *in situ* hybridization to brain sections (Boulter *et al.*, 1986c; Goldman *et al.*, 1986). Figure 22 shows an autoradiogram of a coronal section of a rat brain after hybridization to a ^{32}P-labeled RNA probe made by transcription of the neural cDNA. Labeling is seen most heavily in the medial habenula, ventral tegmental area, and the substantia nigra pars compacta. These regions are all areas that have been implicated by numerous criteria to contain cholinergic innervation. In addition, the strong hybridization seen in the medial habenula is consistent with the idea that

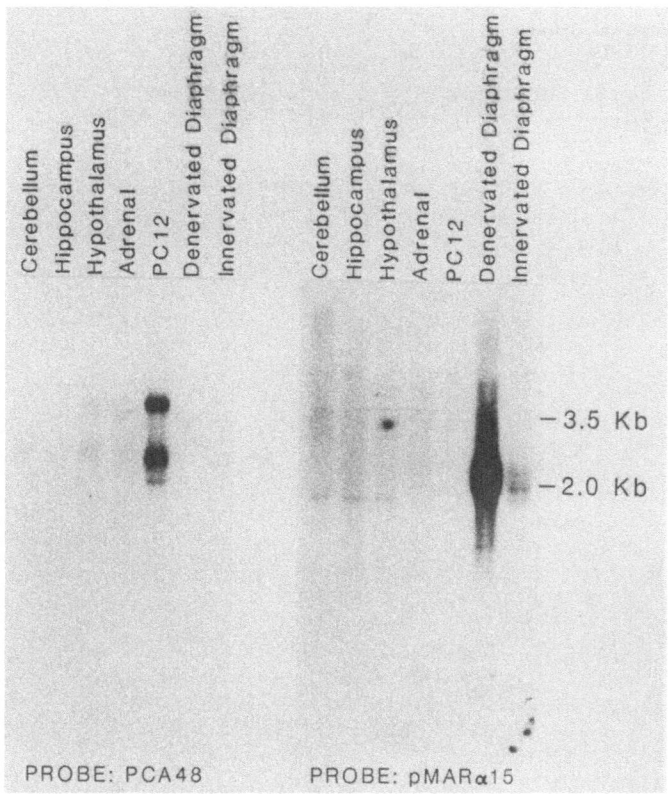

Figure 20. Northern blot analysis of RNA isolated from Sprague-Dawley rats. Indicated regions of brain or muscle tissue were dissected, poly (A)$^+$ RNA was prepared, fractionated on 1.4% agarose gels, and transferred to Gene Screen Plus. Filters were hybridized to probes prepared by nick translation of λPCA48 or pMARα15. Hybridization was carried out in 5 × SSPE, 50% formamide at 49°C and the filters were washed in 0.2 × SSPE at 65°C. Neuronal clone = PCA48; muscle clone = pMARα15.

the protein encoded by λPCA48 corresponds to the functional nicotinic receptor expressed in the PC12 cell line rather than the α-bungarotoxin site, since this region of the brain binds cholinergic ligands but does not bind α-bungarotoxin (Clark *et al.*, 1985).

An important question is whether the gene encoding the λPCA48 mRNA expressed in the PC12 cell line is the same gene that is expressed in the brain. To answer this question, RNA was extracted from PC12 cells and the habenular region dissected from rat brain. The RNA was

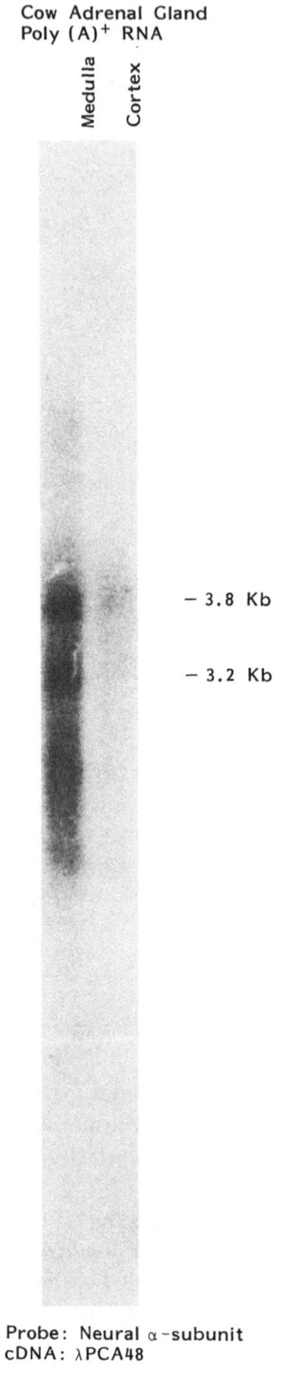

Cow Adrenal Gland
Poly (A)⁺ RNA

Medulla Cortex

— 3.8 Kb

— 3.2 Kb

Probe: Neural α-subunit
cDNA: λPCA48

Figure 21. Northern blot of poly(A)⁺ RNA extracted from the adrenal gland of the cow. The probe was ^{32}P-labeled cDNA prepared by nick translation of the PCA48 clone coding for the neuronal α-subunit. For sequence, see Fig. 8 and Boulter *et al.* (1986).

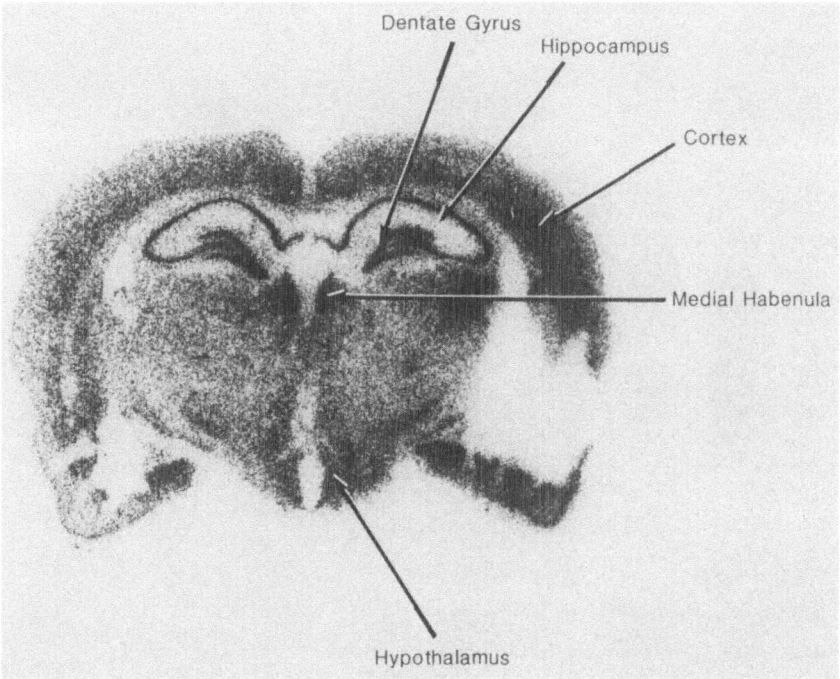

Figure 22. Autoradiograph of *in situ* hybridization in a coronal section of rat brain. Tissue was fixed as described by Swanson *et al.* (1983), and hybridization was performed using a procedure based on that described by Cox *et al.* (1984). Briefly, 20-μm sections were mounted on polylysine-coated slides, air-dried, digested with proteinase K, acetylated with 0.25% acetic anhydride, dehydrated, and hybridized with ^{32}P-radiolabeled single-stranded RNA probe prepared from an SP6 vector containing PCA48 cDNA insert. After hybridization, sections were treated with RNase A and washed at a final stringency of 0.5 × SSC, 42°C. Dehydrated slides were exposed to X-ray film for 4 days.

hybridized to the λPCA48 cDNA clone and the RNA/DNA duplex subjected to S1 nuclease digestion. Figure 23 shows that PC12 and habenular RNA protect the 2000-base λPCA48 clone. Thus S1 analysis demonstrates that the same gene is expressed in the PC12 cell line and the habenula (Boulter *et al.*, 1986c).

Northern blot analysis of RNA purified from various brain regions also demonstrates that there exists RNA homologous to the cDNA clone pMARα15 that encodes the muscle-type α-subunit. This RNA has a different size from the RNA that hybridizes to the neural cDNA clone (see Fig. 20), demonstrating that these species must be encoded by a gene different from the λPCA48 gene. These results suggest that there

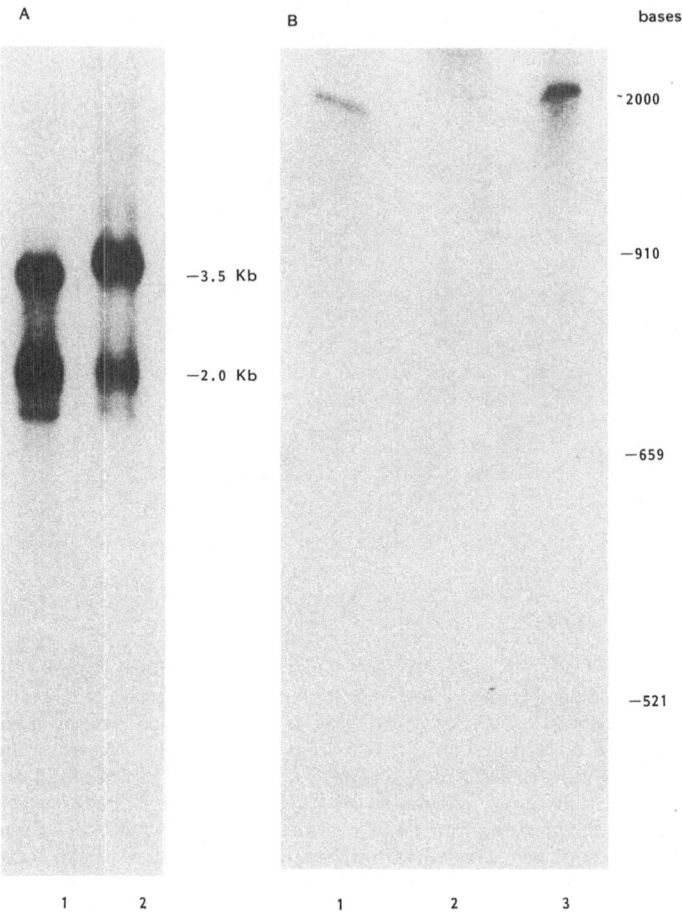

Figure 23. (A) Blot hybridization analysis. Poly(A)⁺ RNA from either the PC12 cell line (lane 1) or a region of the brain containing the habenula (lane 2) was size-fractionated by electrophoresis in denaturing agarose gels, transferred to Gene Screen Plus, and hybridized with ^{32}P-labeled PCA48 insert. Sizes of major hybridizing species are given in kilobases. (B) Nuclease S1 analysis. PCA48 was subcloned in M13 vectors mp18 and mp19. Single-stranded DNA was prepared and used to form heteroduplexes with poly(A)⁺ isolated from the PC12 cell line (lane 1) and a region of the brain containing the habenula (lane 3). Lane 2 represents a control where RNA was omitted. Hybridization reaction mixtures were incubated with nuclease S1 and those molecules surviving digestion were fractionated in denaturing acrylamide gels and electroblotted to Gene Screen Plus. Heteroduplexes surviving digestion were visualized by hybridization with ^{32}P-labeled PCA48. Blots were exposed to X-ray film with an intensifying screen for 18 hr at −70°C. pBR322 restriction fragments run in a parallel lane served as size markers (lengths in base pairs at right).

is a gene family coding for polypeptides with homology to the nicotinic α-subunits.

6.4. Evidence for a Gene Family

The possibility that there is a gene family coding for nicotinic receptor α-subunits was examined directly. Mouse genomic DNA was digested with two different restriction enzymes, Pst 1 and EcoR1. The DNA fragments from each digestion were size-fractionated on agarose gels and the resulting bands transferred to nitrocellulose. The Southern blots were hybridized to ^{32}P-labeled probes made by nick translation of the two α-subunit cDNAs: muscle (pMARα15) and neural (λPCA48). The hybridization was carried out at low stringency (3 × SSC 65°C) and high stringency (0.1 × SSC 65°C). At high stringency the two probes, muscle and neuronal, hybridize to DNA fragments of different lengths consistent with the idea that the mRNAs are products of two separate genes (Fig. 24). At low stringency, the neural probe (λPCA48) hybridizes to a number of additional bands that are not observed when the blots are probed at high stringency with either the neural or muscle probes. The 2.1 Kbase Pst 1 fragment and the 2.3 EcoR1 fragment (Fig. 24) are clear examples of such genomic fragments. These additional fragments, seen only at low stringency, may be parts of additional genes coding for nicotinic ACh receptors (Boulter et al., 1986c). This conclusion is confirmed by further screening of genomic and cDNA libraries. We have identified three additional genes coding for polypeptides with homology to the α-subunit of the muscle nicotinic ACh receptor.

7. CONCLUSION

The molecular cloning of the muscle nicotinic ACh receptor has led to the elucidation of the primary structure of this receptor in numerous species including Torpedo, chick, mouse, calf, and man. Knowledge of the primary structure has led for the first time to testable models relating structure and function. These models are being tested using classical biochemical and electrophysiological techniques as well as the new approach of site-specific mutagenesis.

The availability of cDNA clones makes it possible to study the regulation of the receptor genes in muscle and to answer some long-standing questions about the control of synthesis and the relationship between the receptors found at the synapse, and extrasynaptic receptors synthe-

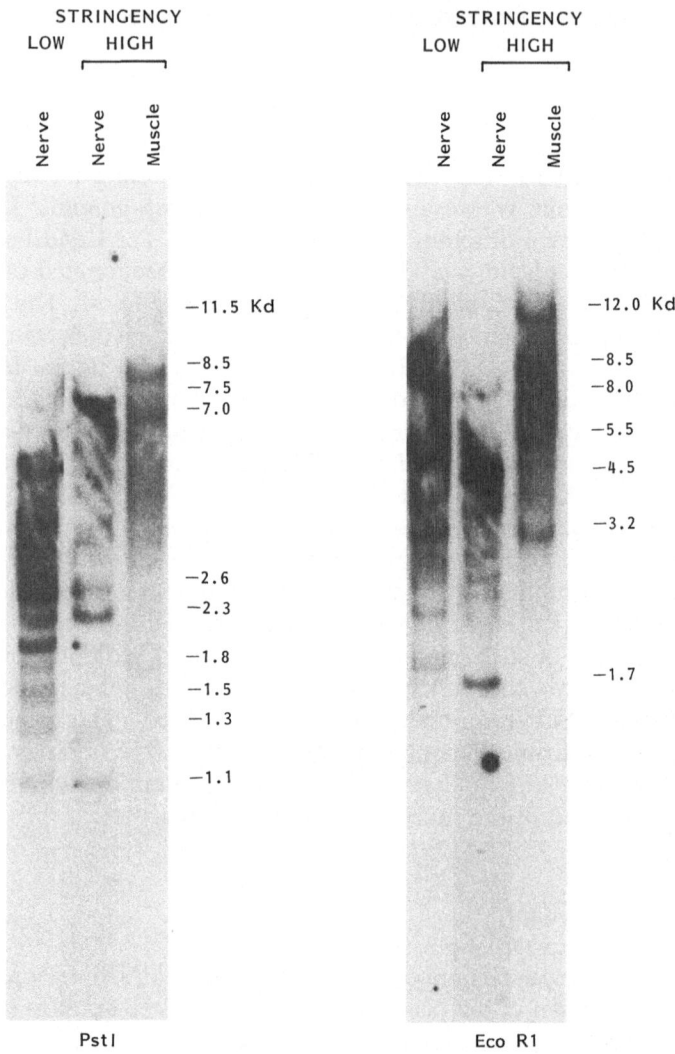

Figure 24. Mouse genomic restriction fragments hybridizing to PCA48 or pMARα15. Genomic DNA was purified from liver of adult female C3H mice and digested with EcoRl or Pst l. The digestion products were resolved on 1.4% agarose gels, transferred to nitrocellulose, and hybridized to probes prepared by nick translation of PCA48 or pMARα15. Hybridization was carried out in 3 × SSC, 1 × Denhardt's, and 100 μg/ml carrier DNA at 65°C. The filters were washed at 65°C in 5 × SSC (low stringency) or 0.1 × SSC (high stringency). The size of the hybridizing fragments was determined from the sizes of known restriction fragments. PCA48 codes for the neuronal α-subunit. pMARα15 codes for the skeletal muscle α-subunit.

sized after denervation. The isolation of the genes and analysis of their regulatory elements could lead to an understanding of how electrical activity regulates gene expression in the nervous system.

The recent isolation of cDNA clones coding for receptorlike polypeptides expressed in the brain brings the full power of molecular biology to the study of brain receptors for the first time. The diversity of the structure and function of the ACh receptors in the central nervous system can now be studied through the use of molecular cloning.

ACKNOWLEDGMENTS

This work was supported by grants from the Amoco Foundation, the Weingart Foundation, the Army Medical Research, and Development Command, contract DAMD17-85-C-5198, the National Institutes of Health (NIH), and the Muscular Dystrophy Association. Daniel Goldman and Paul Gardner are supported by NIH postdoctoral fellowships. Walter Luyten is supported by a Ruth and Louis Wolfe Fellowship in Myasthenia Gravis Research.

REFERENCES

Anderson, D. J., and Blobel, G., 1981, *In vitro* synthesis, glycosylation, and membrane insertion of the four subunits of *Torpedo* acetylcholine receptor, *Proc. Natl. Acad. Sci. U.S.A.* **78**:5598–5602.

Anderson, D. J., Walter, P., and Blobel, G. 1982, Signal recognition protein is required for the integration of acetylcholine receptor delta subunit, a transmembrane glycoprotein, into the endoplasmic reticulum membrane, *J. Cell Biol.* **93**:501–506.

Ballivet, M., Patrick, J., Lee, J., and Heinemann, S., 1982, Molecular cloning of cDNA coding for the gamma subunit of *Torpedo* acetylcholine receptor, *Proc. Natl. Acad. Sci. U.S.A.* **79**:4466–4470.

Ballivet, M., Nef, P., Stalder, R., and Fulpius, B., 1983, Genomic sequences encoding the alpha-subunit of acetylcholine receptor are conserved in evolution, *Cold Spring Harbor Symp. Quant. Biol.* **48**:83–87.

Barnard, E. A., Miledi, R., and Sumikawa, K., 1982, Translation of exogenous messenger RNA coding for nicotinic acetylcholine receptors produces functional receptors in *Xenopus* oocytes, *Proc. R. Soc. Lond. B* **215**:241–246.

Berg, D. K., and Hall, Z. W., 1974, Fate of alpha-bungarotoxin bound to acetylcholine receptors of normal and denervated muscle, *Science* **184**:473–474.

Berg, D. K., and Hall, Z. W., 1975, Loss of alpha-bungarotoxin from junctional and extrajunctional acetylcholine receptor in rat diaphragm muscle *in vivo* and in organ culture, *J. Physiol. (Lond.)* **252**:771–789.

Bevan, S., and Steinbach, J. H., 1977, The distribution of alpha-bungarotoxin binding sites on mammalian skeletal muscle developing *in vivo*, *J. Physiol. (Lond.)* **267**:195–213.

Boulter, J., and Patrick, J., 1977, Purification of an acetylcholine receptor from a nonfusing muscle cell line, *Biochemistry* **16**:4900–4908.

Boulter, J., Luyten, W., Evans, K., Mason, P., Ballivet, M., Goldman, D., Stengelin, S., Martin, G., Heinemann, S., and Patrick, J., 1985, Isolation of a clone coding for the alpha-subunit of a mouse acetylcholine receptor, *J.Neurosci.* **5**:2545–2552.

Boulter, J., Evans, K., Mason, P., Martin, G., Heinemann, S., and Patrick, J., 1986a, Isolation of a clone coding for the precursors to the beta-subunit of mouse muscle nicotinic acetylcholine receptor (manuscript in preparation).

Boulter, J., Evans, K., Martin, G., Mason, P., Stengelin, S., Goldman, D., Heinemann, S., and Patrick, J., 1986b, Isolation and sequence of cDNA clones coding for the precursor to the γ-subunit of mouse muscle nicotinic acetylcholine receptor, *J. Neurosci. Res.* **16**: 37–49.

Boulter, J., Evans, K., Goldman, D., Martin, G., Treco, D., Heinemann, S., and Patrick, J., 1986c, Isolation of a cDNA clone coding for a possible neural nicotinic acetylcholine receptor alpha-subunit, *Nature (Lond.)* **319**:368–374.

Brisson, A., and Unwin, P. N. T., 1985, Quaternary structure of the acetylcholine receptor, *Nature (Lond.)* **315**:474–477.

Brockes, J. P., and Hall, Z. W., 1975a, Acetylcholine receptors in normal and denervated rat diaphragm muscle. II. Comparison of junctional and extrajunctional receptors, *Biochemistry* **14**:2100–2106.

Brockes, J. P., and Hall, Z. W., 1975b, Synthesis of acetylcholine receptor by denervated rat diaphragm muscle, *Proc. Natl. Acad. Sci. U.S.A.* **72**:1368–1372.

Brown, J. R., 1976, Structural origins of mammalian albumin, *Fed. Proc.* **32**:2141–2144.

Chang, C. C., and Huang, M. C., 1975, Turnover of junctional and extrajunctional acetylcholine receptors of the rat diaphragm, *Nature (Lond.)* **253**:643–644.

Clarke, P. B. S., Schwartz, R. D., Paul, S. M., Pert, C. B., and Pert, A., 1985, Nicotinic binding in rat brain: Autoradiographic comparison of [^{3}H]acetylcholine, [^{3}H] nicotine, and [^{125}I]-alpha-bungarotoxin, *J. Neurosci.* **5**:1307–1315.

Claudio, T., Ballivet, M., Patrick, J., and Heinemann, S., 1983, Nucleotide and deduced amino acid sequences of *Torpedo californica* acetylcholine receptor gamma-subunit, *Proc. Natl. Acad. Sci. U.S.A.* **80**:1111–1115.

Cohen, S. A., and Fischbach, G. D., 1973, Regulation of muscle acetylcholine sensitivity by muscle activity in cell culture, *Science* **181**:76–78.

Conti-Tronconi, B. M., and Raftery, M. A., 1982, The nicotinic cholinergic receptor: Correlation of molecular structure with functional properties, *Ann. Rev. Biochem.* **51**:491–530.

Cox, K., DeLeon, D., Angerer, L., and Angerer, R., 1984, Detection of mRNAs in sea urchin embryos by *in situ* hybridization using asymmetric RNA probes, *Dev. Biol.* **101**:485–502.

Criado, M., Hochschwender, S., Virender, S., Fox, J. L., and Lindstrom, J., 1985, Evidence for unpredicted transmembrane domains in acetylcholine receptor subunits, *Proc. Natl. Acad. Sci. U.S.A.* **82**:2004–2008.

Devillers-Thiery, A., Giraudat, J., Bentaboulet, M., and Changeux, J.-P., 1983, Complete mRNA coding sequence of the acetylcholine binding α-subunit of *Torpedo marmorata* acetylcholine receptor: A model for the transmembrane organization of the polypeptide chain, *Proc. Natl. Acad. Sci. U.S.A.* **80**:2067–2071.

Fayhey, R. C., Hunt, J. S., and Windham, G. C., 1977, On the cysteine and cystine content of proteins. Differences between intracellular and extracellular proteins, *J. Mol. Evol.* **10**:155–160.

Finer-Moore, J., and Stroud, R. M., 1984, Amphipathic analysis and possible conformation of the ion channel in an acetylcholine receptor, *Proc. Natl. Acad. Sci. U.S.A.* **81**:155–159.

Goldman, D., Boulter, J., Heinemann, S., and Patrick, J., 1985, Muscle denervation increases the levels of two mRNAs coding for the acetylcholine receptor alpha-subunit, *J. Neurosci.* **5**:2553–2558.

Goldman, D., Simmons, D., Swanson, L., Patrick, J., and Heinemann, S., 1986, Mapping brain areas expressing RNA homologous to two different acetylcholine receptor alpha subunit cDNAs, *Proc. Natl. Acad. Sci. U.S.A.* **83**:4076–4080.

Gordon, A., Davis, C., Milfay, D., and Diamond, I., 1977, Phosphorylation of acetylcholine receptor by endogenous membrane protein kinase in receptor-enriched membranes of *Torpedo californica*, *Nature (Lond.)* **267**:539–540.

Guy, H. R., 1984, A structural model of the acetylcholine receptor channel based on partition energy and helix packing calculations. *Biophys. J.* **45**:249–261.

Hall, Z. W., and Reiness, C. G., 1977, Electrical stimulation of denervated muscles reduces incorporation of methionine into the ACh receptor, *Nature (Lond.)* **268**:655–657.

Hall, Z. W., Roisin, M. P., Gu, Y., and Gorin, P. D., 1983, A developmental change in the immunological properties of acetylcholine receptors at the rat neuromuscular junction, *Cold Spring Harbor Symp. Quant. Biol.* **48**:101–108.

Heinemann, S., Merlie, J., and Lindstrom, J., 1978, Modulation of acetylcholine receptor in rat diaphragm by anti-receptor sera, *Nature (Lond.)* **274**:65–68.

Huganir, R. L., Miles, K., and Greengard, P., 1984, Phosphorylation of the nicotinic acetylcholine receptor by an endogenous tyrosine-specific protein kinase, *Biochemistry* **81**:6968–6972.

Jackson, R. C., and Blobel, G., 1977, Post-translational cleavage of presecretory proteins with an extract of rough microsomes from dog pancreas containing signal peptidase activity, *Proc. Natl. Acad. Sci. U.S.A.* **74**:5598–5602.

Kao, P. N., and Karlin, A., 1986, Acetylcholine receptor binding site contains a disulfide crosslink between adjacent half-cystinyl residues, *J. Biol. Chem.* **261**:8085–8088.

Kao, P. N., Dwork, A. J., Kaldany, R. J., Silver, M. L., Wideman, J., Stein, S., and Karlin, A., 1984, Identification of two alpha-subunit half-cystines specifically labeled by an affinity reagent for the acetylcholine binding site, *J. Biol. Chem.* **259**:1162–1165.

Karlin, A., DiPaola, M., Kao, P. N., and Lobel, P., 1986, Functional sites and transient states of the nicotinic acetylcholine receptor, in: *Proteins of Excitable Membrane* (B. Hille and D. M. Fambrough, eds.), in press, Wiley, New York.

Karlin, A., Weill, C. L., McNamee, M. G., and Valderrama, R., 1976, Facets of the structures of acetylcholine receptors from *Electrophorus* and *Torpedo*, *Cold Spring Harbor Symp. Quant. Biol.* **40**:203–210.

Klarsfeld, A., and Changeux, J-P., 1985, Activity regulates the levels of acetylcholine receptor alpha-subunit mRNA in cultured chicken myotubes, *Proc. Natl. Acad. Sci. U.S.A.* **82**:4558–4562.

Klymkowsky, M. W., and Stroud, R. M., 1979, Immunospecific identification and three-dimensional structure of a membrane-bound acetylcholine receptor from *Torpedo californica*, *J. Mol. Biol.* **128**:319–334.

Kyte, J., and Doolittle, R. F., 1982, A simple method for displaying the hydropathic character of a protein, *J. Mol. Biol.* **157**:105–132.

La Polla, R. J., Mixter-Mayne, K., and Davidson, N., 1984, Isolation and characterization of a cDNA clone for the complete protein coding region of the delta-subunit of the mouse acetylcholine receptor, *Proc. Natl. Acad. Sci. U.S.A.* **81**:7970–7974.

La Rochelle, W. J., Wray, B. E., Sealock, R., and Froehner, S. C., 1985, Immunochemical demonstration that amino acids 360–377 of the acetylcholine receptor gamma-subunit are cytoplasmic, *J. Cell Biol.* **100**:684–691.

Legase, L., Chandra, T., Woo, S. L. C., and Means, A. R., 1983, Identification of multiple species of calmodulin messenger RNA using a full length complementary DNA, *J. Biol. Chem.* **258**:1684–1688.

Lomo, T., and Westgaard, R. H., 1975, Further studies on the control of ACh sensitivity by muscle activity in the rat. *J. Physiol. (Lond.)* **252**:603–626.

Luyten, W., Kellaris, K., Kyte, J., Heinemann, S., and Patrick, J., 1983, A model for the acetylcholine binding site of the acetylcholine receptor, *Neurosci. Abst.* **10**:734.

McCarthy, M. P., Earnest, J. P., Young, E. F., Choe, S., and Stroud, R. M., 1986, The molecular neurobiology of the acetylcholine receptor. *Annu. Rev. Neurosci.* **9**:383–413.

Merlie, J. P., Isenberg, K. E., Russell, S. D., and Sanes, J. R., 1984, Denervation supersensitivity in skeletal muscle: Analysis with a cloned cDNA probe, *J. Cell Biol.* **99**:332–335.

Miledi, R., 1960a, The acetylcholine sensitivity of frog muscle fibers after complete or partial denervation, *J. Physiol. (Lond.)* **151**:1–23.

Miledi, R., 1960b, Junctional and extrajunctional receptors in skeletal muscle fibers, *J. Physiol. (Lond.)* **151**:24–30.

Miledi, R., Parker, I., and Sumikawa, K., 1982, Synthesis of chick brain GABA receptors by frog oocytes, *Proc. R. Soc. (Lond.)* **216**:509–515.

Mishina, M., Kurosaki, T., Tobimatsu, T., Morimoto, Y., Noda, M., Yamamoto, T., Terao, M., Lindstrom, J., Takahashi, T., Kuno, M., and Numa, S., 1984, Expression of functional acetylcholine receptor from cloned cDNAs, *Nature (Lond.)* **307**:604–608.

Nef, P., Mauron, A., Stalder, R., Alliod, C., and Ballivet, M., 1984, Structure, linkage and sequence of the two genes encoding the delta and gamma subunits of the nicotinic acetylcholine receptor, *Proc. Natl. Acad. Sci. U.S.A.* **81**:7975–7979.

Neher, E., and Sakmann, B., 1976, Noise analysis of drug induced voltage clamp currents in denervated frog muscle fibers, *J. Physiol. (Lond.)* **258**:705–729.

Noda, M., Takahashi, H., Tanabe, T., Toyosato, M., Furutani, Y., Hirose, T., Asai, M., Inayama, S., Miyata, T., and Numa, S., 1982, Primary structure of alpha-subunit precursor of *Torpedo californica* acetylcholine receptor deduced from cDNA sequence, *Nature (Lond.)* **299**:793–797.

Noda, M., Takahashi, H., Tanabe, T., Toyosato, M., Kikyotani, S., Hirose, T., Asai, M., Takashima, H., Inayama, S., Miyata, T., and Numa, S. 1983, Primary structures of beta- and delta-subunit precursors of *Torpedo californica* acetylcholine receptor deduced from cDNA sequences, *Nature (Lond.)* **301**:251–255.

Okayama, H., and Berg, P., 1982, High efficiency cloning of full length cDNA, *Mol. Cell. Biol.* **2**:161–170.

Parnes, J. R., Robinson, R. R., and Seidman, J. B., 1983, Multiple mRNA species with distinct 3' termini are transcribed from the $beta_2$-microglobulin gene, *Nature (Lond.)* **302**:449–452.

Patrick, J., and Stallcup, W., 1977a, Immunological distinction between acetylcholine receptor and the alpha-bungarotoxin-binding component on sympathetic neurons, *Proc. Natl. Acad. Sci. U.S.A.* **74**:4689–4692.

Patrick, J., and Stallcup, W., 1977b, Alpha-bungarotoxin binding and cholinergic receptor function on a rat sympathetic nerve line, *J. Biol. Chem.* **252**:8629–8633.

Patrick, J., Heinemann, S. F., Lindstrom, J., Schubert, D., and Steinback, J. H., 1972, Appearance of acetylcholine receptors during differentiation of a myogenic cell line, *Proc. Natl. Acad. Sci., USA* **69**:2762–2766.

Patrick, J., Ballivet, M., Boas, L., Claudio, T., Forrest, J., Ingraham, H., Mason, P., Stengelin, S., Ueno, S., and Heinemann, S., 1983, Molecular cloning of the acetylcholine receptor, *Cold Spring Harbor Symp. Quant. Biol.* **48**:71–79.

Popot, J.-L., and Changeux, J-P., 1984, The nicotinic receptor of acetylcholine: Structure of an oligomeric integral membrane protein, *Physiol. Rev.* **64**:1162–1239.

Porter, S., and Froehner, S., 1983, Characterization and localization of the M_r = 43,000 proteins associated with acetylcholine receptor-rich membranes, *J. Biol. Chem.* **258**:10034–10040.

Porter, S., and Froehner, S., 1985, Interaction of the 43K protein with components of *Torpedo* postsynaptic membranes, *Biochemistry* **24**:425–432.

Raftery, M. A., Hunkapiller, M. W., Strader, C. D., and Hood, L. E., 1980, Acetylcholine receptor: Complex of homologous subunits, *Science* **208**:1454–1457.

Ratnam, M., and Lindstrom, J., 1984, Structural features of the nicotinic acetylcholine receptor revealed by antibodies to synthetic peptides, *Biochem. Biophys. Res. Commun.* **122**:1225–1233.

Ross, M. J., Klymkowsky, M. W., Agard, D. A., and Stroud, R. M., 1977, Structural studies of a membrane-bound acetylcholine receptor from *Torpedo california*, *J. Mol. Biol.* **116**:635–659.

Sakmann, B., 1978, Acetylcholine-induced ionic channels in rat skeletal muscle, *Fed. Proc.* **37**:2654–2659.

Schiffer, M., and Edmundson, A. B., 1967, Use of helical wheels to represent the structures of proteins and to identify segments with helical potentials, *Biophys. J.* **7**:121–135.

Schubert, D., Harris, D. J., Devine, C., and Heinemann, S., 1974, Characterization of a unique muscle cell line, *J. Cell Biol.* **61**:398–413.

Schulz, G. E., and Schirmer, R. H., 1979, *Principles of Protein Structure*, Springer-Verlag, New York.

Shibahara, S., Kubo, T., Perski, H. J., Takahashi, H., Noda, M., and Numa, S., 1985, Cloning and sequence analysis of human genomic DNA encoding gamma subunit precursor of muscle acetylcholine receptor, *Eur. J. Biochem.* **146**:15–22.

Sine, S., and Taylor, P., 1980, The relationship between agonist occupation and the permeability response of the cholinergic receptor revealed by bound cobra alpha-toxin, *J. Biol. Chem.* **255**:10144–10156.

Sine, S. M., and Steinbach, J. H., 1984*a*, Activation of a nicotinic acetylcholine receptor, *Biophys. J.* **45**:175–185.

Sine, S. M., and Steinbach,J. H., 1984*b*, Agonists block currents through acetylcholine receptor channels, *Biophys. J.* **46**:277–284.

Sine, S. M., and Steinbach, J.H., 1985, Activation of acetylcholine receptors on clonal mammalian BC_3H-l cells by low concentrations of agonist, *J. Physiol. (Lond.)* **358**:91–108.

Sine, S. M., and Steinbach, J. H., 1986, Acetylcholine receptor activation by a site-selective ligand: Nature of brief open and closed states in BC_3H-l cells, *J. Physiol.* **370**:357–379.

Sobel, A., Heidmann, T., Hofler, J., and Changeux, J-P., 1978, Distinct protein components from *Torpedo marmorata* membranes carry the acetylcholine receptor site and the binding site for local anesthetics and histrionicotoxin, *Proc. Natl. Acad. Sci. U.S.A.* **75**:510–514.

Steinbach, J. H., 1981, Developmental changes in acetylcholine receptor aggregates at rat skeletal neuromuscular junctions, *Dev. Biol.* **84**:267–276.

Steinbach, J. H., Merlie, J., Heinemann, S., and Bloch, R., 1979, Degradation of junctional and extrajunctional acetylcholine receptors by developing rat skeletal muscle, *Proc. Natl. Acad. Sci. U.S.A.* **76**:3547–3551.

Stroud, R. M., and Finer-Moore, J., 1985, Acetylcholine receptor structure, function and evolution, *Annu. Rev. Cell Biol.* **1**:369–401.

Swanson, L. W., Sawchenko, P. E., Rivier, J., and Vale, W. W., 1983, Organization of ovine cortiocoptropin-releasing factor immunoreactive cells and fibers in the rat brain: An immunohistochemical study, *Neuroendocrinology* **36**:165–186.

Takai, T., Noda, M., Furutani, Y., Takahashi, H., Notake, M., Shimizu, S., Kayano, T., Tanabe, T., Tanaka, K., Hirose, T., Inayama, S., and Numa, S., 1984, Primary structure of gamma subunit of calf-muscle acetylcholine receptor deduced from the cDNA sequence, *Eur. J. Biochem.* **143:**109–115.

Tanabe, T., Noda, M., Furutani, Y., Takai, T., Takahashi, H., Tanaka, K-I., Hirose, T., Inayama, S., and Numa, S., 1984, Primary structure of beta subunit precursor of calf muscle acetylcholine receptor deduced from cDNA sequence, *Eur. J. Biochem.* **144:**11–17.

Teichberg, V., Sobel, A., and Changeux, J.-P., 1977, *In vitro* phosphorylation of the acetylcholine receptor, *Nature (Lond.)* **267:**540–542.

Thornton, J. M., 1981, Disulphide bridges in globular proteins, *J. Mol. Biol.* **151:**261–287.

Wennogle, L. P., and Changeux, J.-P., 1980, Transmembrane orientation of proteins present in acetylcholine receptor-rich membranes from *Torpedo marmorata* studied by selective proteolysis, *Eur. J. Biochem.* **106:**381–393.

Wennogle, L. P., Oswald, R., Saitoh, T., and Changeux, J-P., 1981, Dissection of the 66000-Dalton subunit of the acetylcholine receptor, *Biochemistry* **20:**2492–2497.

White, M. M., Mayne, K. M., Lester, H. A., and Davidson, N., 1985, Mouse–*Torpedo* hybrid acetylcholine receptors: Functional homology does not equal sequence homology, *Proc. Natl. Acad. Sci. U.S.A.* **82:**4852–4856.

Wilson, P. T., Lentz, T. L., and Hawrot, E., 1985, Determination of the primary amino acid sequence specifying the alpha-bungarotoxin binding site on the alpha-subunit of the acetylcholine receptor for *Torpedo californica, Proc. Natl. Acad. Sci. U.S.A.* **82:**8790–8794.

Wise, D. S., Schoenborn, B. P., and Karlin, A., 1981, Structure of acetylcholine receptor dimer determined by neutron scattering and electron microscopy, *J. Biol. Chem.* **256:**4124–4126.

Young, E. F., Ralston, E., Blake, J., Ramachandran, J., Hall, Z. W., and Stroud, R. M., 1984, Topological mapping of the acetylcholine receptor: Evidence for a model with five transmembrane segments and a cytoplasmic C-terminal peptide, *Proc. Natl. Acad. Sci. U.S.A.* **82:**622–630.

Zehner, Z. E., and Paterson, B. M., 1983, Vimentin gene expression during myogenesis: Two functional transcripts from a single copy gene, *Nucl. Acid Res.* **23:**8317–8332.

4

Molecular Biology of Muscle Development

The Myosin Gene Family of *Caenorhabditis elegans*

JONATHAN KARN, NICK J. DIBB, DAVID M. MILLER, and E. JANE MITCHELL

1. INTRODUCTION

The small soil nematode *Caenorhabditis elegans* is an attractive organism in which to apply biochemical and genetic approaches to the study of muscle development and function. Only two major muscles are present in the organism—the pharyngeal muscle and the body wall muscle—and the anatomy and lineage of all the muscle cells in the organism throughout development is known (Sulston *et al.*, 1983; Sulston and Horvitz, 1977). Because a large fraction of the nematode tissue mass is muscle, the major contractile proteins can be isolated in milligram quantities from a few liters of nematode culture (Epstein *et al.*, 1974; Waterston *et al.*, 1977; Harris and Epstein, 1977; MacLeod *et al.*, 1977a,b; Zengel and Epstein, 1980c). *Caenorhabditis elegans* exhibits a characteristic swimming pattern on the surface of agar plates. Genetic analysis (Brenner, 1974) defined mutations in more than 100 different genes that produce

JONATHAN KARN, NICK J. DIBB, DAVID M. MILLER, and E. JANE MITCHELL • MRC Laboratory of Molecular Biology, University Postgraduate Medical School, Cambridge CB2 2QH, England.

animals with defective motility (the uncoordinated or *unc* phenotype). Mutations in 25 of these *unc* genes produce gross abnormalities in muscle ultrastructure (Waterston *et al.*, 1980; Zengel and Epstein, 1980b).

This chapter reviews our studies of the myosin heavy-chain genes of the nematode. Much of our work is based on studies of mutants of the *unc*-54 gene. The *unc*-54 gene encodes the major myosin heavy chain present in the body wall musculature (Epstein *et al.*, 1974; MacLeod *et al.*, 1977a,b, 1979, 1981; Waterston *et al.*, 1982a,b; Karn *et al.*, 1982, 1983a). Mutations of the *unc*-54 gene can paralyze the animal but are not lethal, as the pharyngeal muscle is unaffected (Brenner, 1974; Epstein *et al.*, 1974). Mutations in the *unc*-54 gene generally result in disorganized muscle with a reduced complement of thick filaments. This assembly defect occurs in mutants that accumulate defective myosin heavy chain as well as mutants that do not produce any detectable *unc*-54 product (MacLeod *et al.*, 1977b; Waterston *et al.*, 1982a,b). A few *unc*-54 mutants show relatively normal muscle structure but have impaired movement (Moerman *et al.*, 1982). This extensive genetic information provides a basis for defining amino acid residues in the myosin heavy chain that are of importance to thick-filament assembly and tension generation.

Biochemical and immunological studies have shown that three other myosin heavy chains in addition to *unc*-54 are expressed in *Caenorhabditis elegans* (Epstein *et al.*, 1974; 1982a,b; MacLeod *et al.*, 1977b; Miller *et al.*, 1983, 1986; Schachat *et al.*, 1977, 1978). One myosin heavy-chain isoform (MHC A) is expressed in the body wall musculature coordinately with the *unc*-54 heavy chain (MHC B) (Garcea *et al.*, 1978; MacKenzie *et al.*, 1978b). Decoration of thick filaments with monoclonal antibodies has indicated that MHC A is exclusively incorporated within a central segment of the body wall thick filaments, while MHC B forms the remaining distal portions of the filament (Miller *et al.*, 1983). The other two myosin isoforms (MHC C and D) are expressed exclusively in the pharynx (Epstein *et al.*, 1974, 1982a; Waterston *et al.*, 1982b; Waterston, personal communication).

This chapter describes the genetics of the *unc*-54 gene, as well as the immunological identification and localization of the myosin isoforms. The molecular cloning of the four genes in the nematode sarcomeric myosin heavy-chain gene family is then described, and a structural interpretation of the deduced heavy-chain protein sequences presented. Finally, the way in which sequence alterations in a representative set of *unc*-54 alleles account for these mutant phenotypes and preliminary results of *in vitro* mutagenesis studies of myosin fragments expressed in *Escherichia coli* are discussed. These studies are contributing to an understanding of the molecular basis for differential gene expression during muscle development, the assembly of contractile proteins into filament lattices, the functions of contractile protein isoforms in different muscle tissues, and structure–function relationships in the myosin molecule.

2. GENETICS OF THE *UNC*-54 LOCUS

Strains harboring homozygous *unc*-54 mutations are paralyzed due to defects in body wall muscles. The movement of *unc*-54 mutants is slow and uncoordinated. The muscle disorganization is readily seen with polarized light and electron microscopy (Epstein *et al.*, 1974; MacLeod *et al.*, 1979). In *unc*-54 strains, the characteristic birefringence associated with ordered arrays of thick filaments in the wild-type body wall muscle is dispersed throughout the cell. Figure 1 compares the distribution of thick filaments in cross sections of muscle from the wild-type strain (N2) and a strain homozygous for the *unc*-54 allele *e190*. In addition to the disruption of the highly ordered lattice, it is also evident in these images that the number of thick filaments in the *e190* sarcomere is significantly reduced compared with the wild type. This aspect of the mutant phenotype has been correlated with the absence of *unc*-54 heavy chains in *e190* (MacLeod *et al.* 1977*b*; 1979).

The *unc*-54 mutations have been assigned to four classes on the basis of dominance behavior, muscle ultrastructure, and protein chemistry (MacLeod *et al.*, 1977*b*; Waterston *et al.*, 1982*b*). The mutants are defective for either synthesis of MHC B (class I) or assembly of MHC B into thick filaments (classes II and III), or they produce MHC B that assembles normally but results in a muscle with impaired movement (class IV).

Most *unc*-54 mutations are recessive (class 1). Strains homozygous for these mutations (e.g., *e190*) are paralyzed, have disorganized muscle with greatly reduced numbers of thick filaments, and do not accumulate detectable amounts of the MHC B (Epstein *et al.*, 1974; MacLeod *et al.*, 1977*b*; Waterston *et al.*, 1982*b*). These mutations are usually nonsense mutations (see Section 8).

A second class of mutations is represented by two *unc*-54 alleles, *e675* and *s291*, which produce normal levels of shortened myosin heavy chains lacking internal segments of the rod (MacLeod *et al.*, 1977*a*; Dibb *et al.*, 1985). Animals homozygous for these mutations are, like the class I mutations, paralyzed with disorganized muscle and reduced numbers of thick filaments. However these animals exhibit a subtle twitching of the body wall muscles, not seen in other *unc*-54 strains.

The third class of mutants is dominant and produces normal levels of myosin (MacLeod *et al.*, 1977*b*). More than 30 mutations of this class have now been isolated (R. H. Waterston and D. H. Moerman, personal

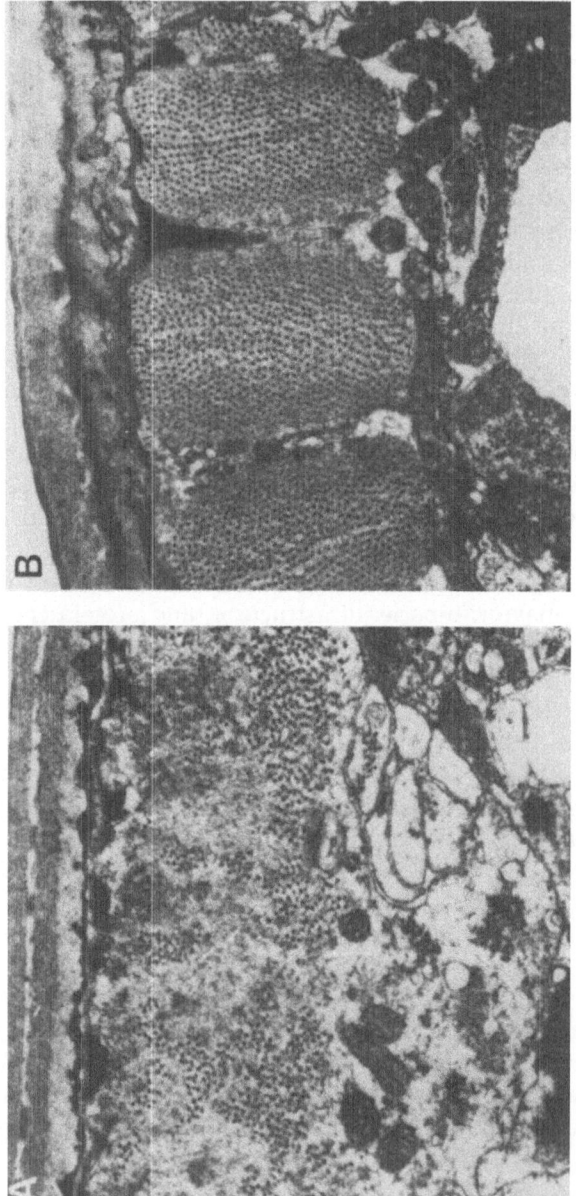

Figure 1. Electron micrographs of cross sections of the body wall muscle of adult *Caenorhabditis elegans*. ×7500. (A) Strain homozygous for *e*190. (B) Wild-type (N2). Note that the regular lattice of thick and thin filaments seen in the wild-type muscle is absent as a result of the *e*190 mutation, and that there is a drastic reduction in the thick filament number. (After MacLeod *et al.*, 1977*b*.)

communication). Animals heterozygous for these mutations show a mild reduction in the number of thick filaments. These filaments fail to assemble into ordered A bands. When homozygous, the effects of these mutations are more pronounced. The phenotypes of some alleles resemble those of class 1 recessive homozygotes, but others are homozygous lethal and arrest in the early larval stages.

Moerman *et al.* (1982) described a fourth class of *unc*-54 mutants that have nearly normal levels of *unc*-54 myosin synthesis and well-ordered thick-filament lattice structures. These mutants were selected as suppressors of the "twitching" phenotype of *unc*-22 mutants (see Sections 2.5 and 8). These strains exhibit slow movement due to sequence alterations in the myosin head (Dibb *et al.*, 1985; R. H. Waterston, personal communication).

The phenotypes and sequence alterations of 19 representative *unc*-54 mutations are described in Section 8.

2.1. Selection of unc-54 Mutations

Anderson and Brenner (1984) developed a rapid screening technique for isolation of recessive alleles of *unc*-54. Paralyzed heterozygotes were prepared with the dominant allele *e*1152 (*e*1152/+). Subsequent mutations that eliminated the dominant character in *e*1152 were easily detected because these animals were fully motile. A balanced-lethal system was required for this selection. An X-ray-induced balancer chromosome was isolated (*let*-209) containing a wild-type copy of *unc*-54. Animals homozygous for *let*-209 die in early larval stages. In the heterozygous strain *unc*-54(*e*1152) +/+ *let*-209, virtually all the viable progeny are paralyzed. The strain segregates one-third of its progeny as paralyzed *e*1152/*e*1152 homozygotes and two-thirds as paralyzed *e*1152 +/+ *let*-209 heterozygotes. Wild-type progeny were observed at less than 7×10^{-5}. However, after mutagenesis, wild-type progeny arise at a frequency of 8×10^{-4}. Approximately 30% of these wild types were shown to be heterozygous for recessive *unc*-54 alleles because they segregate *unc*-54 paralyzed homozygotes. The remainder did not contain new *unc*-54 mutations and presumably resulted from the low level of recombination permitted by the balancing system.

2.2. Deletions and Duplications of the unc-54 Locus

The screening technique of Anderson and Brenner (1984) was also used by them to prepare a series of deletions surrounding the *unc*-54 gene. A deletion of the *unc*-54 gene plus a nearby essential gene is lethal when homozygous, yet strains heterozygous for such deletions are fully motile. Deletions of this type were obtained following mutagenesis of *e*1152 +/+ *let*-209 with DEO (1,2,7,8,-diepoxyoctane). Eighteen percent of the DEO-induced *unc*-54 mutations were deletions, while the remainder were point mutations.

Anderson and Brenner (1984) also noted genetically unstable mutants whose phenotypes were intermediate between paralyzed *e*1152 heterozygotes and the phenotypically wild-type revertants. These strains contain free extrachromosomal duplications including the wild-type *unc*-54 gene. These results suggested that the dominance properties of mutant polypeptide in *e*1152 could be alleviated by increasing the level of wild-type gene expression.

As shown in Fig. 2, the *unc*-54 gene maps to the extreme right-hand end of the linkage group I. In Fig. 3, the approximate end points of the DEO-induced deletions are mapped against a collection of EMS-induced recessive mutations tightly linked to *unc*-54. Complementation experiments have shown that the deletions fall into three classes. Members of class I (*eDf*5, *eDf*10) delete genes to the right of *unc*-54 and are viable when heterozygous with members of class II (*eDf*3, *eDf*7, *eDf*11, *eDf*12, *eDf*15, *eDf*16). Class III deletions (*eDf*13, *eDf*14, *eDf*4, *eDf*9, *eDf*6) are not viable when heterozygous with all other deletion mutations and appear to delete material on both sides of the *unc*-54 gene.

2.3. Orientation of the unc-54 Gene

Waterston *et al.* (1982*a*) constructed a fine-structure map of the *unc*-54 gene (see Section 8.2) using a levamisole resistance gene *lev*-11 (Lewis *et al.*, 1980*a*,*b*) and a homozygous lethal *let*-50 as outside markers. The outer *unc*-54 alleles *e*1300 and *st*60 map at each end of the gene (see Section 8, Fig. 24). It is now known that *e*1300 is an amber mutation (Wills *et al.*, 1983; Dibb *et al.*, 1985) 60 residues from the carboxy-terminal of the myosin heavy chain (see Section 8.3). It follows that the 3' end of the *unc*-54 gene is closest to *lev*-11 and that the 5' end of the *unc*-54 gene is closest to *let*-50. This orientation can also be deduced from the properties of the deletion mutations isolated by Anderson and Brenner (1984). The deletion break points of *eDf*5 and *eDf*10, which extend from *unc*-54 to the *let*-50 side of the gene have been located in the 5' end of the *unc*-54 gene by Southern transfer-hybridization experiments (P. Anderson, unpublished observations).

2.4. In Situ Hybridization to Chromosomes

Albertson (1984, 1985) located nematode genes on chromosomal squashes using *in situ* hybridization with cloned probes. Her results

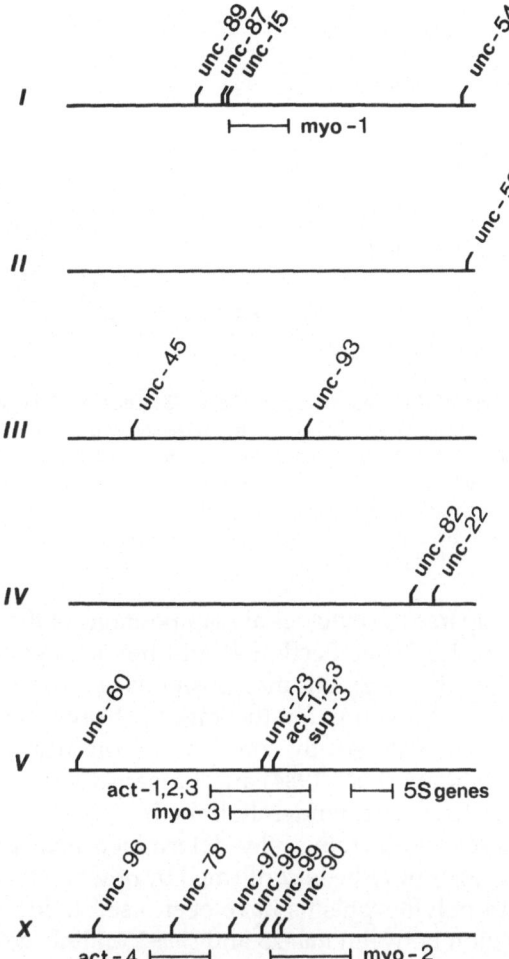

Figure 2. Genetic map of *Caenorhabditis elegans* genes that affect muscle structure. Genetic loci that have been identified by mutations that disrupt muscle structure are shown (Waterston *et al.*, 1980; Zengel and Epstein, 1980c). The approximate positions of the actin, *myo*-1, *myo*-2, and *myo*-3 genes, as determined by *in situ* hybridization, are plotted below the line for each linkage group. (After Alberston, 1985.)

demonstrate that the *let*-209 balancer chromosome (Section 2.2) has a chromosomal break point within the ribosomal DNA cluster of nematode, placing the rDNA genes (Files and Hirsh, 1981; Ellis *et al.*, 1986) on the extreme tip of chromosome I near the *unc*-54 gene. Hybridization to cloned myosin gene probles also permits placement of the *unc*-54 gene on the tip of chromosome I.

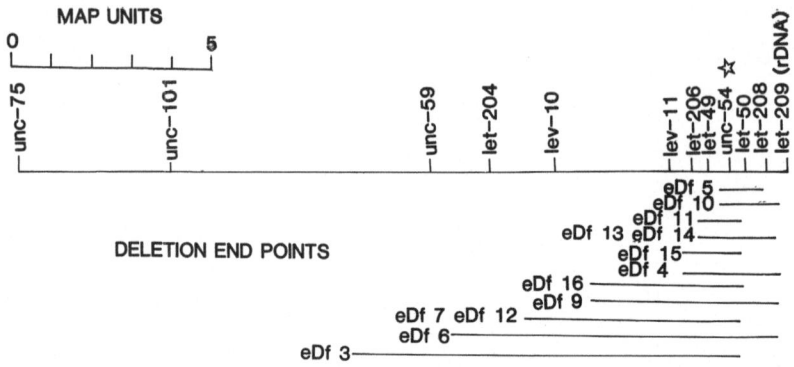

Figure 3. Genetic map of the region around the *unc*-54 gene on linkage group I. The set of overlapping DEO-induced deletions isolated by Anderson and Brenner (1984) (*eDf3–eDf16*, bars) are aligned with a set of lethal and nonlethal point mutations closely linked to *unc*-54 (upper scale). *Let*-209 is a break point within the ribosomal DNA cluster of nematode present in a balancer chromosome isolated by Anderson and Brenner. The position of the *unc*-54 gene is marked by a star. (After Anderson and Brenner, 1984.)

Figure 2 compares the cytological map positions of the myosin heavy-chain genes (*myo*-1,2,3; see Section 4) and the actin genes (*act*-1,2,3,4; Files *et al.*, 1983) with the genetic map positions of other genes specifying muscle structure. The genes for the myosin heavy-chain isoforms in nematodes are not clustered. By contrast, myosin heavy-chain isoform genes in rats (Mahdavi *et al.*, 1984) and humans (Leinwand *et al.*, 1983) appear to be clustered and tandemly linked.

In some cases, the cytologically determined map positions of the muscle protein genes have been confirmed by other methods. Restriction fragment-length polymorphisms have been used to locate the *act*-1,2,3 genes to the region between *unc*-23 and *sma*-1 on linkage group V (Files *et al.*, 1983). Subsequently, mutations at the *act*-1,2,3 locus were correlated with structural alterations in the three closely linked actin genes (Landel *et al.*, 1984). Recently, the *myo*-3 clone (Section 4.4) has been used to identify a break point in a deficiency, *eDf1*, which maps to the center of linkage group V (Miller and Maruyama, 1986). Since *eDf1* displays *sup*-3 activity (see Section 2.5), these experiments place the *myo*-3 gene on the genetic map, closely associated with *sup*-3 (Miller and Maruyama, 1986).

2.5. Interactions with Other Genes

Expression of *unc*-54 phenotypes is dependent on the interactions of the myosin heavy chains with numerous other proteins in the sarcomere. Muscle mutants were among the original set of uncoordinated mutants described by Brenner (1974). Subsequently a large number of additional mutants in more than 25 genes have been isolated (Waterston *et al.*, 1980; Zengel and Epstein, 1980*b*). Reversion analyses of muscle gene mutants have been carried out to identify loci that can partially restore normal muscle structures. These studies yield information about the nature of the original alleles as well as identifying new genes that can compensate for defects in muscle-specific genes.

2.5.1. *unc-15.* Paramyosin is a rodlike protein that forms the core of invertebrate thick filaments (Cohen *et al.*, 1971; Szent-Györgyi *et al.*, 1971). Paramyosin is present at nearly 1 : 1 molar ratios with myosin heavy chains in the body wall muscles, but at a reduced level in pharyngeal muscles (Waterston *et al.*, 1977). Mutations in the paramyosin gene, *unc-15*, disrupt the normal assembly of myosin nematode into thick filaments (Waterston *et al.*, 1977; MacKenzie and Epstein, 1980; Kagawa *et al.*, 1987). The *e*1214 allele of *unc-15* is an amber mutation that produces no detectable paramyosin (Waterston and Brenner, 1978; Willis *et al.*, 1983). Body wall muscle thick filaments isolated from this strain are much shorter than normal thick filaments (1.5 μm in *e*1214, compared with 9.7 μm in wild type). MacKenzie and Epstein (1980) therefore suggested that paramyosin is required for thick filament length determination.

The paramyosin gene has recently been cloned by screening an expression library with polyclonal and monoclonal antibodies (Kagawa *et al.*, 1987). The paramyosin clones have been used as probes in a number of experiments that demonstrate directly that the *unc-15* locus encodes paramyosin. In Southern transfer experiments, only a single paramyosin locus is detected. This maps in *in situ* hybridization experiments to the center of chromosome I as expected for *unc-15*. A. Besjevec and P. Anderson (personal communication) obtained a series of spontaneous mutants in the *unc-15* gene by tagging with the transposable element Tc-1 (Emmons *et al.*, 1983). S. Roux and R. H. Waterston (personal communication) have shown that these Tc-1 insertions generate restriction fragment polymorphisms detectable with the cloned probes.

2.5.2. *sup-3/myo-3.* The gene *sup-3* was identified as yielding suppressors of *unc-54* null alleles (Riddle and Brenner, 1978). The *sup-3* suppressors are of variable strength; in cross-suppression tests, *sup-3* suppresses all *unc-54* null alleles equally well, but *sup-3* poorly suppresses dominant alleles of *unc-54*. *sup-3* also acts on *unc-15* missense alleles and on certain alleles of *unc-87*. In order to explain these effects, Riddle and Brenner (1978) suggested that *sup-3* might affect the gene encoding MHC A (*myo-3*) or regulate its expression. Support for this notion came from observations that MHC A accumulates about twofold in *sup*-3 animals (Waterston *et al.*, 1982*b*; Otsuka, 1987). Recent evidence suggests that *sup*-3 alleles are multiplications of the *myo*-3 gene (Miller and Maruyama, 1986).

Mutations in *sue*-1 partially reverse the improved sarcomere organization and movement conferred by *sup*-3 (S. Brown and D. L. Riddle, unpublished data). The mechanism of action of these two interacting genes is not understood.

2.5.3. sup-5 and sup-7.   The protein products of certain classes of *unc*-54 recessive null alleles may be restored by direct suppression of chain termination during protein synthesis. The loci *sup*-5 and *sup*-7 encode altered RNAs that specifically suppress amber mutations (Waterston, 1981; Waterston and Brenner, 1978; Willis *et al.*, 1983) (see Section 8.4). In addition to the *sup*-5, *sup*-7 suppressors, a large number of other allele-specific multigenic suppressors with different spectra of suppression have been isolated (S. Brenner, unpublished data). It is reasonable to assume that these include altered tRNAs capable of suppressing additional amber mutations or ochre and UGA mutations.

2.5.4. unc-52.   In mutations of the *unc*-52 gene, the normal progressive assembly of body wall muscle sarcomeres is significantly retarded (Zengel and Epstein, 1980*a*; MacKenzie *et al.*, 1978*a*). Zengel and Epstein (1980*a*) presented evidence that *unc*-52 mutations result in decreased amounts of *unc*-54 heavy chains relative to the other myosin isoforms. These investigators have suggested that the *unc*-52 gene product may be required for normal synthesis of *unc*-54 myosin heavy chain. However, the effects of *unc*-52 on *unc*-54 synthesis may be an indirect effect associated with the general disruption of muscle in these animals.

2.5.5. unc-22.   The *unc*-22 gene was among the set of muscle genes originally identified by Brenner (1974). *unc*-22 mutants are uniquely characterized by a pronounced twitch of the body wall muscles. The twitch is caused by a defect in the muscle cells themselves and is not the result of faulty neuronal activity, since it is initiated independently in each muscle cell (Moerman and Billie, 1979; Waterston *et al.*, 1980). The muscle structure of *unc*-22 mutants is variable, but in homozygotes of the amber allele *s*32, the A bands as visualized by polarized light are severely disorganized. Thick filaments are present in near-normal numbers but are not restricted to ordered arrays.

Reversion analysis of *unc*-22 alleles has resulted in the isolation a unique class of *unc*-54 mutants (Moerman *et al.*, 1982). These *unc*-54 suppressors were recovered as dominant suppressors of the weak *unc*-22 allele *s*12. More than 15 *unc*-54 suppressors have been obtained; they map exclusively to the myosin head region (see Section 8.7). *unc*-54 is unique in its ability to suppress *unc*-22; no other loci have been detected in extensive reversion analyses.

The structure of the *unc*-22 gene was recently determined by Moerman *et al.* (1987) and Benian *et al.* (1987). The *unc*-22 locus was tagged by insertion with the transposable element Tc-1. This led to the cloning of the gene and analysis of the *unc*-22 gene products. The *unc*-22 probes hybridize to an mRNA species of about 14 kb, a message sufficient to encode a protein of more than 500,000 M_r. Specific antisera to *unc*-22 were made using fusion proteins between *Escherichia coli* β-galactosidase and restriction fragments from the coding region of *unc*-22. These antisera localize the *unc*-22 protein to the A bands of the body wall muscle (R. H. Waterston, personal communication). These results suggest that *unc*-22 protein is a structural component of the A-band structure. The indirect suppression of the twitching phenotype by *unc*-54 mutations suggests that MHC A and the *unc*-22 product interact directly, however, it is uncertain whether the *unc*-22 protein is a component of the thick filament.

3. IMMUNOLOGICAL IDENTIFICATION AND LOCALIZATION OF MYOSIN ISOFORMS

3.1. Production of Four Sarcomeric Myosin Heavy-Chain Isoforms by Caenorhabditis elegans

Fractionation of myosin heavy chains by sodium dodecyl sulfate–polyacrylamide gel electrophoresis (SDS-PAGE) indicated that four sarcomeric myosin heavy chains are synthesized by *Caenorhabditis elegans* (Figs. 4 and 9A). Many mutations in the *unc*-54 gene show altered patterns of synthesis or mobility of the MHC B heavy chain (Epstein *et al.*, 1974; MacLeod *et al.*, 1977a,b; Waterston *et al.*, 1982b). As shown in Fig. 4, the *unc*-54 allele *e675* produces a shortened MHC B heavy chain. MacLeod *et al.* (1977a) demonstrated by peptide-mapping experiments that the *e675* defect is an internal deletion of 100 amino acids near the COOH terminus of the myosin rod. Nucleotide sequence data (Dibb *et al.*, 1985) have confirmed this result (Section 8.3). Recessive alleles of *unc*-54, such as *e190*, produce a simplified myosin heavy chain pattern on SDS gels due to the absence of detectable MHC B protein (Epstein *et al.*, 1974; MacLeod *et al.*, 1977a, 1979; Schachat *et al.*, 1977; Miller *et al.*, 1983).

The minor body wall myosin heavy chain (MHC A) comigrates with the wild-type MHC B heavy chain but is well resolved when the MHC B heavy chain is altered or removed by the *unc*-54 mutations *e675* and *e190*. Biochemical (MacLeod *et al.*, 1977a,b; Schachat *et al.*, 1977, 1978) and genetic (MacLeod *et al.*, 1981) analyses have shown that separate genes encode each of these heavy chains. Both heavy chains are pro-

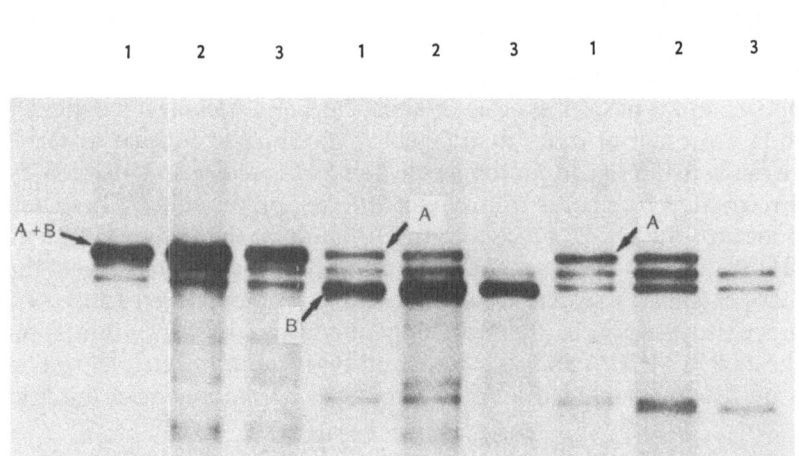

Figure 4. SDS-PAGE of nematode myosin heavy chains. The isozyme patterns of wild-type (N2) and strains homozygous for the mutations *e675* and *e190* are compared. For each nematode strain, lane 1 is an immunoprecipitation of myosin heavy chains labeled *in vivo* with [^{35}S]methionine. Lane 3 is an immunoprecipitation of [^{35}S]methionine-labeled proteins synthesized *in vitro* in a wheat germ cell-free system programmed with RNA isolated from the indicated strain. Lane 2 is a mixture of heavy chains synthesized *in vivo* and *in vitro*. The major protein present in the N2 strain is MHC B. In the *e675* strain, MHC B is also the major myosin heavy chain, but its molecular weight is reduced by 10,000 M_r. This shift in the mobility of MHC B reveals the MHC A chain (which comigrates MHC B in N2). The MHC A chain is not translated efficiently in the *in vitro* system and is absent in lane 3. In the *e190* strain, the MHC B is absent, but the MHC A chain is present. The two other myosin heavy chains seen in this strain correspond to the pharyngeal MHC C and D. These myosins are present in the other strains but are partially obscured by MHC B. The myosin heavy chains are especially well separated in this gel. Typically, one of the pharyngeal myosins is not resolved from MHC A in the *e675* and *e190* gel patterns. Consistent resolution of all four myosin heavy chains in Tris-borate SDS-gels has been reported (Waterston *et al.*, 1982; Otsuka, 1987). (After MacLeod *et al.*, 1979.)

duced in the body wall musculature but not in the pharynx (MacKenzie *et al.*, 1978; Miller *et al.*, 1983). MHC A and MHC B are expressed at a constant ratio throughout larval morphogenesis (Garcea *et al.*, 1978) in adult muscle tissues; the ratio of MHC B to MHC A is approximately 4 : 1 (Waterston *et al.*, 1982*b*). MHC A and MHC B are homodimeric molecules; heterodimers have not been detected (Schachat *et al.*, 1977, 1978).

The remaining two bands in the gel pattern shown in Fig. 4 have been designated MHC C and D. Microdissection experiments (Epstein *et al.*, 1974; Waterston *et al.*, 1982*b*; Waterston, personal communication) and immunofluorescence data (Epstein *et al.*, 1982*a*) indicated that myosins C and D are expressed only in the pharyngeal muscle (see Section 3.2).

3.2. Monoclonal Antibodies Specific for Different Myosin Isoforms

Miller *et al.* (1981, 1983) prepared 24 hybridoma clones that secreted antibodies with high affinity to nematode myosin and paramyosin. Antibodies that preferentially recognize each of the four myosin heavy chains were identified by immunoblotting of myosin heavy chains fractionated by SDS-PAGE (Table I) (Epstein *et al.*, 1982*a*).

Table I. Immunoblotting Reactions of Monoclonal Antibodies with Nematode MHC Isoforms and with Fusion Peptides Derived from Cloned MHC Genes[a]

Monoclonal antibody[b]	MHC specificity[c]	MHC isoform reaction[d]			Epitope[e]	Cloned gene fragment, bp[f]			
		A/D"	B	C		unc-54	myo-1	myo-2	myo-3
5–6	A	A	(+)	—	S-2	—	—	—	460
5–2	B	(+)	+	—	S-2	357	—	—	—
5–3		—	+	—	LMM	1054	—	—	—
28.2		(+)	+	—	S-2[g]	357	—	—	(460)
5–13		(+)	+	(+)	LMM[g]	—	—	—	(1464)
12.1		(+)	+	(+)	S-2[g]	117	—	—	—
					LMM	294	—	—	—
9.2.1	C	—	(+)	+	LMM	—	—	2156	—
5–11		(+)	(+)	+	LMM	(1054)	(1577)	2156	(1464)
5–12	B and C	—	+	+	LMM	1054	—	2156	—
25.1	B and A or D	+	+	—	S-2	357	—	—	460
5–25		+	+	—	LMM	1054	1577	—	—
10.2.1		+	+	—	LMM	1054	1577	—	1464

[a] Immunoblotting was as in Miller *et al.* (1983).
[b] Monoclonal antibodies listed that have not been tested with proteins resolved in the Neville gel system.
[c] For MHC isoform reactions: +, strong reaction; (+), weak reaction; —, no reaction.
[d] Monoclonal antibodies that are specific to either MHC A or MHC D cannot be resolved on immunoblots of Tris/glycine (Laemmli) buffered PAGE. Immunoblots of Tris/borate (Neville) buffered PAGE in which MCH A and MCH D are resolved have shown that monoclonal antibody 5–6 is specific for MHC A.
[e] Epitope: myosin regions (S-1, S-2, LMM) reacting with antibody.
[f] For cloned gene fragments: numbers indicate the size of strongly reactive HindIII fragment; numbers in parentheses indicate weakly reactive HindIII fragments; —, no reaction.
[g] Determined from chymotrypic digestions of MHC and from reactivity toward the products of cloned gene fragments.

The antibodies were first used to study the expression of each of the myosin heavy chains in different tissues. Figure 5 shows the reactions of nematode embryo squashes (Gossett and Hecht, 1980; Epstein *et al.*, 1982*a*) with fluorescently labeled antibodies. The MHC C specific monoclonal antibody, 9.2.1, exclusively labeled the pharyngeal muscle in the head of the animal (Fig. 5A). In contrast, the MHC B specific antibody, 28.2, outlined the embryo with fluorescence, indicating reaction with the developing body wall muscle cells extending along the length of the animal (Fig. 5B). A similar pattern of fluorescence labeling was obtained by reaction with the MHC A specific antibody, 5–6 (D. M. Miller, unpublished data). The MHC D specific antibody, 5–17, did not label vermiform embryo squashes, although gel analysis of dissected pharyngeal muscle has shown that MHC D is expressed in this tissue (R. H. Waterston, unpublished data). These experiments have shown that two different pairs of myosin heavy chains are expressed in each of the two major muscle types in *Caenorhabditis elegans*. MHC A and B are expressed in the body wall, while MHC C and D are expressed in the pharynx.

3.3. Two Myosins Required to Make the Body Wall Thick Filament

In the nematode, the body wall musculature is composed of four elongated groups of interlocking spindle-shaped muscle cells that extend along the entire length of the animal. In the adult, each muscle cell contains 8–10 complete sarcomeres that appear as alternating light and dark stripes under polarized light (Epstein *et al.*, 1974; MacKenzie and Epstein, 1980). The thick filaments are restricted to the light stripes (A bands), while the dark stripes (I bands) contain the thin filaments. Nematode muscle is obliquely striated; myofilaments cross the sarcomere on a diagonal axis with respect to the junctions between sarcomeres rather than at right angles as in other striated muscles (MacKenzie and Epstein, 1980; Miller *et al.*, 1983).

The expression of both the MHC A and MHC B heavy chains in the same cells of the body wall muscle is consistent with two alternative models for thick-filament structure. The myosins could either coassemble in the same thick filament, or alternatively, form two distinct thick-filament types. Miller *et al.* (1983) studied this question by using MHC A and MHC B specific monoclonal antibodies to locate these heavy chains in muscle tissue and in isolated thick filaments.

Squash preparations of adult nematodes, simultaneously stained with MHC A and MHC B specific antibodies, showed distinctive patterns of labeling for each antibody in the A bands of body wall muscle cells. Reactions with MHC A specific antibodies (e.g., 5–6) produced a bright

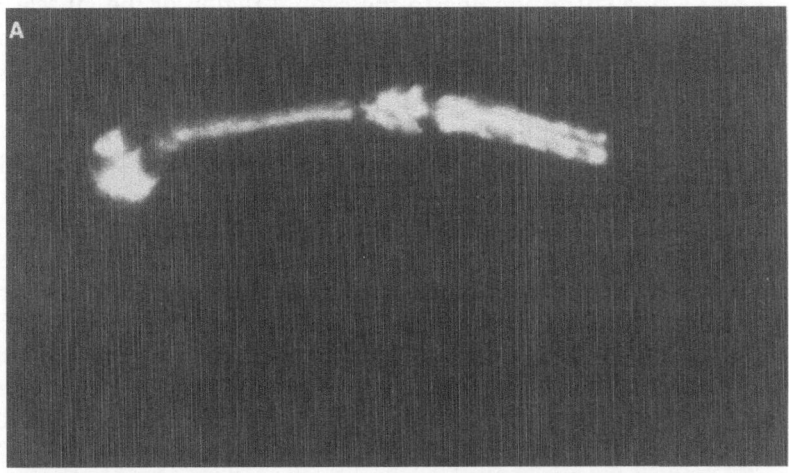

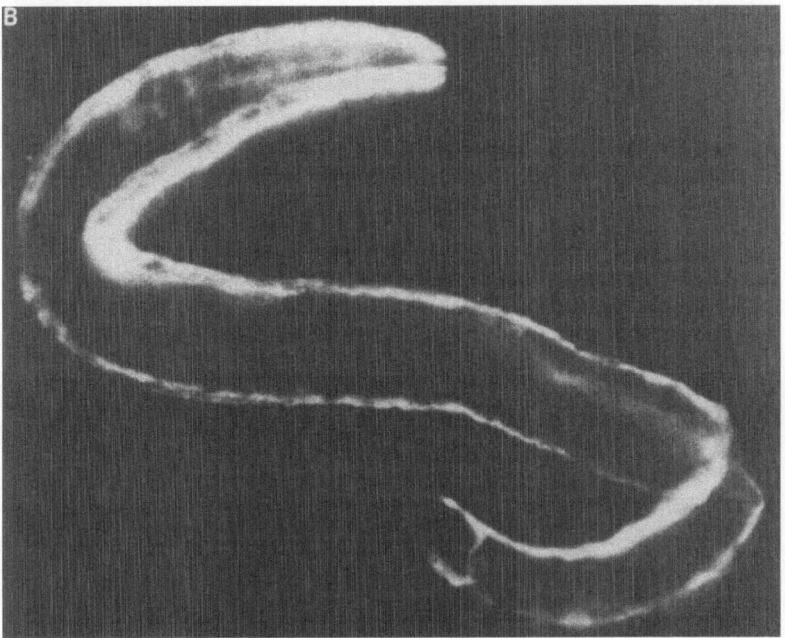

Figure 5. Staining of vermiform embryo squashes with monoclonal antibodies. (A) Re-action of antibody 9.2.1. (MHC C specific). Note the localization of fluorescence to the pharynx. (B) Antibody 28.2 (MHC B specific) reacts with the developing body wall muscle extending along the entire length of the animal. A very slight reaction with the pharynx may be seen near the head. This is probably due to weak affinity of the antibody 28.2 with a pharyngeal myosin. (After Epstein *et al.*, 1982a.)

fluorescent stripe corresponding to the central region of the muscle A band (Fig. 6A). MHC B specific antibodies (e.g., 28.2) produced a complementary pattern in which all of the A band except for the medial region was stained (Fig. 6B).

The simplest interpretation of these results is that MHC A and MHC B are incorporated into different regions of each thick filament. This conclusion has been confirmed by electron microscopy of antibody-decorated thick filaments (MacKenzie and Epstein, 1980; Miller *et al.*, 1983). Antibody 5–6 reacts exclusively with a short central region of the thick filament, whereas antibody 28.2 labels the distal portions of the structure (Fig. 7). Since all thick filaments gave a similar pattern of labeling with these antibodies, it was assumed that only one type of thick filament containing both MHC A and B is produced in the body wall muscle. A schematic diagram of myosin packing was constructed from measurements of the reacted zones of antibody-labeled thick filaments (Fig. 7). In the proposed model, MHC A is restricted to the central 1.8 μm of the 9.7-μm-long native thick filament. MHC B is incorporated into the polar regions exclusive of a 0.9-μm zone in the middle of the filament. Two bilaterally placed regions, approximately 0.45 μm in length, are composed of both kinds of myosin.

The locations of the myosins in the nematode filament suggest distinct roles for the isoforms in thick-filament assembly. Huxley (1969) proposed that thick-filament polymerization occurs in two steps. Myosin molecules first pack into a short bipolar structure in which the myosin rods are arrayed in antiparallel orientations. Filament elongation then proceeds through the addition of parallel-oriented myosins to both ends

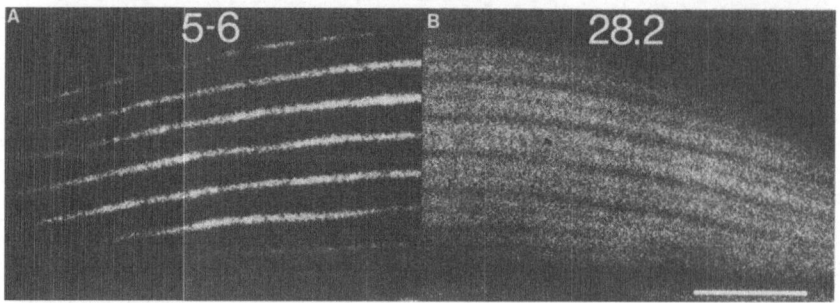

Figure 6. Simultaneous immunofluorescence labeling of a single body wall muscle cell with antibody 28.2 (MHC B specific) and antibody 5–6 (MHC A specific). Bar: 10 μm. (A) Left-hand side of the cell showing rhodamine coupled antibody 5–6 (1.7 μg/ml). (B) Right-hand side of the cell showing antibody 28.2 (10 μg/ml) plus fluorescein-coupled goat antimouse IgG. (After Miller *et al.*, 1983.)

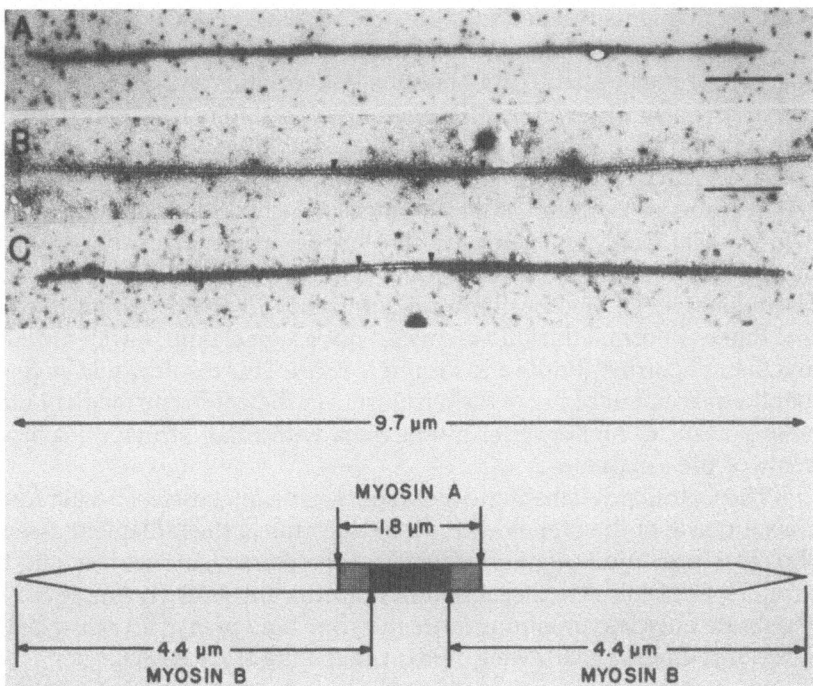

Figure 7. Reaction of isolated nematode thick filaments with antibodies. Bars are 1 μm. Pointers indicate the boundary of antibody reaction. (A) Negatively stained only. (B) Reaction with 1 μg/ml antibody 5–6 (MHC A specific) plus goat antimouse IgG (20 μg/ml). Note binding to central region of the filament. (C) Reaction with 10 μg/ml antibody 28.2 (MHC B specific) plus goat antimouse IgG (20 μg/ml). The distal segments of the filament are stained with antibody 28.2. (Bottom) Interpretive drawing of the antibody labeling patterns derived from measurements of antibody-reacted zones in randomly selected filaments. The diameter of the thick filament has been exaggerated. (After Miller *et al.*, 1983.)

of the bare zone assembly (see Section 7.2). Recent experiments with synthetic and native filaments have substantiated this hypothesis (Reisler *et al.*, 1980; Davis, 1981; Niederman and Peters, 1982). Miller *et al.* (1983) proposed that the nematode MHC A and B participate in different phases of filament assembly. MHC A forms the antiparallel packed nucleation center and adjacent parallel packed regions; MHC B participates exclusively in the polymerization reactions that elongate the filament.

Although normal body wall thick filaments contain both the MHC B and MHC A heavy chains, residual thick filaments are seen when MHC B has been removed by *unc*-54 mutations (Epstein *et al.*, 1974) (Fig.

1). Thick filaments isolated from the *unc*-54 allele *e*190 are indistinguishable from wild-type filaments with respect to length and morphology (MacKenzie and Epstein, 1980). Antibody-decoration experiments (Miller *et al.*, 1983) have shown that *e*190 filaments react along their entire length with antibody 5–6. Thus, MHC A will assemble into a thick filament of wild-type length when MHC B, which normally participates in the elongation reaction, is absent from the muscle. Increased amounts of MHC A in *sup*-3/*e*190 strains result in more of these thick filaments and improved movement (Waterston *et al.*, 1982b; Otsuka, 1987; Miller and Maruyama, 1986) (see Sections 2.5.2 and 4.5). It is interesting to note that these abnormal thick filaments do not assemble into a well-ordered myofilament array (Riddle and Brenner, 1978). This result would suggest that the placement of the myosin isoforms in the wild-type filament may be important to higher-order interactions with other structural components of the sarcomere.

These antibody-labeling experiments have suggested specific functions for each of the nematode myosin isoforms in thick-filament assembly, structure, and integration into the myofilament lattice. It would be of interest to learn whether different myosins are similarly employed in vertebrate muscles containing more than one kind of myosin heavy chain (Starr and Offer, 1973; Zweig, 1981; Chizzonite *et al.*, 1982).

4. MOLECULAR CLONING OF THE *UNC*-54 AND *MYO*-1,2,3 MHC GENES

4.1. Cell-Free Translation of MHC mRNA

Total RNA from nematodes programs the synthesis of nematode myosins in cell-free systems derived from both wheat germ and rabbit reticulocytes (MacLeod *et al.*, 1981) (Figs. 4, 8, and 9). The heavy chains synthesized *in vitro* comigrate with the heavy chains synthesized *in vivo* and may be precipitated with antibodies directed against myosin. Cell-free synthesis of MHC B was demonstrated because RNA from *unc*-54 mutant strains, including *e*675 and *e*190 (Figs. 4 and 9), programs the synthesis of appropriately altered proteins. The cell-free system also produced the MHC C and D chains but for unknown reasons failed to synthesize detectable MHC A (Figs. 4 and 9) (MacLeod *et al.*, 1981).

4.2. unc-54 cDNA Clones

The cell-free translation assay was used to monitor the purification of the myosin heavy-chain mRNAs (MacLeod *et al.*, 1981). A considerable

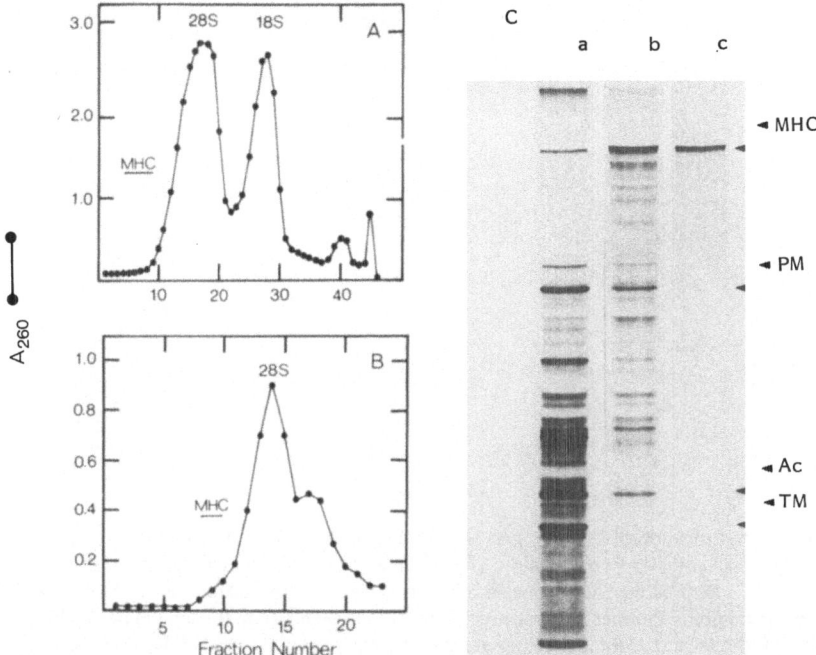

Figure 8. Purification of *unc*-54 mRNA. (A) Thirty mg of nematode RNA was fractionated on a 10–40% sucrose gradient generated in a MSE type XIV zonal rotor. After centrifugation for 15 hr fractions were collected and assayed for optical density (●——●) and myosin mRNA activity in the wheat germ cell-free translation system. The bar indicates the position of myosin mRNA activity. (B) Enriched myosin mRNA from the zonal centrifugation step was denatured with methylmercury hydroxide and applied to a 13-ml 5–20% sucrose gradient. After centrifugation in a Beckman SW40 rotor at 40,000 rpm for 5 hr, fractions were collected and assayed for optical density (●——●) and myosin mRNA activity. (C) SDS-PAGE gel electrophoresis of the products of the wheat germ cell-free system programmed with a, total RNA from nematode larvae; b, myosin mRNA purified by zonal centrifugation; c, myosin mRNA repurified by denaturing sucrose gradient centrifugation. The positions of myosin heavy chains (MHC), paramyosin (PM), actin (Ac), and tropomyosin (TM) are indicated. (After MacLeod *et al.*, 1981.)

purification of these large mRNAs can be achieved by successive sedimentation through neutral and denaturing sucrose gradients (Fig. 8). The partially purified myosin heavy-chain mRNA was used as a template to construct recombinant plasmids containing *unc*-54 specific sequences. As expected, the large size of the *unc*-54 mRNA, 6000 base pairs (bp), prevented the synthesis of full-length cDNA clones. However, short cDNA fragments cleaved by the restriction endonuclease *Mbo*I (which cleaves at the sequence GATC) were successfully cloned (MacLeod *et al.*, 1981).

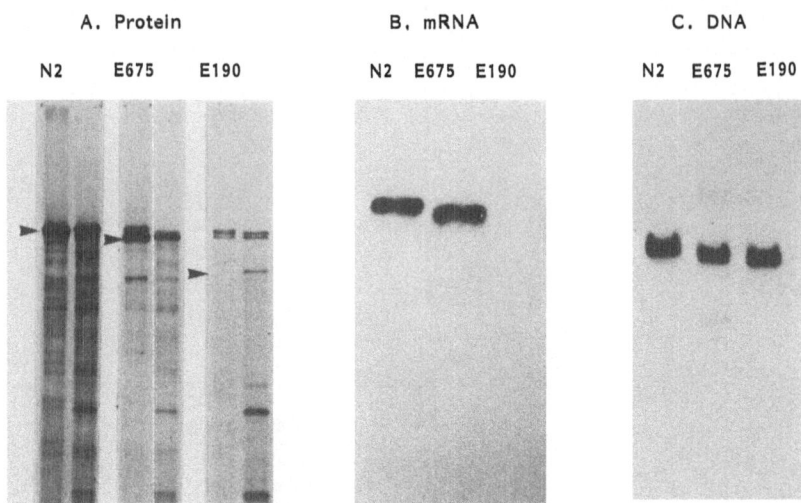

Figure 9. Identification of MHC B, *unc*-54 mRNA, and *unc*-54 DNA sequences by cell-free translation and transfer-hybridization. (A) ^{35}S-labeled myosin heavy chains, synthesized *in vivo* and *in vitro* using a wheat germ cell-free system, were fractionated by SDS-PAGE after purification by indirect immunoprecipitation. For each strain, the left gel panel shows *in vivo* products and right gel panel shows *in vitro* products. (▶) MHC B or MHC B fragments. Note the absence of any immunoprecipitable MHC B fragments in the *in vivo* myosin fraction isolated from *e*190. (B) RNAs from N2, *e*190, and *e*675 were fractionated on a 1.5% agarose gel, transferred to diazotized paper, and hybridized to a ^{32}P-labeled cDNA plasmid probe. Note that the *unc*-54 mRNA is reduced in size in the *e*675 strain and is not detected in the *e*190 strain. (C) *Bam*H1-digested DNAs from N2, *e*675 and *e*190 were fractionated on a 0.8% agarose gel, denatured with alkali, transferred to nitrocellulose, and hybridized to a ^{32}P-labeled cDNA plasmid probe. The probe specifically hybridizes to a 5547-bp restriction fragment containing 3' end of the *unc*-54 gene (see Fig. 10). This fragment is reduced in size in both *e*675 and *e*190 due to deletion of part of the *unc*-54 gene sequence. (After MacLeod *et al.*, 1981; Karn *et al.*, 1982.)

Because the presence of multiple MHC genes might confuse the identification of *unc*-54 specific clones, we devised assays based on the properties of *e*190 and *e*675 to assist in the detection of specific clones (MacLeod *et al.*, 1981). We reasoned both the *unc*-54 mRNA itself and restriction fragments from the *unc*-54 gene should be of altered length in *e*675 strains, since this allele was an internal deletion mutant. In *e*190 a reduced level of mRNA should be observed, as the translation polypeptide is absent. Thus, clones that specifically hybridize to *unc*-54 sequences should be identifiable as clones that detected strain differences between *unc*-54 mutants in hybridization experiments. Figure 9B shows the hybridization pattern obtained when a plasmid containing *unc*-54 sequences was used as a probe against N2, *e*675 and *e*190 mRNA fractionated by agarose gel electrophoresis. A single mRNA species of approximately 6 kb was detected in the wild-type fraction, and this species showed the expected reduction in size in *e*675. A corresponding mRNA was absent in *e*190. Because the genetic evidence unambigously assigned *e*675 and *e*190 to a single locus, this hybridization experiment demonstrated we had obtained an *unc*-54 specific probe. The original cDNA clones (MacLeod *et al.*, 1981) are now known to contain sets of contiguous *Mbo*I fragments from the extreme 3' end of the *unc*-54 sequence (Karn *et al.*, 1980, 1983*a*).

4.3. Cloning the unc-54 Gene

The short cDNA clones were used to identify restriction fragments from the nematode genome that contained *unc*-54 sequences. Figure 9C shows a Southern transfer hybridization experiment that detected *unc*-54 specific sequences in N2, *e*675 and *e*190 DNA digested with *Bam*H1 (MacLeod *et al.*, 1981). A single restriction fragment of 5547 bp (Figs. 9 and 10) was detected in the wild-type DNA. It is reduced by 270 bp in *e*675 and, unexpectedly, also reduced by 401 bp in *e*190. This experiment demonstrated that the cDNA probes hybridize specifically to *unc*-54 genomic sequences and indicated that *e*190 was also a small deletion.

These cDNA probes were also used to isolate the *unc*-54 structural gene from a bacteriophage λ genome library that contained nematode DNA fragments of 15–20 kb in size (Karn *et al.*, 1980, 1983*a,b*; MacLeod *et al.*, 1981). Twenty-two recombinant λ phages that included sequences of the *unc*-54 gene were picked. These bacteriophages contained a series of overlapping restriction fragments that could be ordered to generate a map extending more than 30 kb. Subsequent DNA sequencing studies have confirmed the map and the placement of the *unc*-54 coding sequence (Fig. 10).

The nematode genome library was prepared with the bacteriophage λ1059 cloning vector (Karn *et al.*, 1980; 1983*b*). This vector allows positive genetic selection for inserts. High-molecular-weight DNA fragments prepared by partial digestion with the restriction enzyme *Sau*3a may be directly ligated to the vector arms cleaved with *Bam*H1. Because *Sau*3a has a 4-bp recognition sequence, *Sau*3a sites should occur once every 256 bp in DNA with 50% G + C. Only 1/80th of these sites need to be cleaved to produce 20-kb DNA fragments. In practice, fragments prepared in this way show a random distribution of cut sites. A permanent collection of recombinant phages was then established by permitting the phages harboring nematode inserts to amplify through several genera-

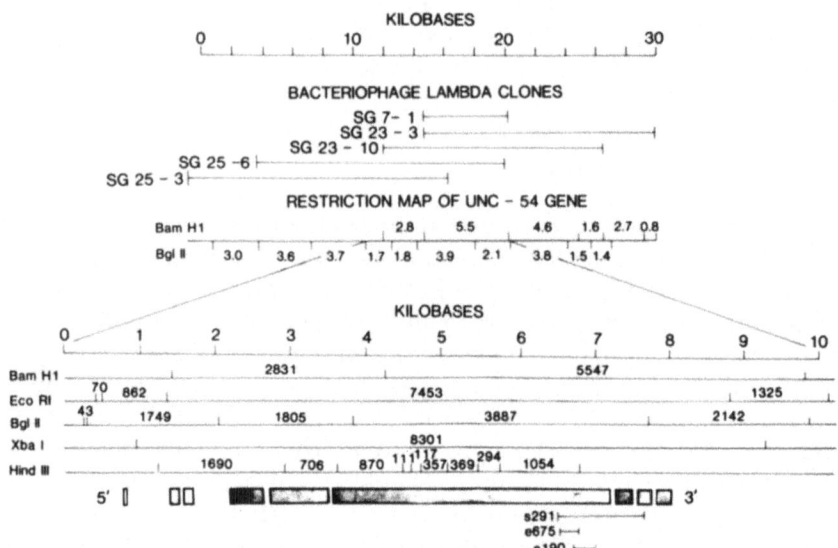

Figure 10. Restriction endonuclease maps of clones containing the *unc*-54 myosin heavy-chain gene and flanking sequences. Representative clones in bacteriophage λ1059 are shown in the top panel. The maps are drawn without the bacteriophage vector sequences. SG7-1, SG23-3, and SG23-10 contained nematode fragments prepared by partial digestion with *Bam*H1, while SG25-6 and SG25-3 contained nematode fragments prepared by partial digestion with *Sau*3a. Below the bacteriophage maps is a 30-kb map showing the *Bam*H1 fragments and *Bgl*II fragments contained in the clones. The sizes of the fragments is given in kilobases (kb). The central 10 kb of the restriction map is enlarged in the bottom of the panel to show restriction maps for *Bam*H1, *Eco*R1, *Bgl*II, *Xba*I, and *Hind*III. The sizes of the fragments were determined by DNA sequencing and are expressed as base pairs (bp). The positions of the *unc*-54 coding sequences are aligned with the restriction map (shaded boxes). The coding sequence is discontinuous due to the presence of introns. The endpoints for the deletions *s*291, *e*675, and *e*190 are also included. (After MacLeod *et al.*, 1981; Karn *et al.*, 1983*a*.)

tions on a strain that restricted the growth of the original vector phage. Clones corresponding to nematode actin (Files *et al.*, 1983; Hirsh *et al.*, 1982) and yolk proteins (T. Blumenthal, unpublished data) as well as the myosin heavy chains have been obtained from this collection.

4.4. Identification of the myo-1,2,3 MHC Genes by Homology to unc-54

Comparisons between the amino acid sequence of the *unc*-54 myosin heavy chain and other known myosin heavy-chain sequences, including rabbit skeletal myosin, showed that myosin heavy chains are well conserved (McLachlan and Karn, 1982; Capony and Elzinga, 1981; Elzinga and Collins, 1977; Elzinga and Trus, 1980; Kavinsky *et al.*, 1983; Strehler *et al.*, 1986) (see Fig. 17; Section 6.1). It was therefore assumed that other nematode myosin heavy-chain genes would include homologous regions that could be detected with *unc*-54 probes.

The λ1059 genome library was screened with a *Bam*H1-*Bgl*II fragment (residues 3989–4401) (Karn *et al.*, 1983*a*) that covered the highly conserved active thiol region (Capony and Elzinga, 1981). Fifty-two positive clones were isolated and classified by restriction mapping. Although this collection included a number of spurious clones, such as some that contained rDNA sequence, three groups of new MHC clones were recovered. Further hybridization experiments were undertaken to determine if each of these clones represented separate muscle myosin heavy chains. The λ clones were digested with restriction enzymes, fractionated by agarose gel electrophoresis, and hybridized to a series of *unc*-54 probes spanning the head and rod sequences. Figure 11 shows the hybridization pattern of representative clones from *unc*-54 and the three new MHC genes, *myo*-1,2,3, to a head-specific probe. In general, there is less sequence homology between rod sequences than between head sequences. Very little hybridization to rod-region probes could be detected after stringent washes. DNA-sequencing studies (Karn *et al.*, 1983*a*) (see Section 6) have shown that the *myo*-1,2,3 clones include the sequences of three genes from a highly conserved gene family.

5. IMMUNOLOGICAL IDENTIFICATION OF THE PRODUCTS OF THE MYO-1,2,3 MHC GENES

The genetic and biochemical studies outlined in Sections 2 and 4 clearly show that the *unc*-54 locus encodes the MHC B heavy chain. The three other cloned myosin heavy-chain genes, *myo*-1, *myo*-2, and *myo*-3, were found not to be initially linked to myosin heavy-chain isoforms

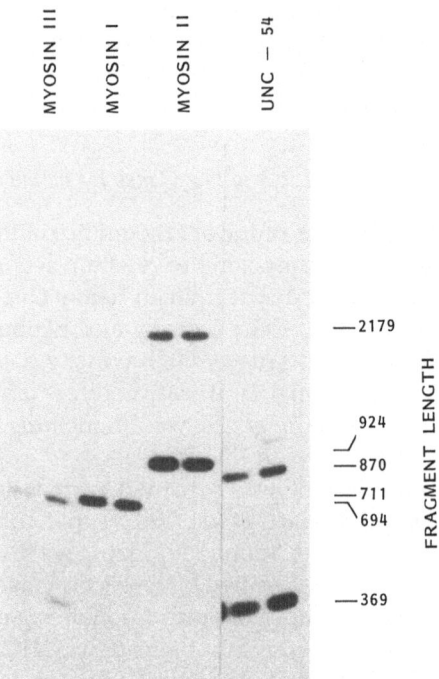

Figure 11. Identification of nematode myosin heavy-chain isoforms by hybridization to *unc*-54 sequences. Bacteriophage λ clones that hybridized to *unc*-54 DNA sequences were digested with *Hind*III, fractionated by agarose gel electrophoresis, and hybridized at low stringency to a cloned probe containing two *Hind*III fragments from *unc*-54; the 870-bp head fragment and the 369-bp rod fragment (Fig. 10). The fragment lengths were determined by DNA sequencing and are expressed as base pairs (bp).

by mutations. Therefore, other approaches were needed to determine whether these genes are transcribed and, if so, which of the additional isoforms (MHC A, C, or D) they produce. Miller *et al.* (1986) prepared fusion peptides composed of *E. coli* β-galactosidase and myosin rod fragments as outlined in Fig. 12. The hybrid proteins were screened with a panel of specific monoclonal antibodies to identify epitopes corresponding to each myosin heavy-chain isoform.

The construction of suitable clones was greatly simplified by our observation that the rod regions of the myosin heavy-chain genes each contained numerous short *Hind*III fragments (Figs. 11 and 13). All the *Hind*III sites showed the same translational phasing because each derived from the frequent appearance of the dipeptide sequence Lys-Leu

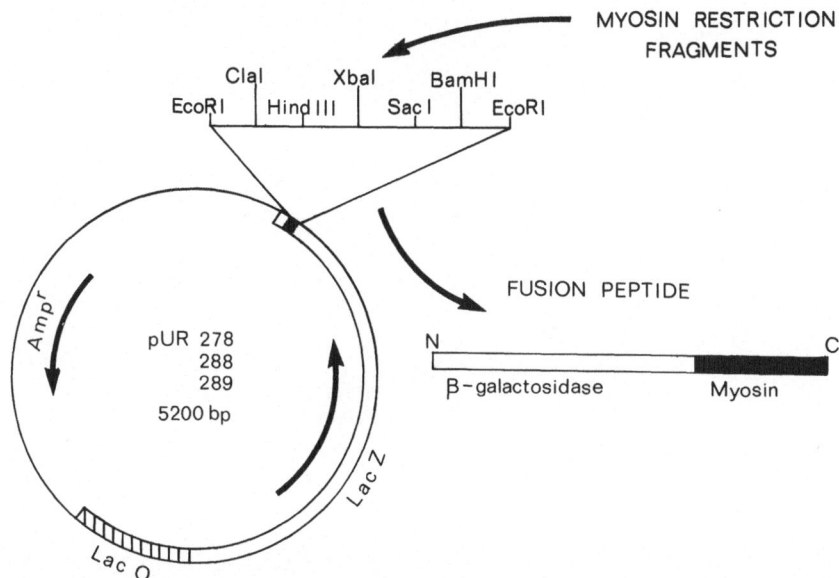

Figure 12. Vectors used to produce fusion peptides containing myosin fragments inserted at the carboxy-terminal of β-galactosidase. The plasmids pUR278, pUR288, and pUR289, carry an intact *lac-z* gene and specify ampicillin resistance (Amp^r). A polylinker sequence with unique restriction endonuclease cleavage sites has been inserted near the COOH terminus of the β-galactosidase protein. The polylinker is shifted by one nucleotide in each of the three plasmids so that fragments can be inserted into the vectors in each of three possible phases. (After Rüther and Müller-Hill, 1983.)

(AGG-CTT). Rüther *et al.* (1982) and Rüther and Müller-Hill (1983) prepared a series of plasmid vectors with restriction endonuclease cloning sites adjacent to the COOH-terminus of β-galactosidase. The positions of the restriction sites in each of the vectors differs by 1 bp, permitting DNA from a single-restriction digest to be inserted into a different reading frame in each of the plasmids. We were able to obtain a complete set of fusion peptides using the pUR288 vector, which had appropriately phased *Hind*III sites.

Colonies expressing fusion proteins were identified by immunoassays (Miller *et al.*, 1986) of nitrocellulose filter replicas reacted either with rabbit antiserum to nematode myosin (MacLeod *et al.*, 1981) or with monoclonal antibodies (Miller *et al.*, 1983). Clones expressing fusion peptides encoded by most of the *Hind*III fragments from the rod region of each myosin heavy chain gene were isolated in this manner (see Table I and Fig. 13).

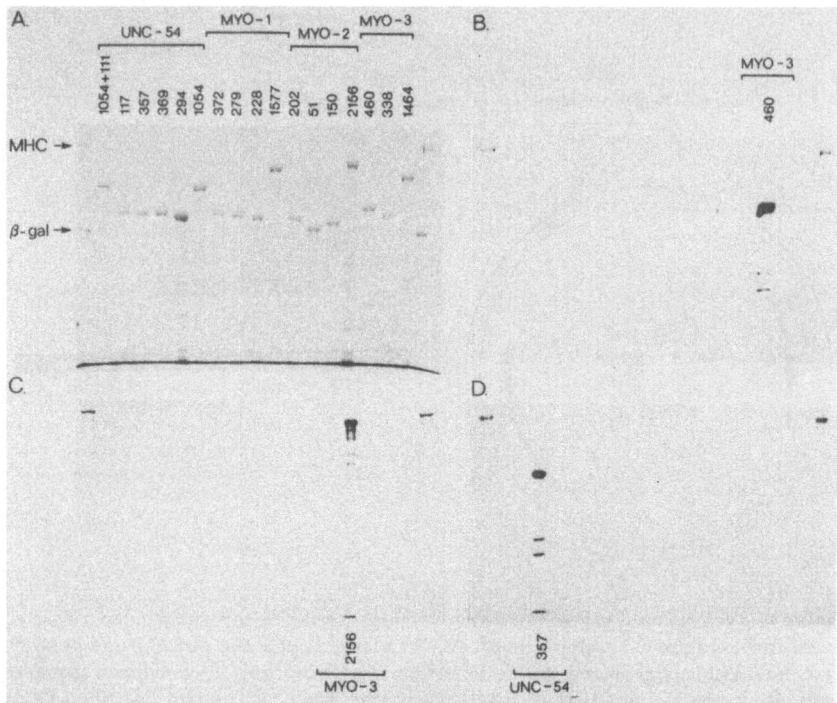

Figure 13. Reaction of fusion proteins with specific monoclonal antibodies. Clones containing *Hind*III fragments from each of the MHC genes in pUR288 were grown in the presence of IPTG and lysed with SDS. The fusion proteins were separated by SDS-PAGE. (A) Coomassie blue-stained gel. The numbers above each gel lane refer to the size of each *Hind*III insert in bp. Immunoblots of the gel shown in A reacted with antibody 5–6 (B), antibody 9.2.1 (C), and antibody 28.2 (D). Nematode myosin (MHC; 226,000 M_r) and *E. coli* β-galactosidase (β-gal, 116,000 M_r) standards were run in the outside lanes of the gel, and the MHC is visible in each immunoblot.

The induced fusion proteins constituted a major fraction of the bacterial protein and migrated on SDS polyacrylamide gels in the expected high-molecular-weight range between β-galactosidase (116,000 M_r) and nematode myosin (226,000 M_r) standards (Fig. 13A). Most of the clones contained a single inserted *Hind*III fragment, and the relative sizes of the hybrid proteins are directly related to the length of the *Hind*III insert.

Immunoblotting was used to define the reactivity of the fusion proteins with each of the monoclonal antibodies. Some examples are shown in Fig. 13; complete results are presented in Table I. The patterns of reactivity establish that each of the four myosin heavy chain genes encodes one of the four myosin heavy chain isoforms. For example, antibody 5-6 is highly specific for an epitope in the S-2 region of MHC A. This antibody reacts only with the hybrid protein encoded by the 460-bp *Hin*dIII fragment from the S2 domain of the *myo*-3 gene and does not recognize epitopes encoded by homologous fragments from the other genes (Fig. 13B). Similarly, antibody 9.2.1, which is specific to MHC C, preferentially recognizes the large fusion peptide encoded by the 2156-bp *Hin*dIII fragment from the *myo*-2 rod domain (Fig. 13C). A number of antibodies tested showed multiple reactivity toward the different myosin heavy chains. For example, antibody 28.2 reacts strongly with MHC B and weakly with MHC A. This antibody reacts strongly with the 257-bp fragment of the *unc*-54 gene (Fig. 13D) and weakly with the 260-bp fragment of *myo*-3. The evidence linking *myo*-1 to MHC D is less direct than the evidence linking *myo*-3 to MHC A and *myo*-2 to MHC C. MHC A and D migrated as a single band on the original immunoblots used to classify the reactivity of the monoclonal antibodies. However, since antibody 5–25 reacts strongly with the A + D band from immunoblots and with epitope encoded by the 1577-bp fragment of *myo*-1 (Table I), and *myo*-3 is clearly assigned to MHC A by antibody 5-6, it follows that *myo*-1 must encode MHC D.

Taken together, these data suggest the following assignments: *myo*-1 encodes MHC D, *myo*-2 encodes MHC C; *myo*-3 encodes MHC A, and *unc*-54 encodes MHC B. Our results indicate that each of the nematode MHC isoforms is encoded by a discrete gene. It is likely that these four genes represent the entire nematode gene family. No other MHC-like sequences were detected in exhaustive screens of genome banks from *Caenorhabditis elegans* or on Southern blots. The reactivities of the monoclonal antibodies with the fusion peptides are also in agreement with this conclusion. We tested 12 different monoclonal antibodies that recognize epitopes encoded by fragments from all four genes. The patterns of reactivity we observed are fully consistent with a one-to-one correspondence between the cloned MHC genes and the known myosin isoforms.

By cloning smaller fragments of the myosin gene sequence, it should be possible to localize the recognition sites of monoclonal antibodies to within a few amino acid residues. Knowledge of the precise sequences recognized by the monoclonal antibodies could be used to relate the results of thick filament antibody-decoration experiments (Miller *et al.*, 1983) to molecular models of the myosin rod (McLachlan and Karn, 1982, 1983) and to distinguish between proposed models of myosin packing in the thick filament (Wray, 1979; Squire, 1981; Vibert and Craig, 1983).

Binding of monoclonal antibodies to subregions of the myosin molecule has been used to identify regions involved in filament assembly, ATPase activity, and the ability of myosin to catalyze movement *in vitro* (Flicker *et al.*, 1985; Kiehart and Pollard, 1984).

6. STRUCTURAL ORGANIZATION OF THE NEMATODE MHC GENES

The complete DNA sequence of the *unc*-54 gene and *myo*-1,2,3 have now been determined by cloning small DNA fragments into M13 vectors (Heidecker *et al.*, 1980; Messing *et al.*, 1981; Sanger *et al.*, 1980) for sequence analysis by the dideoxy method (Biggin *et al.*, 1983; Sanger *et al.*, 1977, 1980, 1982; Staden, 1980; Messing *et al.*, 1981).

The complete *unc*-54 DNA sequence is given in Karn *et al.* (1983*a*). We have followed the nucleotide numbering given in that paper throughout the text. Figure 14 shows the relative lengths of the introns and exons in the *unc*-54 gene and the *myo*-1,2,3 genes (Dibb *et al.*, 1987). Sequence comparisons are presented in Figures 15, 16, and 17. Figure 17 compares the deduced amino acid sequences of *unc*-54, *myo*-1,2,3 with the deduced amino acid sequence of a rat cardiac myosin heavy chain determined by Strehler *et al.* (1986).

6.1. Highly Conserved Head Sequences and Variable Rod Sequences

The protein-coding sequences of the genes in the nematode myosin heavy-chain family are extremely similar, and have clearly arisen by a series of gene duplication events (Karn *et al.*, 1983*a;* McLachlan and Karn, 1982, 1983). The hybridization results described in Section 4.4 suggested that the globular head regions of the proteins are better conserved than the α-helical rod regions, which appear more tolerant of amino acid substitutions (Section 7.2). The sequencing data confirm and extend this finding. In the myosin head region, MHC C (*myo*-1) and MHC D (*myo*-2) show 659 out of 859 matching amino acids (76.7%). By contrast, in the rod region, MHC C and MHC D show only 505 of 1091 matching amino acids (46.3%). Similar patterns of sequence conservation can be detected between each pair of the nematode MHC sequences.

Most of the amino acid substitutions in the head and the rod are conservative and preserve the chemical characteristics of the sequences (Karn *et al.*, 1983*a;* Kavinsky *et al.* 1983; MacLachlan and Karn, 1982, 1983). However, throughout the MHC sequence, restricted regions show unusually high levels of divergence. In the head region these are located at the junctions between the major proteolytic domains (see Section 7.1.4, Table III) and in the 25,000-M_r region (amino acids 50–72) as well as the 23,000-M_r region (amino acids 749–778 and 820–835). In the rod the S-2 region shows the most variation, while the LMM region from amino acids 1555–1670 is especially well conserved. This region is likely to participate in the strongest interactions involved in filament assembly (Section 7.2.3). These sequence comparisons highlight essential residues in the myosin molecule. The extremely high conservation of sequence in the myosin head suggests that most residues in the sequence are essential.

Sequence conservation within the nematode gene family is higher than sequence conservation between nematode and rat embryonic MHC (Strehler *et al.*, 1986) or rabbit skeletal sequences (Elzinga and Collins, 1977; Capony and Elzinga, 1981; Lu, 1980; Kavinsky *et al.*, 1983; M. Elzinga, unpublished data). However, the pattern of high sequence conservation in S-1 remains. The rat embryonic heavy chain is 1939 residues long and aligns well with the nematode sequences. The few length differences between each of the sequences are attributable to changes at both termini and in the loop structures separating the proteolytic domains of the myosin head (Section 7.1.4). Strehler *et al.* (1986) calculated the sequence homology between the rat MHC and *unc*-54 MHC using a program that scores for the relatedness of amino acids. They found an overall homology of 48.7% between rat and *unc*-54 MHC as compared with a 81.6% homology between rabbit skeletal MHC and rat embryonic MHC. Shorter stretches of amino acid sequence, mostly from the rod portion, are available for MHCs from other species. The values for the degree of homology to rat MHC are: *Drosophila* muscle (Rozek and Davidson, 1983) 64.4%; chicken embryonic skeletal muscle (Kavinsky *et al.*, 1983) 81.2%; rat cardiac MHC (Mahdavi *et al.*, 1984); 73.9%. In contrast to the high degree of overall amino acid sequence conservation among sarcomeric MHCs from nematodes to mammals, no homology is detected between nonmuscle MHC sequences from *Dictyostelium discoideum* (L. DeLozanne and J. Spudich, personal communication), *Acanthamoeba* myosin II (G. Jung, B. M. Paterson, and E. D. Korn, personal communication), and *Drosophila* nonmuscle MHC (D. P. Kiehart, personal communication).

6.2. Selective Loss of Introns from the Nematode MHC Genes

Each of the nematode MHC genes is split by a series of short introns (Figs. 14, 15, and 16 and Table II) (Dibb *et al.*, 1987). The *unc*-54 gene is split by eight short introns that vary in size from 561 to 37 bp. Five of

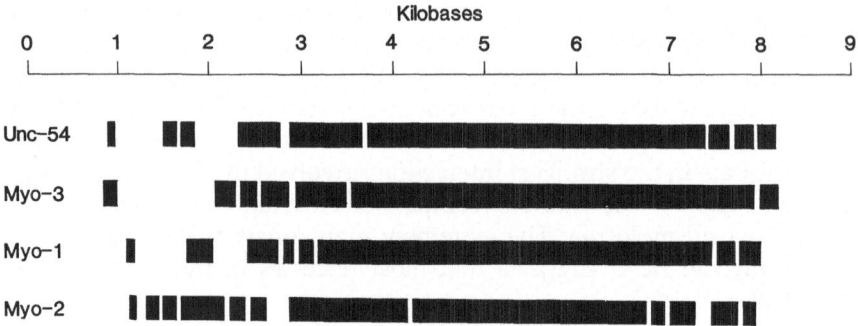

Figure 14. Alignment of coding sequences from the *unc*-54, *myo*-1,2,3 genes. Coding sequences are shown as shaded bars; introns are the variable length gaps that separate the coding sequences.

the eight introns are in the myosin head region. The organization of the other MHC genes is similar (Fig. 14 and Table II). The *myo*-2 gene has 11 introns, *myo*-1 has 7 introns, and *myo*-3 only 6 introns. As in *unc*-54, the introns in the other myosin heavy chain genes are unusually short; most are approximately 50 bp, the longest is 1069 bp. Twenty-two of the 32 mapped introns are in the head regions. By contrast, in the rat embryonic MHC (Strehler *et al.*, 1986) 40 introns are present and the gene extends over 24 kb. The introns are not clustered in the head region; 19 introns interrupt the rod sequences.

The introns show interesting patterns of conservation throughout the gene family (Table II). Three introns (after amino acids 119, 278, and 1909) are shared throughout the gene family. However, there are examples of introns present in only two or three of the genes (for example, after amino acid 26 in *myo*-1,2 and *unc*-54, and after amino acid 1763 in *myo*-3 and *unc*-54). Each gene also has uniquely positioned introns. Remarkably, the positions of most of the introns in the nematode gene family correspond to the positions of introns in the rat embryonic MHC gene (Table II). This high degree of conservation of intron locations between the nematode MHC family and the mammalian sarcomeric MHC gene indicates that each evolved from an ancestral MHC gene containing at least all the common introns. This ancestral sequence predates the separation of the vertebrate and invertebrate lineages 5×10^8 years ago (Dayhoff, 1978). It seems clear that introns have been selectively lost from the nematode MHC genes.

Interpretation of the order of these events is complicated by the possibility that recent gene conversion events have either shifted or removed introns. For example, introns of 47 bp are present within amino acid 334 of the *myo*-1 sequence and after amino acid 338 of *myo*-2 (Figs. 15 and 17; Table II). The coding sequence immediately preceeding the intron junction in *myo*-2 shows a 12-bp inverted repeat. In the *myo*-1 gene the intron has been shifted to divide the repeat.

Introns are frequently found between regions coding for structural units in proteins (Cochet *et al.*, 1979; Dugaiczyk *et al.*, 1978; Gilbert, 1978; Lewin, 1982; Wozney *et al.*, 1981). The myosin heavy-chain genes from nematode do not always follow this pattern. Introns do not separate the three major proteolytic subfragments of the head (Section 7.1), nor do they separate head and rod sequences. The rod sequence is highly periodic (Section 7.2) and shows a more regular and persistent repeat than any other known fibrous protein. It is notable that the introns in the rod sequences do not separate repetitive elements in the rod, are not regularly spaced, and are not obviously associated with any of the special features of the rod sequence (Section 7.2).

6.3. Unusual Features at Intron Junctions

The GT/AG rule is followed at nematode intron junctions, but some deviations from vertebrate consensus sequences are observed (Mount, 1982). The 5' end of the introns is defined by the 8-bp consensus sequence

$$G(20)T(20)A(16)A(15)G(12)T(14)T(15)T(13)$$

in which the numbers in parentheses give the frequency of occurrence of the nucleotide in 20 junction sequences. The 3' ends of the introns are defined by the 9-bp consensus sequence

$$T(11)T(15)T(11)T(18)T(22)T(13)C(17)A(22)G(22)$$

in which the numbers in parentheses give the frequency of occurrence of the nucleotide in 22 junction sequences. Both consensus sequences are longer and better defined than the corresponding vertebrate consensus sequences (Mount, 1982).

Statistical analysis of codon usage within myosin introns and exons (Staden and McLachlan, 1982) has shown that codon-usage bias in coding sequences is not followed in introns that are A + T rich and frequently contain homopolymer tracts. These features are also found in introns in the nematode actin genes (Files *et al.*, 1983; Hirsh *et al.*, 1982),

```
            T   Y   S   G   L   F   C   V   V   I   N   P
ATGTTCCCAATTGCCCATTTTCAGACCTACTCTGGACTTTTCTGCGTTGTCATCAACCCAT
*** *   * *   *** **** ** **  ** ******** ******************
ATGATTTAACATATTTCCTTTTAGACTTATTCGGGTCTTTTCTGTGTTGTCATCAACCCAT
            T   Y   S   G   L   F   C   V   V   I   N   P


    T   E   M   P   P   H   L   F   A   V   S   D   Q   A   Y   R   Y   M   L   Q
ACCGAGATGCCACCTCATTTGTTCGCTGTCTCTGATCAAGCCTACCGTTACATGCTTCAAG
***** *********** ********* ********  ******** *  *********** 
ACCGAAATGCCACCTCACTTGTTCGCCGTCTCTGACGAAGCCTACAGAAACATGCTTCAAA
    T   E   M   P   P   H   L   F   A   V   S   D   E   A   Y   R   M   M   L   Q


    T   K   K   V   I   C   Y   F   A   T   V   G   A   S   Q   K   A   A   L   K
ACAAAGAAGGTTATTTGCTACTTTGCCACCGTCGGAGCTTCACAAAAGGCTGCTTTGAAG-
***** ********* *  ********* * ***** *** * *** * *   * **
ACAAAAAAGGTTATCTCTTACTTTGCCGCTGTCGGTGCTGCTCAACAAGAGACCTTCGGAG
    T   K   K   V   I   S   Y   F   A   A   V   G   A   A   Q   Q   E   T   F   G


    I   V   Q   T   N   P   V   L   E   A   F
ATCGTTCAAACCAATCCAGTATTGGAAGCTTTCGGTACAGACATAAATTAAAAAATTAACA
** **  ************** * ** ******
ATTGTCCAAACCAATCCAGTTCTCGAGGCTTTC----------------------------
    I   V   Q   T   N   P   V   L   E   A   F


    R   F   G   K   F   I   R   I   H   F   N   K   H   G   T   L   A   S   C   D
CCGTTTCGGAAAGTTCATCCGTATTCACTTCAACAAGCACGGAACTCTTGCTTCCTGCGAT
************************************  ******** ** * * ***** ******
CCGTTTCGGAAAGTTCATCCGTATTCACTTCTCCAAGCAAGGACGTGTCGCTTCTTGCGAT
    R   F   G   K   F   I   R   I   H   F   S   K   Q   G   R   V   A   S   C   A


    Y   L   L   E   K   S   R   V   I   R   Q   A   P   G   E   R   C   Y   H
GTTCAGATCTTCTCGAGAAATCTCGTGTCATCCGTCAAGCTCCAGGAGAGCGTTGTTACCA
*  ********** ** ** ** ****  ******************************  ******
TTCCAGATCTTCTTGAAAAGTCACGTGTGATCCGTCAAGCTCCAGGAGAGCGTTCTTACCA
    Y   L   L   E   K   S   R   V   I   R   Q   A   P   G   E   R   S   Y   H


    H   P   I   S   N   Y   W   F   V   A   Q   A   E   L   L   I   D
ACCATCCAATTTCCAACTATTGGTTTGTTGCTCAAGCTGAGTTGCTCATTGATGGTATGTT
** * *** *    **** ****** *  *********************  ******
ACAAGCCAGTCAAGGACTACTGGTTTATCGCTCAAGCTGAGTTGATCATTGATGGTATCAA
    K   P   V   K   D   Y   W   F   I   A   Q   A   E   L   I   I   D   G   I   N


    F   Q   L   T   D   E   A   F   D   V   L   K   F   S   P   T   E   K   M   D
AGTTCCAATTGACCGACGAGGCCTTTGATGTTCTCAAATTCTCGCCCACCGAGAAGATGGA
**   ** ******** ************ * ***** **  * ** ** ** ********
AGCATCAGTTGACCGATGAGGCCTTTGATATCCTCAAGTTTACCCCAACTGAAAAGATGGA
    H   Q   L   T   D   E   A   F   D   I   L   K   F   T   P   T   E   K   M   E
```

Figure 15. Sequence comparison between *myo*-1 (upper sequence) and *myo*-2 (lower sequence) genes. A region of the head sequence from thr-119 to pro-380 is shown. Stars between the lines indicate the matching nucleotides. The MHC C amino acid sequence is above the *myo*-1 nucleotide sequence, the MHC D amino acid sequence is below the *myo*-

```
Y  K  R  L  P  I  Y  T  D  S  V  A  R  M  F  M  G  K  R  R
ACAAACGTCTCCCAATTTACACTGACTCTGTTGCTCGTATGTTCATGGGCAAGCGTCGT
*****************************  **  *****  *****  **  **  ***************  ***
ACAAACGTCTCCCAATTTATACCGACTCCGTTGCCCGCATTTTCATGGGCAAGCGCCGT
Y  K  R  L  P  I  Y  T  D  S  V  A  R  I  F  M  G  K  R  R


D  H  E  N  Q  S  M  L  I  T  G  E  S  G  A  G  K  T  E  N
ACCATGAGAATCAGTCTATGCTTATTACCGGAGAATCTGGAGCCGGAAAGACTGAAAAC
****  **  **  **********  ****  **************  **  *************  ***
ACCACGAAAACCAGTCTATGCTCATTACCGGAGAATCTGGTGCAGAAAGACTGAGAAC
M  H  E  N  Q  S  M  L  I  T  G  E  S  G  A  G  K  T  E  N


                           E  G  E  K  E  V  T  L  E  D  Q
----------------------------GAGGGAGAGAAGGAAGTTACCCTTGAAGACCAG
                           **  *****  ****  **  *****  ******
CCAAGAAGGCCGCCACCGAAGAAGACAAGAACAAGAAGAAAGTCACACTTGAGGACCAG
A  K  K  A  A  T  E  E  D  K  N  K  K  K  V  T  L  E  D  Q


                           G  N  A  K  T  V  R  N  N  N  S  S
CTTGTCTGATAATTTGGTTGTTTAGGTAACGCCAAGACTGTCCGTAACAACAATTCTTC
                           *****  *************************  **  **  **
----------------------------GGTAATGCCAAGACTGTCCGTAACAATAACTCCTC
                           G  N  A  K  T  V  R  N  N  N  S  S


I  E  H
ATTGAACACTGTAAGTTGATAATAATGATAAGAAGTTA---TTACGAAAATCGTAAATT
********  ********  ***  **  *  **  *  *  *  *****
ATTGAACATTGTAAGTTTGAAATCATTTCGAATGTTTGACCTGTATATATTTAAAAATT
I  E  H


   I  F  Y  Q  I  Y  S  D  F  K  P  Q  L  R  D  E  L  L  L  N
TATCTTCTACCAAATCTACTCAGATTTTAAGCCACAACTCCGTGACGAGCTTCTTCTCA
   **  *********  ***  ***  ****  ****  *  **  *  **  **  *****  *
CATTTTCTACCAAGTCTTCTCCGATTACTTGCCAAAACCTTAAAAAGGACCTCCTTCTTA
   I  F  Y  Q  V  F  S  D  Y  L  P  N  L  K  K  D  L  L  L  N


                              G  I  D  D  T  E  E
TTGTATTTCTGGAGAATGAGATTAACTTCGTTTTATTCCAGGAATCGATGACACTGAGG
*  *  *  *  *  **  *  *  **  *
TGACAAGGTAATACTGATCAAAACATGACGAAATGCAGATATTGTTTTCTTGTAGGAAG
   D  K                                          E  E


C  Y  R  L  M  S  A  H  M  H  M  G  N  M  K  F  K  Q  R  P
TTGTTATAGACTCATGTCGGCTCACATGCACATGGGTAACATGAAATTCAAGCAACGCC
**  **  **  *  *  *  **  ********************  ***********  *
GTGCTACCGATTGGTTGCCGCCATGATGCACATGGGTAACATGAAGTTCAAGCAACGTC
C  Y  R  L  V  A  A  M  H  M  G  N  M  K  F  K  Q  R  P
```

2 nucleotide sequence. In order to align the sequences hyphens(---) have been inserted. The original sequences are continuous. Note the presence of shared and unique introns, and the high degree of nucleotide matching in the coding regions.

```
    R  S  S  M  D  N  L  S  E  Q  I  E  T  L  R  R  E  N  K
TTCGTTCATCCATGGACAACCTCAGCGAGCAAATCGAGACCTTGAGACGCGAGAACAAGA
* ***  *     ** * *      ** *** *  **      *  * **********
TCCGTGGACAGCACGATACCTTGGCTGATCAAGTTGAAGGACTTCGTCGCGAGAACAAAT
    R  G  Q  H  D  T  L  A  D  Q  V  E  G  L  R  R  E  N  K


    E  V  H  K  S  V  R  R  L  E  Q  E  K  D  E  L  Q  H  A
AAGAGGTTCACAAGTCTGTCCGCCGTCTCGAACAAGAAAAGGACGAGCTTCAACATGCTC
* *    *   ****    * ** *** * ***    ** ***** ** ********* **
ACGCTCTCTCCAAGAACCTTCGTCGTTTGGAAATGGAGAAGGAAGAACTTCAACGTGGAT
    A  L  S  K  N  L  R  R  L  E  M  E  K  E  E  L  Q  R  G


    V  Q  Q  I  R  S  E  I  E  K  R  I  Q  E  K  E  E  E  F
AGGTTCAACAGATCCGTTCAGAAATCGAGAAGAGAATTCAAGAGAAGGAGGAGGAATTCG
*****    ** ****** *  ***** ***********      ****************
AGGTTTCCCAAATCCGTGCTGAAATTGAGAAGAGAATCGCCGAGAAGGAGGAGGAATTCG
    V  S  Q  I  R  A  E  I  E  K  R  I  A  E  K  E  E  E  F


    A  K  S  K  A  E  L  A  R  A  K  K  K  L  E  T  D  I  N
AAGCTAAGAGCAAGGCCGAGCTCGCCAGAGCCAAGAAGAAGCTTGAGACTGACATCAATC
*   * ***  **** *  *****    *  *  *  *********** *  **   **  *
AGACCAAGGCCAAGTCAGAGCTTTTTCCGTGTCAAGAAGAAGTTGGAAGCCGATATCAACG
    T  K  A  K  S  E  L  F  R  V  K  K  K  L  E  A  D  I  N


                                         K  L  F  D
A------------------------------------------GAAACTTTTCGAC
                                         ** ***
GGTAATTTATAAGTTTTTTTTTTGTGTACACCAACTTTTTCATTCAGACGCTACTTGGAC
                                         R  Y  L  D


    E  N  Y  L  A  A  E  K  R  L  A  I  A  L  S  E  S  E  D  L
GAGAACTATTTGGCTGCTGAGAAGAGACTTGCTATTGCTCTTTCTGAAAGCGAGGATCTT
*** **  ********** ****  *  *** *** **** ***      ** ** **  **
GAGCACCTTTTGGCTGCCGAGAGAAAACTCGCTGTTGCCAAACAGCAGCAAGAAGAGCTT
    E  H  L  L  A  A  E  R  K  L  A  V  A  K  C  E  Q  E  F  L


       S  D  K  H  K  K  Q  L  E  I  E  Q  A  E  L  K  S  S  M
----TCTGATAAGCATAAGAAACAACTCGAAATCGAACAGGCTGAACTCAAGTCTTCCAA
       ***      ****** *    *    ****  *       **
TCAGCTTGAGCGTGCCCGTCGTGTTGTCGAAAGCTCGGTGAAGGAACATCAAGAGCATAA
       L  E  R  A  R  R  V  V  E  S  S  V  K  E  H  Q  E  H  N


    V  Q  I  A  R
GGTTCAGATTGCCAGAGTGAGTTTATTAAACGGATTGTCAGATTTTAAAAAATAAAACTT
**     **    **
AATTGCCCTTCTCAAC--------------------------------------------
    I  A  L  L  N
```

Figure 16. Sequence comparison between the *myo*-1 (upper sequence) and *myo*-2 (lower sequence). A region of the rod from arg-1509 to lys-1781 (*myo*-1) or arg-1781 (*myo*-2) is shown. Stars between the sequences indicate matching nucleotides. The MHC C amino acid sequence is above the *myo*-1 nucleotide sequence; the MHC D amino acid sequence

```
  I  F  S  Q  E  I  R  D  I  N  E  Q  I  T  Q  G  G  R  T  Y  Q
TCTTTTCGCAAGAGATTCGCGACATTAACGAGCAAATCACCCAAGGAGGACGTACATACC
   *     *  ****  **  *** *  *  *** *  *  **  *  ********* *   **
CTCTCAGCGACGAGACCCGTGACCTCACCGAATCACTTTCCGAGGGAGGACGTGCTACCC
  S  L  S  D  E  T  R  D  L  T  E  S  L  S  E  G  G  R  A  T  H


  L  D  E  A  E  A  A  L  E  A  E  E  S  K  V  L  R  L  Q  I  E
TTGATGAGGCCGAGGCTGCACTTGAAGCCGAAGAGAGCAAGGTTCTCCGTCTTCAAATCG
* **  *****  **  ***** ** ***  *  **  ** ***** ** *  ***   ** ****
TGGACGAGGCTGAAGCTGCCCTCGAATCTGAGGAAAGCAAGGCTCTCCGTTGCCAGATCG
  L  D  E  A  E  A  A  L  E  S  E  E  S  K  A  L  R  C  Q  I  E


  E  N  T  R  K  N  H  Q  R  A  L  E  S  I  Q  A  S  L  E  T  E
AGAACACTCGCAAGAACCATCAACGTGCCCTCGAGTCCATCCAAGCTTCCCTCGAAACCG
*****   ***** **  ****   **  ****   ********** *   *  **   **
AGAACCACCGCAAGGTTCACCAACAAACCATCGACAGCATCCAAGCTACTTTGGACTCCG
  E  N  H  R  K  V  H  Q  Q  T  I  D  S  I  Q  A  T  L  D  S  E


  Q  L  E  I  A  L  D  H  A  N  K  A  N  V  D  A  Q  K  N  L  K
AACTCGAGATTGCTTTGGATCATGCCAACAAGGCTAACGTTGACGCCCAGAAGAACTTGA
* **  *****  **  *  * **  *****  ***** **  ****   **  ***********  *  *
AGCTGGAGATCGCCCTCGACCATGCGAACAAAGCGAACGAAGATGCCCAGAAGAATATCA
  E  L  E  I  A  L  D  H  A  N  K  A  N  E  D  A  Q  K  N  I  R


  Q  V  K  E  L  Q  G  Q  V  D  D  E  Q  R  R  E  E  I  R
CAAGTCAAGGAACTTCAAGGACAAGTCGATGATGAGCAACGCAGACGTGAGGAGATCCGT
***  **    **  **  ****** *     **  *  ***** ** ***   ***** **  **  *****
CAAATCCGCGAGCTTCAACAAACCGTGGATGAGGAACAAAAGAGCGCGCGAAGAATTCCGT
  Q  I  R  E  L  Q  Q  T  V  D  E  E  Q  K  R  R  E  E  F  R


  A  H  R  I  E  A
GCTCACCGCATCGAGGCA--------------------------------------------
   *        *  *****
ATTGTTAAGCTTGAGGCTGTAAGAATTTGTTTCGGAAATCCCCTTGTATGTGATTTTTTT
  I  V  K  L  E  A


  T  E  L  I  G  N  N  A  A  L  S  A  M  K  R  K  V  E  N  E
CACTGAGCTTATTGGAAACAATGCTGCTCTTTCTGCCATGAAGAGAAAGGTTGAGAACGA
**  **  **  *      *  **** *** **  ****     *****  *  *  **  *****
CAACGAACTCAACTCCCAAAATGTTGCCCTCGCTGCTGCCAAGAGCCAACTCGACAACGA
  H  E  L  N  S  Q  N  V  A  L  A  A  A  K  S  Q  L  D  N  E


  N  E  L  D  E  Y  L  N  E  L  K  A  S  E  E  R  A  R  K
TCAGAACGAATTGGACGAGTACTTGAACGAGCTTAAGGCTTCAGAGGAGCGTGCCCGCAA
* ***  **  *  ***  *   **  ***   **  **  **  **** **
----AGCGACATCGCTGAGGCCCACACTGAACTTTCCGCATCCGAAGACCGTGGACGTCG
  S  D  I  A  E  A  H  T  E  L  S  A  S  E  D  R  G  R  R
```

is below the *myo*-2 nucleotide sequence. In order to align the sequences hyphens (---) have been inserted. The original sequences are continuous. Note the presence of uniquely placed introns, and the lower number of nucleotide and amino acid matches than in the head region (Fig. 15).

Table II. Positions and Lengths of Introns in the Nematode MHC Gene Family and Rat Embryonic MHC

Amino acid residue[a,b]	myo-1	myo-2	myo-3	unc-54	Rat
26/27	581	110		562	
46		1069			
68/69					1954
69				38	
69/70		55			
119/120	369	49	69	480	514
172			47		373
181					296
217/218					496
243	52				
256/257					768
278	52	55	98	79	84
311					2056
334	47				
338/339		48			
345/346					87
392		260			587
431/432					85
454			64		
482/483					963
539/540				53	85
642/643					116
673/674					872
703					84
742					360
786/787					93
822/823		52			
832					492
917/918					193
998/999					72
1057/1058					126
1106					88
1136/1137					731
1266/1267					90
1308/1309					861
1348/1349					166
1414					431
1475/1476					583
1531					282
1572/1573					486
1668		46			
1675/1776					111
1718/1719		48			
1743/1744					72

Table II. (Continued)

Amino acid residue[a,b]	myo-1	myo-2	myo-3	unc-54	Rat
1762/1763	48			57	
1785/1786					766
1815/186		189			
1831	49				
1834/1835				48	
18842/1843					103
1877/1878					144
1909/1910	450	46	76	52	138
1955/1956					429

[a] The amino acid residue number indicates the amino acid split by the intron (Fig. 17).
[b] When introns follow an amino acid codon, this is indicated by giving the residue number of the amino acids preceding and following the intron, separated by a slash (/).
[c] The lengths of introns in the *unc-54*, *myo*-1,2,3, and rat MHC gene (Strehler *et al.*, 1986) are given in base pairs.

collagenlike genes (Kramer *et al.*, 1982), and yolk protein genes from *Caenorhabditis elegans* (T. Blumenthal, personal communication).

6.4. Terminal Sequences

Figure 18 compares the *unc*-54 gene sequence and the 5'-terminal sequence of *myo*-2. Two blocks of homology, separated by 27 bp, are found upstream of the initiator methionine. It seems likely that these sequences represent the promoter sequences for the genes and the capping site for the mRNA. Neither shows a TATA box sequence (Benoist and Chambon, 1981) near the beginning of the gene. Similar homologies are detected in pairwise comparisons between each of the nematode MHC genes. Studies of signals for differential gene expression in nematodes will probably make use of the MHC gene family. Although these genes are highly conserved, the *myo*-1 and *myo*-2 genes are exclusively expressed in the pharynx, while the *unc*-54 and *myo*-3 genes are exclusively expressed in the body wall muscle. A useful approach may be to look for DNA-binding proteins that can specifically recognize sequences in the promoter regions of these genes. The nematode actin and collagenlike genes have normal TATA box sequences (Files *et al.*, 1983; Kramer *et al.*, 1982).

Typical polyadenylation signals, AATAAA, are present shortly after the termination codons of each of the nematode genes. In some cases,

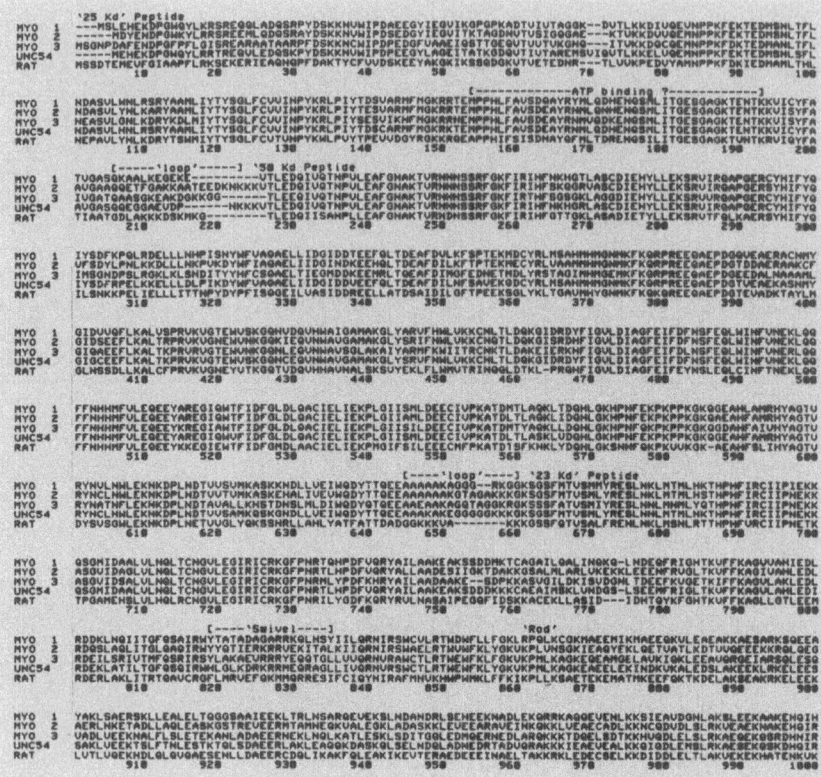

Figure 17. Comparison between the amino acid sequences of MHC C (*myo-1*); MHC D (*myo-2*); MHC A (*myo-3*); MHC B (*unc-54*); rat embryonic sarcomeric MHC (rat) (Strehler *et al.*, 1986). Positions of major sequence features are indicated in the diagram. The se-

there are several tandem AATAAA signals. We have not yet determined the exact termini for the MHC mRNAs.

7. MOLECULAR ANATOMY OF THE MYOSIN MOLECULE

The native myosin molecule is a multimer composed of two heavy chains (each of 230,000 M_r) and two pairs of light chains (16,000–23,000 M_r) (Lowey *et al.*, 1969; Elliot and Offer, 1978). Electron microscopic examination exhibits individual myosin molecules as long thin tails approximately 20 Å in diameter and 1500 Å long (myosin rods), with two

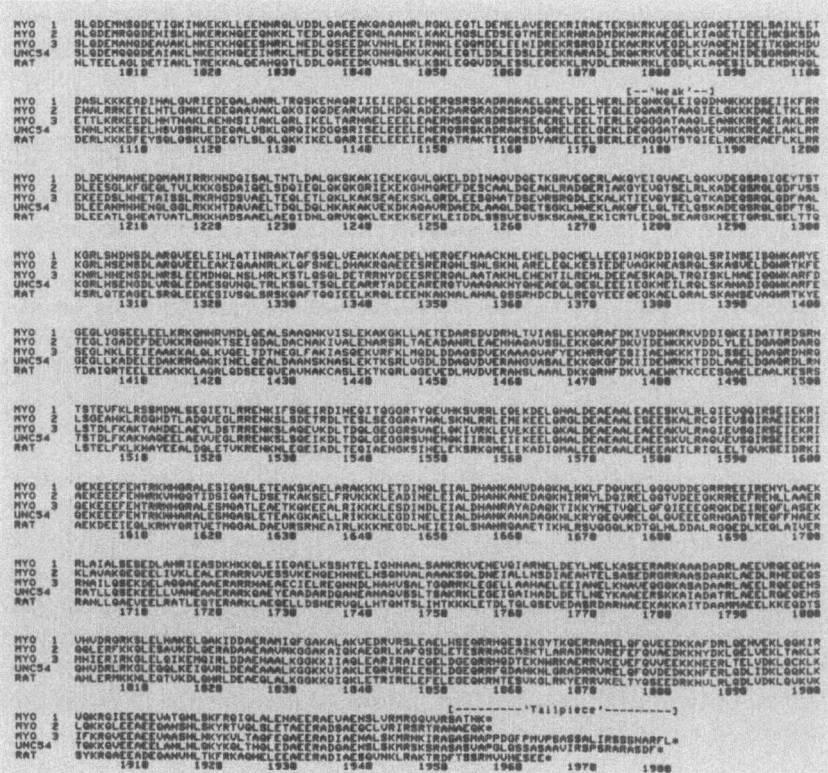

quences have been aligned using hyphens (---) as padding characters. The original sequences are continuous. Throughout the text and figure legends, amino acids are numbered according to this diagram.

globular heads about 70 Å in diameter and 200 Å long emerging at one
end. Each myosin heavy-chain subunit consists of one complete globular
head at the NH₂ end and a long fibrous α-helical tail at the COOH end.
In a native myosin molecule, helices from each heavy chain in the dimer
lock together to form a single coiled coil, while the two heads diverge.
The light chains bind to the head near the neck of the molecule near
the head/rod junction (Vibert and Craig, 1982; Flicker *et al.* 1983; Winkelmann *et al.*, 1984). Although light chains are not required for the
ATPase activity of the head (Wagner and Giniger, 1981; Sivaramakrishnan and Burke, 1982), recent NMR work has shown that the vertebrate
skeletal A1 light chain participates directly in actin binding (Henry *et*

```
                                        [---------- Homolog
TCGTAGAATCCCCAAAGAGGGGCGTGGCTTGCGGGTGCCCAACATCCTCCTGCCGAGGAAG
 *    *  *                    *   * * *  *   *   **** **  ** *** *
CACTTTCAGGCTACCTAGATACATGGATATCCCCGCCTCCCAATCCACCCACCCAGGGAA

                                          M  E  H
AGGTAGAGAA-----------------ACCATTTGAAAGAAGCGAGAAATCATGGAGCAC
 *                          * *   * *  **  *        *   ***** **
TGACTTCTCTCCACCACTTTTCATTTTAACCCTCGATCGTCAGACACAGAAATGGATTAC
                                                    M  D  Y

ATTTATTTTTTATTTATTTATTTTTCTTTCATTTGTCTTTATTGTTATGCTCATTTCTAT
 *  **      *  ** *** * *        ** *      *  ***   * *      *
TTATAAAGCATGGTTTTTTTGTCTAAAGAATGTTATAAGCTTGCTTAGTTTAAAAAAAAA
```

Figure 18. Comparison between the nucleotide sequences of the *unc*-54 gene (upper sequence) and the *myo*-2 gene (lower sequence) in the region of the 5' exon. Stars between the sequences indicate identical nucleotides. In order to align the sequences hyphens (---) have been added. The original sequences are continuous. The MHC B amino acid

al., 1985). One class of light chains, the regulatory light chain, are involved in Ca^{2+} regulation in smooth muscle and molluscan muscle but their role in vertebrate muscle is not understood (Adelstein and Eisenberg, 1980; Frank and Weeds, 1974; Kendrick-Jones *et al.*, 1976; Weeds and Lowey, 1971).

The complete amino acid sequence of the 1966 residue *unc*-54 myosin heavy chain and the *myo*-1,2,3 genes are aligned in Fig. 17. In this section we describe how features of these sequences account for many of the chemical properties of myosin. This information forms the basis for the interpretation of the myosin mutants described in Section 8.

7.1. Topography of the Head

Proteolytic enzymes such as trypsin and papain divide the myosin heavy chain into characteristic structural fragments (Lowey *et al.*, 1969; Lu, 1980; Weeds and Pope, 1977). The rod is cut at a specific site about one-third of the way along its length, separating the COOH-end portion (light meromyosin or LMM) from two intact heads attached to a truncated rod (heavy meromyosin or HMM). Further digestion of HMM detaches the globular head (subfragment-1, or S-1) from the amino terminal of the rod (subfragment-2, or S-2) (Weeds and Pope, 1977). The head itself contains three major proteolytic segments that are probably structural domains (Balant *et al.*, 1978; Cardinaud, 1979; Mornet *et al.*, 1979; 1981*a,b*). The positions of the major proteolytic cuts are known in rabbit (M. Elzinga, personal communication) and have been identified in the nematode sequences by homology (Figs. 17 and 24). The calculated

```
U 1 -----------]                           [Homology 2]
AAGCAGGCACTCATCACTCGCATCATCAACCTCGACTAAGGGAGCACCACATTAGTTTTG
** ** * ***  *              *            * *** * **
AAAGAAGGGCTCGCCGAAAAATCAAAGTTATCTCCAGGCTCGCGCATCCCACCGAGCGGT

E   K   D   P   G   W   Q   Y   L   R   R   T   R   E   Q   V   L   E
GAGAAGGACCCAGGATGGCAATATCTCCGCCGTACCAGAGAGCAGGTTTTGGAGGTGAGT
** ** ************** * ** ** ***** *  * *** *** * * **** *
GAAAACGACCCAGGATGGAAGTACCTTCGCCGCAGTCGCGGGAGATGCTCCAGGTACTT
E   N   D   P   G   W   K   Y   L   R   R   S   R   E   E   M   L   Q

AGAGTAAATCGTATAGTACTCCATAAGTATCTAATTTTAACAAAACTGAGCATTTATTTT
*       *  ** * *
AATCATCTTTATAAATTTTC-----------------------------------------
```

sequence is shown above the *unc*-54 nucleotide sequence; the MHC C amino acid sequence is shown below the *myo*-2 nucleotide sequence. Note the two blocks of 5'-terminal homology. (After Karn *et al.*, 1983a.)

molecular weights of the major domains of MHC B are given in Table III. Throughout the text we refer to the NH$_2$-terminal domain as 25,000 M_r, the central domain as 50,000 M_r, the COOH-terminal domain as 23,000 M_r.

7.1.1. ATP-binding Sites. The amino-terminal segment 25,000-M_r fragment contains important components of the ATP-binding site. A photoreactive analogue of ATP reacts with a peptide from the 25,000-M_r segment and binds to tryptophan-134 (trp-134) (Okamoto and Yount,1983; Nakamaye *et al.*, 1985). ATPase activity can also be selectively inhibited by the reaction of lys-89 with 2,4,6-trinitrobenzene sulfonate (Fabian and Muhlrad, 1968; Hozumi and Muhlrad, 1981; Kubo *et al.*, 1960, 1965; Takahashi *et al.*, 1982). Lys-133 is trimethylated in rabbit myosin (M. Elzinga, personal communication). This ensures that the nitrogen atom remains positively charged within the protein even when close to a phosphate group. Recently, the ATPase site was visualized at about 140 Å from the head/rod junction by electron microscopy using a biotinylated ATP analogue and avidin (Sutoh *et al.*, 1986).

Walker *et al.* (1982) noted that a wide range of ATP-binding proteins share related sequences that may be involved in nucleotide binding. Homologies have been noted between the α- and β-subunits of membrane ATP synthetase complex, adenylate kinase, *E. coli recA*, phosphofructokinase, and myosin. The region of homology in myosin includes amino acids 154–194. In the adenylate kinase crystal structure (Pai *et al.*, 1977), the adenine nucleotide-binding site is a hydrophobic pocket formed between a β-sheet, a glycine-rich loop, and an α-helix. The myosin sequence can be folded into an analogous structure with amino acids 176–181 forming the β-sheet, 192–189 the loop, and 190–203 the α-helix. In this model, ser-184 and lys-188 would also participate in ATP binding.

7.1.2. Actin Binding. Chemical cross-linking studies indicate that actin probably binds to both the 50,000- and 23,000-M_r regions (Mornet *et al.*, 1979, 1981*a,b*; Labbe *et al.*, 1982; Sutoh, 1983). The actin to S-1 ratio in the complex is equimolar (Heaphy and Treager, 1984; Greene, 1984; Chen *et al.*, 1985). Chaussepied *et al.* (1986*a,b*) recently showed that cleavage by thrombin of DTNB-treated S-1 specifically cuts within the 50,000-M_r region and abolished actin-activated ATPase activity. This indicates that a major role of the 50,000-M_r segment is for transmission of signals between the 23,000-M_r regulatory domain, containing the light chain and actin-binding sites, and the ATP-binding site.

The 23,000-M_r segment contains two reactive SH groups called SH-1 (cys-726) and SH-2 (cys-716) (Burke and Reisler, 1977; Elzinga and Collins, 1977; Wells and Young, 1979, 1980). Wells and Young (1979) cross-linked the two SH groups with a variety of cross-linking agents with spacings of 2–14 Å. Separation of the active thiol residues by the spacer groups of the crosslinking agents is necessary for the ATP-photoaffinity analogue to bind near trimethyllysine-133 (Okamoto and Yount, 1983). These experiments suggest that there is a cleft between the 25,000- and 23,000-M_r segments, which can be opened and closed by the movement of cys-726 and cys-716 or the binding of actin.

The actin-binding region in the 23,000-M_r segment appears to be near SH-1 and SH-2 (Katoh *et al.*, 1984; Katoh and Morita, 1984). Recent localization studies place the actin-binding site between SH-1 and the essential light-chain-binding site (Katoh *et al.*, 1984; Mitchell *et al.*, 1986). Electron microscopic visualization of SH-1 with an avidin-biotin system indicates that it is about 130 Å from the head/rod junction (Sutoh *et al.*, 1984).

7.1.3. Light-Chain Binding. The light-chain binding sites (both regulatory and essential) are also in the 23,000-M_r segment of S-1 (Burke *et al.*, 1983; Burke and Kamalakannan, 1985; Sellers and Harvey, 1984; Szentkiralyi, 1984; Bennet *et al.*, 1984; Mitchell *et al.*, 1986). The light chains are localized in the neck of the molecule (Vibert and Craig, 1982; Flicker *et al.*, 1983; Winkelmann *et al.*, 1984). The binding region for the two kinds of light chains is located in the COOH-terminal half of the 23,000-M_r segment constituting a regulatory domain (Szentkiralyi, 1984; Mitchell *et al.*, 1986). The regulatory light chain binds closer to the COOH-terminus of the S-1 heavy chain than the essential light chain (Szentkiralyi, 1984; Bennett *et al.*, 1984; Mitchell, *et al.*, 1986). Immunoelectron microscopy has been used to place the N-terminal region of the regulatory light chain near the S-1/rod junction (Winkelmann *et al.*, 1983) and the N-terminal region of the essential light chain further into the S-

1 head (Waller and Lowey, 1985). Chemical cross-linking studies with scallop myosin have shown that the regulatory and essential light chains are in close proximity for much of their length and demonstrated relative movements of the light chains upon Ca^{2+} activation of myosin (Hardwicke et al., 1982, 1983; Hardwicke and Szent-Györgyi, 1985). The N-terminal segments of the two regulatory light chains on each of the myosin heads are in close proximity (Hardwicke and Szent-Györgyi, 1985; Vibert et al., 1985). It is postulated that the movements of light chains affect the structure of the regulatory domain and that interdomain motions regulate the actin- and ATP-binding sites (Vibert et al., 1986).

7.1.4. Proteolytic Domains Separated by Variable Sequences. The sequences from the head region of myosin are extremely well conserved (see Section 6.1). It is notable that the sites of proteolysis in the head correspond to regions of sequence variability that are rich in charged amino acids and glycine (Karn et al., 1983a). These sequences are likely to be flexible surface loops. In contrast to the rest of the MHC molecule, the loop regions tolerate changes in length and amino acid composition. The loop region between the 50,000- and 23,000-M_r fragments is much better conserved than is the comparable region between the 25,000- and 50,000-M_r fragments. The loop begins with a hydrophobic sequence of polyalanine, and ends after a glycine-rich linker in a lysine-rich sequence. Three other regions of extensive sequence variability are seen in myosin. In the 25,000-M_r peptide there is a glycine and proline-rich sequence from amino acids 50–72 that is highly variable. In the 23,000-M_r domain there are two regions of sequence diversity, amino acids 749–778 and amino acids 820–835. This latter region is the site of proteolysis that separates S-1 from the rod and is believed to be a region of high flexibility (Hvidt et al., 1982; Mendelson et al., 1973; Thomas et al., 1975, 1980). It is not known whether the variable region in the 25,000-M_r peptide is also flexible.

7.2. Rod Sequences

Interactions between the rod segments of the myosin molecules are necessary for the assembly of thick filaments, while the heads act as movable cross-bridges between the thick and thin filaments and perform the enzymatic reactions required to generate tension. Thick filaments are bipolar and contain myosin molecules assembled in an antiparallel arrangement in the central bare zone which is about 1600 Å wide (Craig, 1977; Sjöström and Squire, 1977; Squire, 1981). Throughout the rest of the thick filament, the rods form parallel arrays in which the heads project away from the surface. The myosin heads are helically arrayed with a repeat of 429 Å and an axial stagger of 143–146 Å (Huxley and Brown, 1967; Miller and Treager, 1972; Wray, 1979). These structural features are found in all thick filaments, although different species show considerable variation in filament length and number of myosin molecules per cross section (Squire, 1981; Kensler and Levine, 1982; Kensler and Stewart, 1983; Levine *et al.*, 1983; Vibert and Craig, 1983).

    7.2.1. 28-Residue Repeat Units. The myosin rod is highly repetitive and shows features typical of α-helical coiled-coil proteins (McLachlan and Karn, 1982, 1983; Karn *et al.*, 1983a; Parry, 1981; Kavinsky *et al.*, 1983). The helical region appears to extend from the invariant pro-862

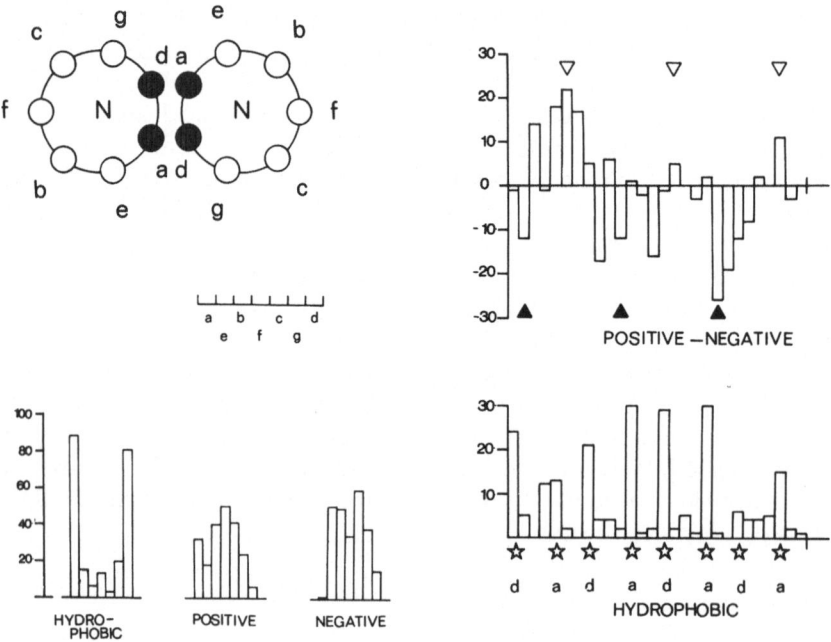

Figure 19. Histograms of amino acid distributions in the MHC B rod sequence. The cross section of coiled coil (upper left) shows the positions of the *a, b, c, d, e, f*, and *g* residues when viewed from the amino end of parallel helices. The histograms in the lower left show the distribution of amino acids in the 7-residue structural repeat of the coil. The histograms show the numbers of amino acids at each type of position in 152 complete units. The histograms at the right show the amino acid distributions along the 28-amino acid repeat. The triangles show the three peaks for the sets of residues with positive and negative charges. Asterisks mark the hydrophobic core positions. The peak heights show the numbers of amino acids at each position (1*d* to 4*c*) in 38 complete cycles. (After MacLachlan and Karn, 1982, 1983a.)

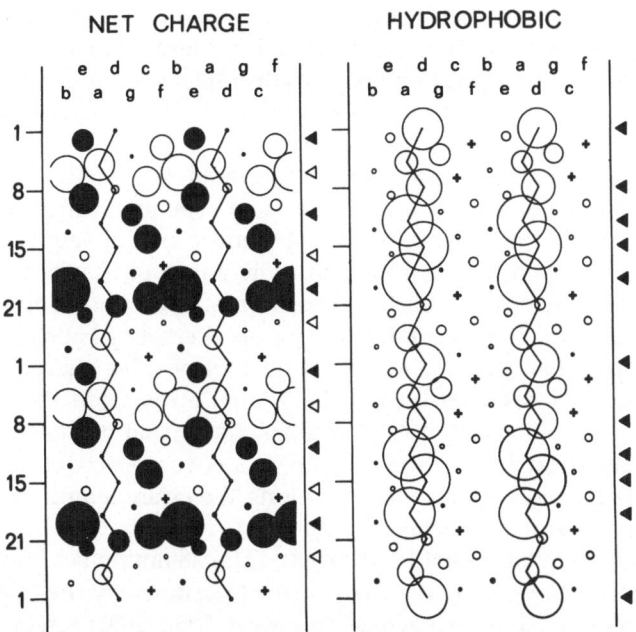

Figure 20. Distribution of charged (left) and hydrophobic (right) residues plotted on the supercoil surface of the myosin molecule. Each panel is a cylindrical projection of the outside surface of two α-helical chains in contact. The positions of residues *a* to *g* are labeled above the diagram. The sequence reads from right to left and is drawn with a pitch of 3.5 residues per turn. The radii of the circles is proportional to the peak heights in the histograms shown in Fig. 19. Two cycles of the 28-amino acid repeat are shown. (A) White and black circles represent positive and negative charges. A (+) sign represents a point of zero net charge. Triangles at the side mark the six bands of charge in a typical 28-residue zone. (B) Hydrophobic residues are plotted in a similar way with triangles to mark the major sites and plus (+) signs to mark forbidden positions. (After MacLachlan and Karn, 1982.)

to amino acid 1951, where a short, apparently irregular tailpiece containing prolines is present in MHC B (*unc*-54) and MHC A (*myo*-3). Vertebrate myosins and the pharanygeal myosins MHC C (*myo*-2) and MHC D (*myo*-1) do not have the tailpiece sequences. α-Helical coiled-coil proteins (Crick, 1952; Fraser and MacRae, 1973), including tropomyosin (McLachlan and Stewart, 1976), fibrinogen (Parry, 1978; Doolittle *et al.*, 1978), and α-keratin (Parry *et al.*, 1977), have a characteristic regular 7-residue pattern (*a, b, c, d, e, f, g*) in which hydrophobic residues are concentrated in the *a* and *d* positions. This configuration permits two chains of α-helix to interact along a hydrophobic interface on one surface of each helix. Figure 19 shows the positions of the *a, b, c, d, e, f*, and *g* residues through two turns of each α-helix, when viewed from the amino end. The myosin rod sequences follow this pattern and have nearly 80% occupancy of *a* and *d* positions with hydrophilic residues (Fig. 19). The helix surface is highly charged, and acidic and basic residues are clustered in the outer positions *b*, *c*, and *f*.

In addition to the 7-amino acid hydrophobic repeat, the myosin rod shows strong evidence for a 28-amino acid repetitive unit (McLachlan and Karn, 1982, 1983; Parry, 1981). The entire helical sequence of the rod can be mapped out as a pattern of 40 zones of the 28 amino acid repeat. Each zone has a characteristic pattern of hydrophobic and charged residues. A typical zone is subdivided into six alternating bands of charged residues spaced so that the strongest peak of positive charge (position 1b) is exactly 14 residues (or one half-zone) away from the strongest peak of negative charge (Figs. 19 and 20). As a result of this zone structure, the outer surface of the myosin rod is composed of alternating bands of charge in a regular pattern along the entire length of the molecule (Figs. 20 and 21, Section 7.2.3).

7.2.2. Skip Residues and Weak Spots. The 28-amino acid repeat pattern is interrupted at four regularly spaced positions by the insertion of one extra skip residue (McLachlan and Karn, 1982, 1983; Kavinsky *et al.*, 1983). We have assigned the skip residues to amino acids 1212, 1409, 1606, and 1831. The positions of skip residues are invariant in all known myosin sequences. The major effect of the skip residues is to break the 7-residue pattern of the hydrophobic seam. In Fig. 21, the sequence of the myosin rod near skip residue 1212 is drawn on a helical net. The seam of hydrophobic residues shifts across the surface over a stretch of 14–21 residues before resuming a regular pattern. This flattens the twist of the supercoil by approximately 30° and makes the helices run parallel to one another for a few turns. Skip residues in the other positions have analogous effects. These local distortions could have important effects on the way myosin molecules pack into thick filaments.

A second form of distortion of the hydrophobic seam arises from placement of polar or charged groups in the *a* or *d* positions. Lysine and arginine often appear in the *a* core positions. Model building suggests that these residues are well tolerated in these positions because the hydrophobic side chains can cross the center of the core and place the charged amino groups on the outer surface of the rod. The negatively charged side chains of glutamic and aspartic acid are, however, totally excluded from position *a* and extremely rare in position *d*. There are a number of short apparently weak sections of the rod in which several successive polar, negatively charged, or glycine residues are placed in *a* or *d* positions. The weakest of these sections corresponds to amino

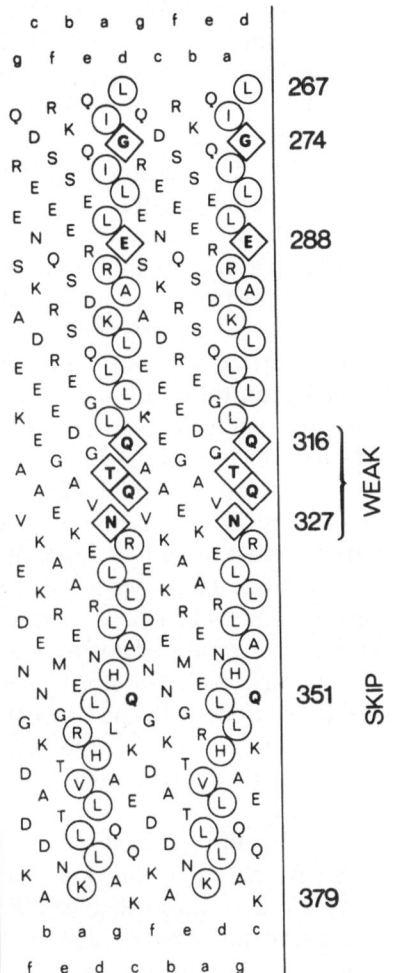

Figure 21. Coiled-coil surface diagram of the MHC B sequence from leu-1128 (leu-267 in the rod sequence) to lys-1240 (379). The sequence is drawn, as in Fig. 20, as a cylindrical projection of the outside surface of two chains in contact. The sequences read from right to left with a slight downward slant and are drawn with 3.5 residues per turn. Residues with hydrophobic side chains located in the core positions *a* and *d* have been ringed. Diamonds enclose *a*- or *d*-position residues having acidic, polar, or glycine side chains. These residues are likely to weaken the core. At skip residue glu-1212 (351), the hydrophobic seams shifts across the helix surface. Residues glu-1177 (316) to asp-1188 (327) form an unusually weak part of the coil and lie near the short S2–LMM junction. (After MacLachlan and Karn, 1982.)

acids 1177–1187, with four consecutive defects, and amino acids 1038–1045 and 1685–1692, with three each.

It has been suggested that myosin contains flexible hinges in the rod that permit enhanced movement of the cross-bridge from the thick-filament backbone (Huxley and Brown, 1967; Huxley, 1969). Myosin often shows localized sharp bends when displayed in the electron microscope (Elliot and Offer, 1978; Takahashi, 1978). Furthermore, proteolysis of myosins from a variety of species tends to cleave the rods at narrowly localized points (Lowey *et al.*, 1969; Weeds and Pope, 1977;

Lu and Wong, 1982). It is interesting to note that the site of cleavage between short S-2 and the COOH terminus of the myosin rod (Lu, 1980; Lu and Wong, 1982) is homologous to the major weak spot at amino acids 1177–1187 of the sequence (Fig. 21). Each of the nematode genes shows a similar pattern of hydrophobic and charged groups in this region (Figs. 17 and 21), suggesting that this sequence has evolved to fold into a specialized structure.

7.2.3. *Electrostatic Interactions between Rods.* The high density of charged groups on the outer surface of the myosin rod suggests that ionic interactions are involved in the assembly of thick filaments (McLachlan and Karn, 1982). The 28-residue (or 41.6 Å long) repeat units should attract one another strongly when myosin molecules are staggered by multiples of 14 residues. These potential interactions have been analyzed using a linear model of the *unc*-54 sequence in which the charge distribution is averaged around the circumference of the helix surface (McLachlan and Karn, 1982; Hulmes *et al.*, 1973).

Figure 22 shows how the net electrostatic interaction between two parallel rod sequences depends on axial stagger. A pair of molecules were treated as arrays of charged and uncharged amino acids shifted successively by 0, 1, 2, . . . *n* residues. Two charges were considered to interact if they are opposite one another within a range of ± 2 residues. The energy oscillates strongly with stagger, showing the expected 28-residue period, with the highest peaks close to odd multiples of 14. The amplitude decreases as the molecular overlap is reduced, but not uniformly, and several peaks at odd multiple staggers of 98 (98, 294, 490) stand out.

A stagger of 98 residues would correspond to a distance of 145.5 Å (146.3 Å, including the skip residues) assuming a helical rise of 1.485 Å per residue in the coiled coil (Fraser and MacRae, 1973). This distance is close to the cross-bridge stagger of 143–146 Å observed in muscle. The favorable interactions between ideal straight molecules translated by 294 residues could account for the 430-Å helical repeat seen in thick filaments.

This overlap calculation does not show which portions of the rod interact most strongly. A more detailed picture was obtained by calculating local interactions between segments of staggered rods. In the Fig. 22B, the interaction density between two rods summed over a span of 99 residues is plotted along the length of the rod for staggers of 294, 100, 98, and 96 residues. The strongest interactions are seen for the staggers of 98 residues. Note that the interactions are not uniform along the length of the rod. Pairing begins around amino acid 320 of the rod

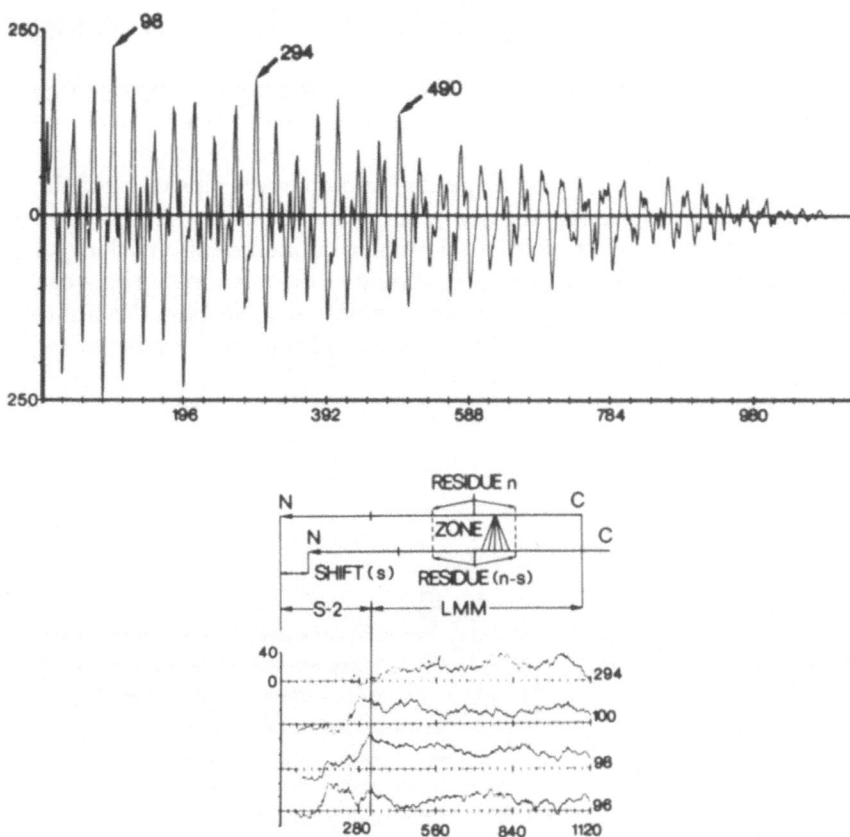

Figure 22. Charge interactions between MHC B rod sequences. The rod sequence of 1094 residues from pro-862 to pro-1956 was used in the calculations. (A) Calculation of optimal staggers. The stagger, s, is plotted horizontally and the net number of interactions (attractions or repulsions) are plotted vertically. Amino acids are counted within an axial range of ± 2 residues, taking each rod as a single chain (because only one side can interact with its neighbor). Note the peaks at staggers which are odd multiples of 98. (B) Distribution of charge interactions along the length of a pair of rods at selected staggers. The upper rod is fixed and the lower rod has been shifted by s = 96, 98, 100, or 294 residues to the right. The height of the graph under residue n of the upper rod is equal to the sum of the attractions minus repulsions of a 99-residue section centered on residue n − s. When residue n nears the end of the sequence, the 99-residue zone becomes shortened and the interactions gradually decrease at each end. The interactions within each 99-residue zone are scored as in the top panel with charge pairs counted within an axial range of ±2 residues. The relationship between the zone and scoring is shown schematically in the diagram above the graphs (not to scale). The S2–LMM junction has been placed at residue 322 of the rod (ala-1184) and refers to the upper rod. (After McLachlan and Karn, 1982.)

(ala-1183) and persists strongly down the rod to amino acid 880 (ala-1742).

    7.2.4. Thick-Filament Assembly.   We have seen how periodic charge distributions in the myosin rod amino acid sequence correlate with the known cross-bridge spacings in muscle. The structurally important features of the sequence, the hydrophobic and charged repeats, and the placement of skip residues and weak spots are highly conserved in all the myosin sequences studied. A simplified model for the packing of myosin molecules diagrammed in Fig. 23 shows the relationship between various molecular staggers, the 28 amino acid zones, and the 197 amino acid spacings between skip residues. Up to 12 rods can participate in these interactions.

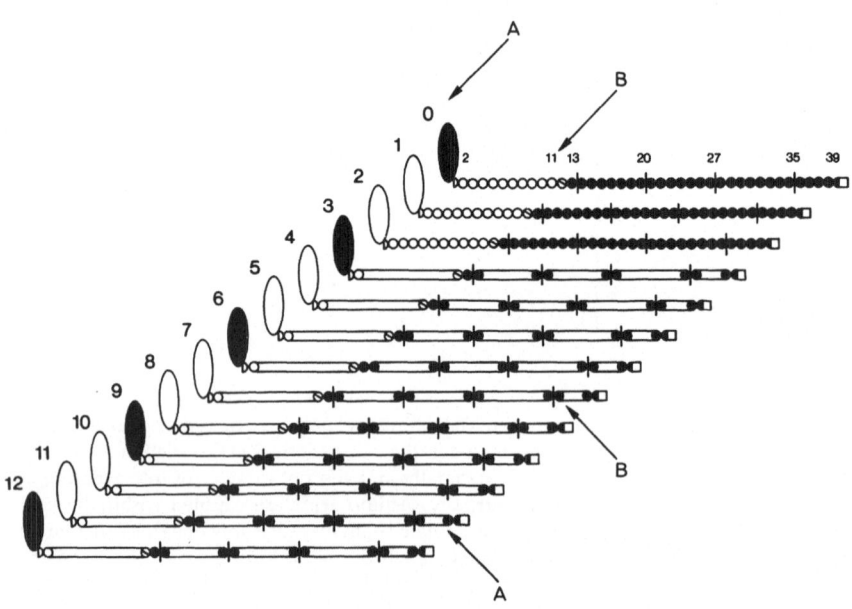

Figure 23. Parallel array of myosin rods drawn to show the relationships between the 98-residue (or 143-Å) and 294-residue (or 430-Å) displacements and the periodic structure of the amino acid sequence. In rods numbered 0–2, ○ represents the 28-residue zones in short S-2, ending within zone 12, and ● represents 28-residue zones in the remainder of the rod. Vertical bars mark the positions of the skip residues, and the open square beyond zone 40 represents the nonhelical tailpiece. Rods 3–12 emphasize the half-staggered arrangement of the skip residues. The diagram is a two-dimensional view and should not be taken literally. The arrows labeled A and B indicate the plane of two different cross sections of arrays, including 7 or 11 rods. (After McLachlan and Karn, 1982.)

In order to extend these packing models to three dimensions, it will be necessary to establish the pitch of the supercoil (Squire, 1981). Estimation of the pitch is complicated by the presence of skip residues that introduce local distortions. But it is important to note that the special features of the amino acid sequence of myosin restrict packing models to those in which zones of opposite charge are in adjacent positions (McLachlan and Karn, 1983).

It is clear, however, that a complete thick-filament structure cannot be built up from a consideration of the myosin rod sequence alone. Nematode thick filaments are composed of two distinct myosin heavy-chain isoforms and several additional proteins (see Section 3.3). MHC B (the *unc*-54 gene product) participates only in parallel interactions, whereas MHC A (the *myo*-3 gene product) interacts with itself in both parallel and antiparallel configurations. There is a short region in the filament in which both myosins are found in parallel assays (Miller *et al.*, 1983). Thus far an analysis of the complete *myo*-3 rod sequence has not helped to explain these complex interactions. The charge distribution in the *unc*-54 and *myo*-3 rod sequences are extremely similar. Only 51 of 474 positions with charged amino acids show differences. These changes are largely conservative and are located in the S-2 region. Perhaps subtle differences in the pitch of the rod, or interactions with other proteins, are responsible for the differences in behavior of the two molecules. Genetic evidence using *sup*-3 alleles to overproduce MHC A in *unc*-54 null backgrounds shows clearly that MHC A can substitute for MHC B (Section 2.5.2). However, MHC B cannot form a thick filament by itself and *myo*-3 hypomorphic mutations are homozygous lethals (R. H. Waterston, personal communication).

Paramyosin is also required for the assembly of nematode thick filaments (MacKenzie and Epstein, 1980) (Section 2.5.1). The sequence of paramyosin is not known, although the gene has recently been cloned by Kagawa *et al.* (1987). Studies of paracrystals (Cohen *et al.*, 1971) indicate that paramyosin may also contain 28-residue charge zones (McLachlan and Karn, 1982), which could interact with myosin rods on the surface of the thick filament (Cohen, 1982). Recent experiments have indicated that the core of the nematode contains a third major structural component in addition to paramyosin (Epstein and Miller, 1983). At least three different accessory proteins, C protein (Offer *et al.*, 1973), H protein, and X protein (Starr and Offer, 1983), are also incorporated into vertebrate thick filaments. It is likely that similar proteins are minor components of nematode thick filaments.

The packing of the myosin molecules is also likely to vary along the length of the filament (Vibert and Craig, 1983; Craig, 1977) to produce tapering toward the tips of the structure (Huxley, 1969; Szent-Györgyi *et al.*, 1971). In vertebrate thick filaments, nonmyosin components, including C protein, are restricted to locations closely matching the 430-Å myosin helical repeat (Sjöström and Squire, 1977). If all myosins were packed into equivalent environments, one would expect the C-protein repeat to follow the cross-bridge stagger of adjacent myosins (143 Å) rather than every third level (430Å) (Craig and Offer, 1976).

8. SEQUENCES AND MOLECULAR INTERPRETATION OF *UNC*-54 MUTATIONS

More than 100 separate *unc*-54 mutants have been isolated. Together with the laboratory of R. H. Waterston, we have cloned and sequenced 17 representative *unc*-54 mutations in order to determine precisely the genetic lesions and define amino acid sequences critical for myosin function.

8.1. *Rapid Cloning and Sequencing of Mutations*

The *unc*-54 gene sequence is greater than 8.5 kb (Fig. 14) and locating point mutations within this sequence created considerable technical problems. The approximate end points of the three deletion mutants in the rod (*e675*, *e190*, and *s291*) were determined by restriction endonuclease mapping (Fig. 10). The appropriate fragments were then cloned into M13 vectors for sequencing. Each of the point mutations we sequenced had been ordered with respect to one or more of the deletions by genetic fine-structure mapping (Waterston *et al.*, 1982a; Moerman *et al.*, 1982). We found that we could reliably restrict searches for the mutations to regions of 1 kb or less on the basis of this information. In most cases, we sequenced large portions of the *unc*-54 gene in order to confirm that only single base-pair alterations were present. Two technical advances greatly improved the speed and accuracy of our sequencing efforts.

We first devised a method that improved the recovery of *unc*-54-specific clones in bacteriophage λ. The entire coding sequence of the *unc*-54 gene, except the short 5' exon, is contained on an *XbaI* fragment of 8301 bp (Fig. 10). The *XbaI* fragments in a limit digest of nematode DNA average only 4 kb, and such short fragments are inefficiently cloned in the bacteriophage λ *XbaI* vector, 2149, which has a minimum insert length requirement of 7 kb (Karn *et al.*, 1983b). When unfractionated

*Xba*I digests of nematode DNA were cloned into this vector, four to five recombinants carrying the entire *unc*-54 gene were typically recovered from a bank of only 20,000 clones. The *unc*-54 clones could be recovered from as few as 10^5 nematodes.

In order to sequence the mutations more rapidly, a series of short synthetic oligonucleotides primers were prepared that were complementary to 17-bp sequences distributed at 250–300-bp intervals along the *unc*-54 sequence. Large restriction fragments such as the 5004 bp *Bam*H1 (residue 4401) to *Xba*I (residue 9405) fragment covering most of the myosin gene, including the entire myosin rod (Fig. 10), were subcloned into M13 vectors. Regions of interest were then sequenced by priming in separate reactions with each of the synthetic oligonucleotide primers. Using these two methods in conjunction, it was possible to clone and sequence a new *unc*-54 mutation in approximately 3 weeks.

8.2. Correlation of Physical and Genetic Fine-Structure Map

The relative positions of more than 30 mutations in *unc*-54 have been ordered by Waterston *et al.* (1982*a*). Crosses were performed between a strain containing a *lev*-11 mutation and one *unc*-54 allele and a second strain containing a *let*-50 mutation and a second *unc*-54 allele. Rare recombinants that are wild type for *unc*-54 but that incorporate one or both of the two flanking markers were collected. If the *unc*-54 allele coupled to *lev*-11 is to the left of the *unc*-54 allele in the *let*-50 background then wild-type recombinants may be recovered over either parental chromosome. These strains will then be fully wild type. However, if the *lev*-11 linked allele is to the right of the second *unc*-54 wild type, the only recombinants recovered will be *lev*-11 + + *let*-50/*lev*-11 + *unc*-54 + because *let*-50 homozygotes are inviable. These strains will express the levamisole-resistance phenotype. This method allows accurate placement of mutations to the left and to the right of one another, but insufficient numbers of recombinants are obtained to permit accurate estimation of the physical distances between the markers. The fine structure map determined by this method is shown in Fig. 24A.

Sixteen point mutations and three deletions have now been sequenced (Dibb *et al.*, 1985; R. H. Waterston, personal communication) (Fig. 24; Table III). The positions of the mutations as determined by DNA sequencing agrees with the relative positions of the mutations as determined by genetic fine-structure mapping (Waterston *et al.*, 1982*a*; Dibb *et al.*, 1985). Approximately 5 kb of sequence within the *unc*-54 gene corresponds to 0.01-map unit. The fine-structure mapping was able to order the deletion *e*190 and nonsense mutation *e*1115, which as shown

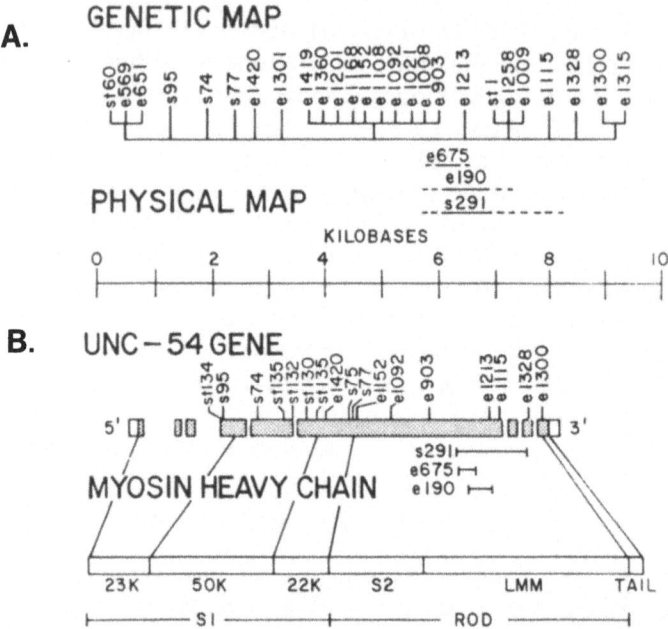

Figure 24. Fine-structure map of the *unc*-54 gene and location of *unc*-54 mutations on the gene sequence. (A) The relative positions of 30 independently derived mutations ordered by genetic crosses (Waterston *et al.*, 1982a). Point mutations are shown above the line, deletions below. (B) The positions of sequenced *unc*-54 mutations. The coding sequences are drawn as shaded bars, introns as gaps. The open boxes indicate the positions of 5′ and 3′ noncoding regions. See Table IV for exact positions of the mutations. A diagram of the protein structural domains of the encoded MHC B is shown below the gene map. Note that the positions of introns do not correspond to the major proteolytic domains in the protein. (After Dibb *et al.*, 1985.)

by our results are separated by only 213 bp. The deletion s291 was separable from the nonsense mutation e1328 (Waterston *et al.*, 1982a). These two mutations are separated on the physical map by only 24 bp. Although the genetic fine-structure mapping proved a remarkably sensitive method for ordering mutations, small sample sizes of recovered recombinants and possible variations of recombination frequencies along the map and between alleles prevented a precise estimate of physical distance from the recombination data.

8.3. Deletions in the Rod

The size and positions of the deletions in e190, e, and s291 are shown in Figs. 10 and 24 and Table III. Each pair of deletions overlaps and is located within the region of *unc*-54, which encodes the LMM portion of

Table III. Molecular Weights of Major Structural Domains in MHC B

Fragment	Residue no.[a]	M_r^b
Total	5–1977	229,486
S-1	6–862	95,392
25,000 M_r	6–225	22,690
50,000 M_r	26–658	49,840
23,000 M_r	659–861	22,898
Rod	862–1958	125,836
S-2	862–1179	6,858
LMM	1179–1955	88,996
Tailpiece	1956–1977	8,422

[a] Residue numbers refer to amino acids in Fig. 17.
[b] Fragment molecular weights were calculated from the amino acid compositions of regions of the MHC B amino acid sequence (*unc*-54).

the rod. The allele *e675* is an in-phase deletion of 270 bp (residues 6548–6817). The allele *s291* is also an in-phase deletion of 1275 bp (residues 6512–7786 or 6513–7787). The exact position of the *s291* deletion is ambiguous because of a duplicated G : C bp. The *e675* and *s291* deletions are predicted to give rise to the truncated myosin heavy-chain fragments that have been detected *in vivo*.

Both the *e675* and *s291* alleles are semidominant, and both mutants make normal amounts of shortened myosin heavy chains that fail to assemble into thick filaments (Epstein *et al.*, 1974; MacLeod *et al.*, 1977*b*; Waterston *et al.*, 1982*b*). From the DNA sequencing results, it is clear that the mutations *e675* and *s291* disrupt both the hydrophobic repeat of 7 amino acids and the charge repeat of 28 amino acids of the myosin rod (Section 7.2). Consequently, deletion is expected to introduce a localized distortion of the rod that could prevent normal assembly into thick filaments. Indeed, it has been noted that the deletion junction region of *e675* is sensitive to proteolysis (MacLeod *et al.*, 1977*a*).

Unlike *e675* and *s291*, the mutant *e190* is recessive and accumulates no detectable myosin heavy chain *in vivo*. The DNA sequencing results show *e190* is an out-of-phase deletion of 401 bp (residues 6746–7146). The deletion introduces many in-phase nonsense codons, the nearest a TGA at residues 7160–7162. A truncated product of the predicted size has been detected in cell-free translation experiments (Fig. 9). The absence of the *e190* heavy chain *in vivo* appears to be analogous to the absence of heavy-chain products from other nonsense mutations (see Section 8.4). This suggests that the stability of *unc*-54 heavy chains *in vivo* appears to be dependent on the presence of the normal COOH-terminus of the protein and not on the length of the myosin rod.

8.4.   Nonsense Mutations and Suppressors

The effects of some *unc*-54 mutations can be partially overcome by genetically distinct suppressor mutations. The loci *sup*-5 III and *sup*-7 X specify suppressors which are able to restore functional gene products to a class of null alleles of nematode muscle proteins (Waterston and Brenner, 1978; Waterston, 1981). Six alleles of *unc*-54 are suppressible by *sup*-5 and *sup*-7. These alleles (*e*576, *e*1008, *e*1021, *e*1108, *e*1300, and *e*1419) are distributed throughout the intragenic map (Waterston *et al.*, 1982*a*). Of these alleles, only *e*1300 is located near the extreme 3' end of the map, to the right of the deletion mutants *e*675 and *s*291. Examination of the pattern of myosin heavy chains in strains homozygous for *e*1300 demonstrated the presence of a small amount of a truncated protein product. The truncated myosin contained *unc*-54 specific cyanogen bromide peptides. The COOH-terminal 37,460-M_r cyanogen bromide peptide (residues 1609–1931) was replaced by a peptide of approximately 34,000 M_r (Willis *et al.*, 1983; Waterston, *et al.*, 1982*b*).

Sequencing studies of cloned *e*1300 (Wills *et al.*, 1983) showed that a T is substituted for a C at position 8013 of the *unc*-54 sequence. This substitution changes glu-902 (60 amino acids upstream from the normal TAA terminator) to a TAG amber terminator. The position of the mutation is in agreement with the position predicted from the size of the *e*1300 polypeptide, and strongly suggested that *sup*-5 and *sup*-7 act to promote readthrough of amber terminators *in vivo*. Wills *et al.* (1983) showed that both *sup*-5 and *sup*-7 are altered tRNAs that specifically suppress UAG (amber) terminators. Thus alleles in any nematode gene that are suppressible by *sup*-5 and *sup*-7 are almost certainly amber mutations. This may be of use in the identification of the protein products of new genes affecting muscle. Bolten *et al.* (1984) cloned the *sup*-7 tRNA and demonstrated that this gene encodes a tryptophan tRNA with an altered anticodon.

Recessive alleles *e*1213, *e*1115, *e*1328, and *e*1092 are not suppressed by *sup*-5 and *sup*-7. Sequencing studies (Dibb *et al.*, 1985) have shown that each is a C to T transition that converts at glutamine codon (CAA) to an ochre terminator (TAA) (Table IV). The identification of ochre mutations in the *unc*-54 gene should prove useful in the search for ochre suppressors in the nematode. These known ochre mutations can be tested for suppression in combination with suppressors with different spectra of suppression from *sup*-5 and *sup*-7. In addition, the *sup*-5 or *sup*-7 loci can be altered by EMS mutagenesis to create potential ochre suppressors by a method analogous to that used in previous studies with yeast and *E. coli* (Hawthorne and Leupold, 1974). Our observations that most point mutations isolated following EMS mutagenesis of *Cae-*

Table IV. Sequenced Mutations in the unc-54 MHC

A. Nonsense mutations

Allele	Mutation	Amino acid change[a]
e1092	C → T	Q → ochre (1087)
e1213	C → T	Q → ochre (1694)
e1115	C → T	Q → ochre (1795)
e1328	C → T	Q → ochre (1875)
e1300	C → T	Q → amber (1918)

B. Missense mutations

Allele	Mutation	Amino acid change	Phenotype
st134	C → T	S → F (122)	Very slow, stiff
st95	G → A	G → R (123)	
st74	C → T	R → C (269)	
st135	G → A	A → T	Slow, stiff
st132	C → T	H → Y (590)	Disorganized muscle, slow
	G → A	E → K (538)	
st130	G → A	C → Y (551)	
s75	G → A	G → R (663)	Slightly slow
s77	G → A	G → R (729)	
el152	G → A	G → R (867)	Disorganized muscle, very
	A → T	K → M (868)	slow, dominant

C. Deletion mutations

Allele	Deletion[b]	Mutagen
e903	5985–5986	X ray
s291	6512–7786 or 6513–7787[a]	Formaldehyde
e675	6548–6817	EMS
e190	6746–7146	EMS

[a] Numbers in parentheses give the amino acid residues altered by the mutation following the numbering of Fig. 17.
[b] Deletions are indicated by the residues removed from the sequence, following the numbering of Karn et al. (1983a). For example, in e675 residues 6547 and 6818 of the wild-type sequence are contiguous.
[c] The exact position of the junction in s291 is ambiguous because of a duplicated G:C base pair.

norhabditis elegans are G : C to A : T transitions (Section 8.7) make this strategy particularly attractive. The availability of both amber and ochre suppressors would enable one to recognize a large fraction of the nonsense alleles in any genetic locus in *Caenorhabditis elegans*. Null mutations that are not suppressed by amber or ochre suppressors would be strong candidates for opal mutations and mutations affecting mRNA processing and translation.

The recessive allele *e*903 was induced by X-ray mutagenesis. It is a deletion of AA at position 5985 and is expected to give rise to a myosin protein product terminating with a TGA, six residues beyond the deletion endpoint. It seems likely that *e*903 arose from excision of a radiation-induced thymidine dimer on the complementary strand.

8.5. Mutations Affecting the Synthesis of MHC

Most *unc*-54 mutations are due to recessive alleles and produce little or no detectable MHC B. All sequenced recessive mutations were either amber or ochre termination mutations or out-of-phase deletions that give rise to nonsense mutations (Table IV).

Stable MHC B is only produced when the carboxy-terminal of the protein is present. The *e*190 deletion mutation and all the other premature termination mutants show drastically reduced levels of MHC B. No detectable MHC B is recovered from mixed populations of adult and larval animals homozygous for the sequenced premature termination mutations *e*190, *e*903, *e*1092, *e*1115, *e*1213, and *e*1328 (MacLeod *et al.*, 1977*b*). However, *e*1300 animals, which have an amber terminator near the extreme 3' end of MHC B, show truncated MHC B at about 20% the normal levels in third stage larvae (Wills *et al.*, 1983) (see Section 8.4). Perhaps because most of the myosin rod is intact, the *e*1300 MHC B may aggregate into myosin bundles that are less susceptible to proteolytic degradation.

We also have noticed a correlation between the position of a terminator and the apparent stability of the *unc*-54 mRNA. Cell-free translation assays have shown that reduced levels of MHC B may be detected (MacLeod *et al.*, 1979), although no MHC B may be recovered from these animals. Approximately three-fold more *e*190 mRNA was detected as compared with the upstream mutation *e*1092. These results suggest that eukaryotic mRNAs are subject to enhanced degradation when the termination codon is placed at long distances from the 3' end of the mRNA. The effect may be analogous to the well-known polarity effect in prokaryotes (Morse *et al.*, 1969).

8.6. A Dominant Assembly-Defective Missense Mutation

Homozygous *e*1152 animals produce normal levels of MHC B but have a greatly reduced number of thick filaments that are not organized into ordered A-bands (MacLeod *et al.*, 1977*b*). We found two adjacent sequence alterations near the amino terminal of the rod in *e*1152. Gly-867 (CGG) was converted to an arginine (AGG) and the adjacent residue lys-868 (AAG) was converted to a methionine (ATG). Of the two alterations, the lysine to methionine change is the easiest to interpret. Amino acid 869 is a glutamic acid in the *d* position in the rod. Lys-868 is in a good position to form neutralizing salt bridges with glu-870. As noted in Section 7.2.2, glutamic or aspartic acids are extremely rare in *a* and *d* positions. There are only 18 such residues in the MHC B sequence (53 in all the myosin sequences). With only two exceptions, these negatively charged groups are neutralized by salt bridges formed with an adjoining positively charged residue (typically lysine, arginine, or glutamine in the *c* or *g* positions). The substitution of methionine for lys-868 removes the neutralizing salt bridge and would be expected to destabilize the hydrophobic core in this region. Gly-867 is conserved throughout the nematode family, and is replaced by alanine (also with a small side chain) in rabbit. The residue is only 5 amino acids from pro-862, which begins the rod. It seems likely that the proline is not well accommodated as a *d* position residue and the small side chain of the glycine permits local distortion of the rod in this region. Replacement of the glycine with the bulky arginine side chain in *e*1152 could produce steric restrictions on the rod backbone. Our analysis suggests that both alterations in *e*1152 are significant and that both contribute to destabilizing the myosin core in this region. The dramatic effects on thick-filament assembly by these localized changes at the tip of S-2 suggest that the overall three dimensional twist of the rod, in addition to the charge distribution, is critical for assembly.

In addition to *e*1152, four other dominant assembly-defective mutations are known (*e*1157, *e*1219, *e*1273, and *e*1274). We have begun sequencing studies on *e*1157 and *e*1274 but have found no changes in the rod sequence from amino acids 862–1480.

All the assembly-defective myosin-producing strains show dominant phenotypes except for a unique allele *e*1301, which is temperature sensitive for assembly and recessive (MacLeod *et al.*, 1977*b*). At the permissive temperature (15°C) a normal lattice of thick filaments is observed; however, at 25°C a full mutant phenotype is observed. This allele is known to map between *s*74 (residue 2885) and *e*1092 (residue 5343). We have sequenced from the start of *e*1301 up to the position of *e*1092 and have not found any substitutions. This raises the intriguing possibility that the assembly phenotype of *e*1301 is due to an alteration in the head sequence and that head sequences may participate in thick-filament assembly.

8.7. Mutations in the Head

Although most *unc*-54 mutations exhibit reduced numbers of thick filaments, Moerman *et al.* (1982) isolated a series of *unc*-54 alleles (*s*74, *s*75, *s*76, *s*77, *s*78, *s*95, *st*130, *st*132, *st*134, *st*135) that, when homozygous, result in animals that are slow but have normal muscle structure except for a slight disorder of the packing of thick filaments in the A band (Section 2.5.5). These animals are also more rigid than wild-type animals, as if these muscles were in rigor. The genetic localization of these mutations to the myosin head suggested that these mutations altered myosin enzymatic activity but not thick-filament assembly.

The alterations in *s*74, *s*75, *s*77, *s*95, *st*130, *st*132, *st*134, and *st*135 have been determined by sequencing studies (Table IV). The mutations are distributed throughout the myosin head; *st*134 and *s*95 affect adjacent residues in the 25,000-M_r peptide near the ATP-binding sites that involve amino acids 154–193 and lys–133, which is homologous to the trimethyl lysine in rabbit skeletal muscle known to be involved in ATP binding. Mutations *s*74, *st*135, *st*132, and *st*130 are distributed throughout the 50,000-M_r proteolytic fragment and are likely candidates for defects in actin binding and the communication system linking actin binding and the ATP-binding site. Two sequence alterations were detected in *st*135, involving amino acids 476 and 590. Mutations *s*75 and *s*77 are both found in the 25,000-M_r peptide. In *s*75 an arginine replaces gly-663, in a flexible loop region between the 50,000- and 23,000-M_r domains. Movement in this loop region is thought to be essential for ATP-dependent actin binding. The mutation in *s*77 converts gly-729 to an arginine. Gly-729 is three amino acids removed from one of the active thiols cys-726. Measurements of myosin functions *in vitro* should help define the enzymatic consequences of these mutational alterations.

Any conclusions reached about the effects of the missense mutations on the function of either the myosin head or the rod could be further tested through the isolation of second-site mutations in *unc*-54 that alleviate the phenotypic effects of the primary mutations. Revertants have already been obtained for several of the dominant assembly-defective mutations (R. H. Waterston and S. Brenner, personal communication). Of five revertants to wild-type *s*95, one *st*136 restored the original glycine at amino acid 126 with an EMS-induced A to G transition, while four others (*st*176, *st*177, and *st*179) converted the arginine to a less bulky serine residue using anamalous A to T transversions (R. H. Waterston, personal communication).

8.8. Mechanisms of Mutagenesis in the Nematode

DNA sequencing studies show that, with one exception, all point mutations isolated following EMS mutagenesis were G : C to A : T transitions. The *unc*-54 point mutations revert at a frequency of less than

10^{-7} (Waterston *et al.*, 1982a). Therefore, the effects of EMS in *Caenorhabditis elegans* and *E. coli* are similar (Coulondre and Miller, 1977). In *e*1152 two alterations are present, a G to A transition and the aberrant A to T transversion. Because two mutations are present it is not possible without further genetic studies to determine whether the A to T transversion was induced by EMS or arose in a second event, possibly as a partial revertant. The single X-ray-induced mutation, *e*903, had deletion of two consecutive A residues, probably the result of excision of a radiation-induced thymidine dimer on the complementary strand.

The origin of the internal deletions in the myosin rod region of the *unc*-54 gene is particularly interesting. Homologous recombination between direct repeats probably generates most deletions in the chromosome of *E. coli* (Albertini *et al.*, 1982). Nematode deletions *e*675 and *e*190 (isolated after EMS treatment) and *s*291 (isolated after formaldehyde treatment) do not follow this pattern. The myosin rod sequence is highly repetitive and includes 187 direct repeats and 106 inverted repeats longer than 11 bp (A. D. McLachlan, personal communication). None of the repeats occurs near the deletion junctions of the mutants. A strong homology has been detected in the sequences flanking the deletion junctions of the mutants *e*190 and *s*291 (Dibb *et al.*, 1985). Each mutant had the sequence AAGATCATC at the 5′ side of the deletion and the sequence GCCGA at the 3′ side of the deletion. The sequence AAGATCATC occurs only three times in the 10-kb sequence of the *unc*-54 gene region while the sequence GCCGA occurs 24 times. Although these sequences occur more often than would be expected in a random sequence of this length, the probability of obtaining two deletion junctions showing these homologies is extremely low. Deletion *e*675 shows only limited homology to the GCCA sequence. As far as we know, the sequence AAGATCATC and GCCA have not been reported as preferred sites for nonhomologous recombination in prokaryotic or eukaryotic systems. The data show that these deletions were not generated by homologous recombination and suggest that they arose by site-directed mechanism.

9. MYOSIN PROTEIN EXPRESSION AND MUTAGENESIS IN *E. COLI*

Studies of MHC mutants should provide insights into structure–function relationships in the myosin molecule. As outlined in Section 8.7, considerable progress has been made studying the myosin head mutations of *unc*-54 arising as suppressors of *unc*-22 twitching. However,

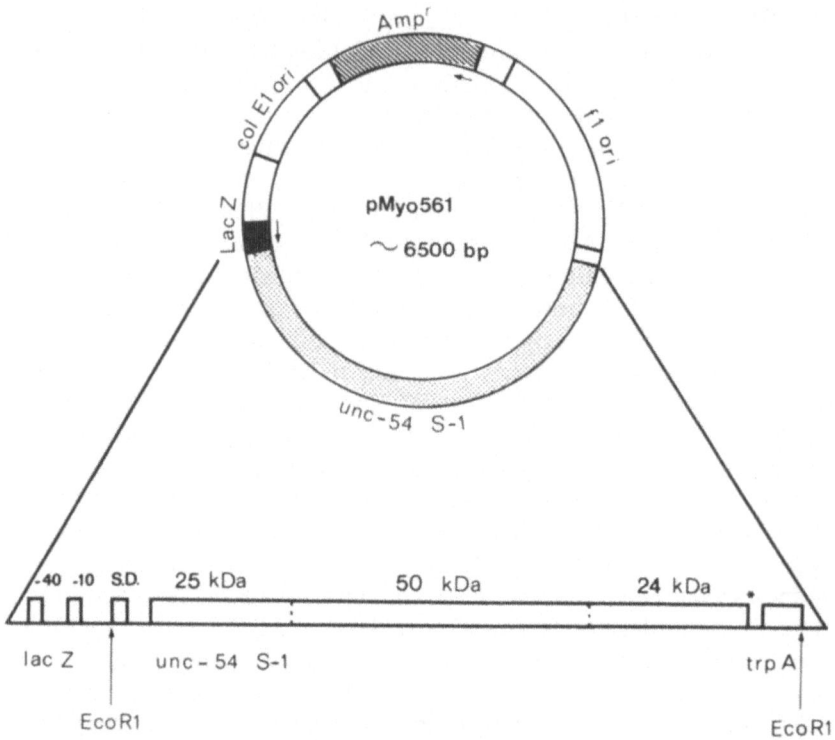

Figure 25. Structure of the pMyo561 S-1 expression plasmid. pMyo561 contains the coding sequence for MHC B from the initator methionine (met-6) to leu-856 [tyr-857 is replaced by an amber codon (*)]. The coding sequence is preceeded by the *lacz* promoter (the −40 and −10 regulatory regions are indicated) and is followed by a *trpA* transcription terminator (*trpA*). EcoR1 sites flank the ribosome binding site (S.D.) and the terminator sequence. The plasmid is derived from *p*EMBL8 (Dente *et al.*, 1983) and contains a ColEl origin of replication (ColEl ori), a β-lactamase gene (Amp^r), and a single strand origin of replication from bacteriophage f1 (f1 ori). The plasmid normally grows as a double-stranded molecule. Upon infection of strains harboring the plasmid with bacteriophage f1, singlestranded copies of the plasmid DNA are produced and packaged. MHC B S-1 synthesis is induced by the addition of IPTG to cultures containing the plasmid.

only a limited number of mutations can be obtained and studied by these *in vivo* mutagenesis techniques. For example, it would not be possible to obtain a complete set of amino acid substitutions at a single critical residue in the myosin sequence. Also, mutations of interest that led to unstable myosin proteins or produced lethal effects could not be obtained.

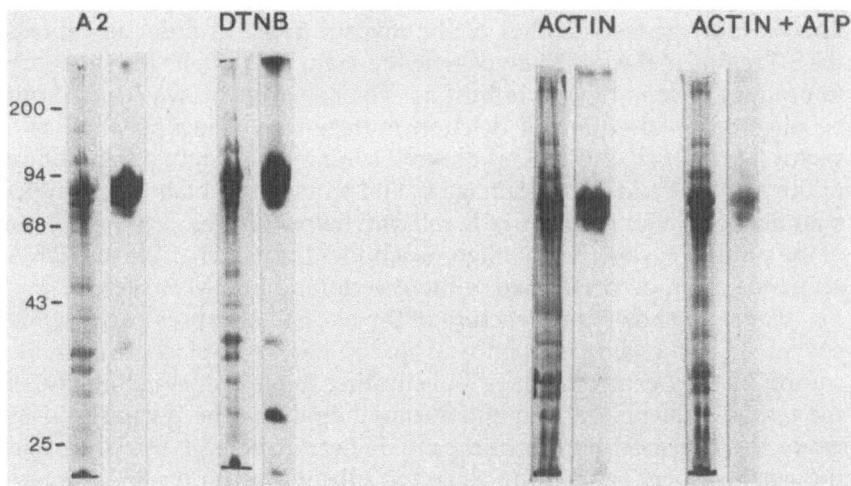

Figure 26. Binding of ^{125}I-labeled myosin light chains and actin to bacterially-synthesized MHC B S-1. Inclusion body proteins from *E. coli* strains harboring the pMyo561 plasmid (Fig. 25) were fractionated by SDS-PAGE and gel overlay assays performed as described by Mitchell *et al.* (1986). The gels were fixed in 10% acetic acid/50% methanol for 30 min at 22°C, washed for 3 hr with three washes of 250 ml of 10% ethanol at 4°C followed by incubation for 30 min at 4°C in buffer A (0.25% gelatin/200 mM NaCl/2 mM MgCl$_2$/20 mM Tris-HCl pH 7.5/10 mM NaH$_2$PO$_4$ pH 7.0/0.25 mM DTT/3 mM Na azide/ ± 2 mM ATP pH 7.0). Gels were incubated overnight at 4°C in 10 ml of buffer A containing ^{125}I-labeled protein. The gels were washed at least four times in buffer A, rinsed in water, fixed, stained, destained, dried, and autoradiographed. Proteins were labeled using ^{125}I-labeled Bolton-Hunter reagent (Amersham International). The specific activities for the ^{125}I-labeled proteins ranged between 4×10^6 to 4×10^7 cpm/nmoles for the light chains and 2×10^6 to 2×10 cpm/nmoles for actin. In each pair of gel lanes (left) Coomassie-stained proteins; (right) Autoradiograph of bound ^{125}I-labeled protein. A2, A2 light chain from chicken skeletal muscle (essential); DTNB, DTNB light chain (regulatory).

A second problem rising from the *in vivo* studies is that biochemical analysis of the mutant myosins has been difficult. This is partly because it has been difficult to isolate completely undenatured myosins from nematodes and partly because it will be necessary to purify the mutationally altered MHC B from the MHC A, C, D chains. Recently R. H. Waterston and M. Sheetz (personal communication) demonstrated that nematode myosins can direct the movement of beads along actin cables from *Nitella* using the method of Sheetz and Spudich (1984).

The cloned *unc*-54 gene permits the possibility of producing defined regions of the myosin molecule in *E. coli*. This could be a potentially valuable source of homogeneous protein of defined sequence for site-directed mutagenesis studies of the myosin head. In order to express the S-1 region of the *unc*-54 myosin heavy chain in *E. coli*, it was necessary to produce a clone devoid of introns. The construction was carried out by oligonucleotide-directed deletion mutagenesis using pEMBL8 as a vector (Dente *et al.*, 1982). This plasmid contains the origin of replication of the single-strand bacteriophage f1 and so may be obtained in single-stranded form after infection of *E. coli* with helper phages. The advantage of the pEMBL system is that oligonucleotide-directed changes and DNA sequencing can be performed without recloning into M13 vectors.

Figure 25 shows the structure of the *unc*-54-S-1-expressing plasmid pMyo561. The coding sequence of *unc*-54 has been placed under the control of the *lacz* promoter by substituting the initiator methionine of the *unc*-54 sequence for the initiator methionine of the β-galactosidase gene. The five short introns of the *unc*-54 head have been removed, and the sequence ends with an amber codon substituting for tyr-857, 5 amino acids before the beginning of the rod at pro-862. A *trp*A transcription terminator sequence has been inserted immediately after the amber codon. The *lacz* ribosome binding site also been slightly modified to improve the translation efficiency of the *unc*-54 sequence in *E. coli*. The entire coding sequence is flanked by unique *Eco*R1 sites to enable us to move the S-1 gene easily into other expression plasmids.

Induction of *E. coli* strains harboring pMyo561 with IPTG leads to the accumulation of the 90,000-M_r S-1 fragment (Fig. 26). The protein reacts specifically with rabbit antinematode MHC antibodies and monoclonal antibodies recognizing the MHC B S-1 region. Unfortunately, like many eukaryotic proteins expressed in *E. coli* (Williams *et al.*, 1982; Reinach *et al.*, 1986; Nagai *et al.*, 1985) the *E. coli* MHC S-1 was found to be an insoluble product accumulating in cytoplasmic inclusion bodies.

Partial refolding of the protein has been obtained using a gel overlay technique. Inclusion body proteins were purified by repeated washing with 0.5% Triton X-100 (Marston *et al.*, 1984). The protein was solubilized with SDS and fractionated by PAGE and the gels were then processed as described by Mitchell *et al.* (1986) to remove SDS and permit refolding of the S-1 molecule. As shown in Fig. 26, the *E. coli* MHC S-1 can specifically bind ^{125}I-labeled myosin regulatory (DTNB) and essential (A2) light chains and ^{125}I-labeled actin. MgATP greatly reduced the actin binding activity. These results are similar to previous results using the gel overlay method to study actin and light-chain binding in rabbit skeletal myosin fragments (Mitchell *et al.*, 1986). We are currently making a series of mutations in the 23,000-M_r region in order to map the amino acid residues of S-1 involved in actin and light-chain binding accurately.

Future studies will require enzymatically active S-1. This will require finding suitable conditions to renature the *E. coli* MHC S-1. Although Wagner and Giniger (1981) demonstrated that both the regulatory and alkali light chains could be removed from chick pectoralis myosin while retaining 30–80% of the ATPase activity of native S-1, it may be necessary to add light chains to the bacterial S-1 before myosin functions can be demonstrated. In nematodes, the light chains are 16,000–18,000 M_r, or approximately 2,000 M_r smaller than their mammalian counterparts (Harris and Epstein, 1977). Cummins and Anderson (1987) recently identified the *Caenorhabditis elegans* light-chain genes using as probes *Drosophila* alkali and regulatory myosin light-chain cDNA clones (Falkenthal *et al.*, 1985). Reinach *et al.* (1986) successfully expressed chick regulatory light chains in *E. coli*. These workers showed that the protein can renature and replace the regulatory light chain in scallop myosin. These experiments, and our experience with the binding of heterologous light chains to bacterial S-1, suggest that it may be possible to complement the cloned *unc-54*-derived-S-1 with cloned *Caenorhabditis elegans* light chains. We are also beginning experiments aimed at expressing the nematode S-1 protein in mammalian tissue culture cells. These cells may provide an alternative source of enzymatically active protein.

Successful expression of enzymatically active S-1 will permit application, the first time, of oligonucleotide-directed mutagenesis methods to the study of structure–function relationships in myosin. It should be possible to produce a wide variety of mutants that could define amino acid residues of importance for actin binding, ATP hydrolysis, light-chain binding, and force generation, and thereby advance our understanding of the molecular basis for muscle contraction.

ACKNOWLEDGMENTS

This contribution represents a complete update of "Cloning Nematode Myosin Genes," which appeared in *Cell and Muscle Motility*, Vol. 6 (J. W. Shay, ed.), pp. 185–237, published by Plenum Press, New York.

The work reported here reflects the close working relationship between our laboratory and the laboratories of Henry Epstein and Bob Waterston. The free exchange of ideas and materials between our groups has greatly speeded work in this field. We are also indebted to our many colleagues who have contributed unpublished information to this chapter. We especially thank Sydney Brenner, Dan Brown, Frank Stockdale, Andrew McLachlan, and Ichiro Maruyama. Jonathan Karn is an Established Investigator of the American Heart Association. David Miller was supported by grants from the Burroughs-Wellcome Foundation and the

European Molecular Biology Organization. Nick Dibb was supported by a research fellowship from the Medical Research Council. E. Jane Mitchell was supported by a Postdoctoral Fellowship from the Medical Research Council of Canada.

REFERENCES

Adelstein, R. S., and Eisenberg, E., 1980, Regulation and kinetics of the actin–myosin–ATP interaction, *Annu. Rev. Biochem.* **49:**921–956.

Albertson, D. G., 1984, Localization of the ribosomal genes in *Caenorhabditis elegans* chromosomes by *in situ* hybridization using biotin-labeled probes. *EMBO J.* **3:**1227–1234.

Alberston, D. G., 1985, Mapping muscle protein genes by *in situ* hybridization using biotin-labeled probes, *EMBO J.* **4:**2493–2498.

Albertini, A. M., Hoter, M., Calos, M. P., and Miller, J. H., 1982, On the formation of spontaneous deletions: The importance of short sequence homologies in the generation of large deletions, *Cell* **29:**319–328.

Anderson, P., and Brenner, S., 1984, A selection for myosin heavy chain mutants in the nematode *Caenorhabditis elegans*, *Proc. Natl. Acad. Sci. USA* **81:**4470–4474.

Balant, M., Wolf, I., Tarcsafalvi, A., Gergely, J., and Streter, A., 1978, Location of SH-1 and SH-2 in the heavy chain segment of heavy meromyosin, *Arch. Biochem. Biophys.* **190:**793–799.

Benian, G. M., Barstead, R., Moerman, D. G., and Waterston, R. H., 1987, The *unc-22* gene of *Caenorhabditis elegans* encodes a 500,000 M_r component of the thick filament, *J. Mol. Biol.*, submitted for publication.

Bennet, A. J., Patel, N., Wells, C., and Bagshaw, C. R., 1984, 8-Anilino-1-naphthalene-sulphonate, a fluorescent probe for the regulatory light chain binding site of scallop myosin. *J. Muscle Res. Cell Motil.,* **5:**165–182.

Benoist, C., and Chambon, P., 1981, In vivo sequence requirements of the SV40 early promoter region, *Nature (Lond.)* **290:**304–310.

Biggin, M. D., Gibson, T. J., and Hong, G. F., 1983, Buffer gradient gels and ^{35}S label as an aid to rapid DNA sequencing, *Proc. Natl. Acad. Sci. USA* **13:**3963–3965.

Bolten, S. L., Powell-Abel, P., Fischoff, D. A., and Waterston, R. H., 1984, The sup-7 (st5) X gene of *Caenorhabditis elegans* encodes a tRNA^{Trp} amber suppressor. *Proc. Natl. Acad. Sci USA* **81:**6784–6788.

Brenner, S., 1974, The genetics of *Caenorhabditis elegans*, *Genetics* **77:**71–94.

Burke, M., and Kamalkannan, V., 1985, Effect of trypic cleavage on the stability of myosin subfragment-1.Isolation and properties of the severed heavy chain subunit, *Biochemistry* **24;**846–852.

Burke, M., and Reisler, E., 1977, Effect of nucleotide binding on the proximity of the essential sulfhydryl groups of myosin. Chemical probing of movement of residues during conformational transitions, *Biochemistry* **16:**5559–5563.

Burke, M., Sivaramakrishnan, M., and Kamalakannan, V., 1983, On the mode of the alkali light chain association to the heavy chain of myosin subfragment-1. Evidence for the involvement of the carboxyl-terminal region of the heavy chain, *Biochemistry* **22:**3046–3053.

Cardinaud, R., 1979, Proteolytic fragmentation of myosin: Location of SH-1 and SH-2 thiols, *Biochemie* **61:**807–821.

Capony, J., and Elzinga, M., 1981, The amino acid sequence of a 34,000 dalton fragment from S-2 of myosin, *Biophys. J.* **33**:148a.

Chaussepied, P., Mornet, D., Andemard, C., Derancourt, J., and Kassab, R., 1986a, Abolition of ATPase activities of skeletal myosin subfragment-1 by a new selective proteolytic cleavage within the 50 kilodalton heavy chain segment, *Biochemistry* **25**:1134–1140.

Chaussepied, P., Mornet, D., Barman, T. E., Travers, F., and Kassab, R., 1986b, Alteration of the ATP hydrolysis and actin binding properties of thrombin-cut myosin subfragment-1; *Biochemistry* **25**:1141–1149.

Chen, T., Appelgate, D., and Reisler, E., 1985, Cross-linking of actin to myosin subfragment-1: Course of reaction and stoichiometry of products, *Biochemistry* **24**:137–144.

Chizzonite, R. A., Everett, A. W., Clark, W. A., Jakovicic, S., Rabinowitz, M., and Zak, R., 1982, Isolation and characterization of two molecular variants of myosin heavy chain from rabbit ventricle, *J. Biol. Chem* **257**:2056–2065.

Cochet, M., Gannon, F., Hen, R., Maroteaux, L., Perrin, F., and Chambon, P., 1979, Organisation and sequence studies of the 17-piece chicken conalbumin gene, *Nature (Lond.)* **282**:567–574.

Cohen, C., 1982, Matching molecules in the catch mechanism, *Proc. Natl. Acad. Sci. USA* **79**:3176–3178.

Cohen, C., Szent-Györgyi, A. G., and Kendrick-Jones, J., 1971, Paramyosin and the filaments of molluscan "catch" muscles, I. Paramyosin: Structure and assembly, *J. Mol. Biol.* **56**:223–237.

Coulondre, C., and Miller, J. H., 1977, Genetic studies of the lac repressor. IV. Mutagenic specificity in the *lacI* gene of *Escherichia coli*, *J. Mol. Biol.* **117**:577–606.

Craig, R. W., 1977, Structure of A-segments from frog and rabbit skeletal muscle, *J. Mol. Biol.* **109**:69–81.

Craig, R. W., and Offer, G., 1976, The location of C-protein in rabbit skeletal muscle, *Proc. R. Soc. Lond. B* **192**:451–461.

Crick, F. H. C., 1952, Is alpha-keratin a coiled coil? *Nature (Lond.)* **170**:882–883.

Cummins, C., and Anderson, P, 1987, The myosin light chain genes of *Caenorhabditis elegans*, *J. Mol. Biol.*, submitted for publication.

Davis, J. S., 1981, pressure-jump studies on the length regulation kinetics of the self-assembly of myosin from vertebrate skeletal muscle into thick filament, *Biochem. J.* **197**:309–314.

Dayhoff, M. O., 1978, *Atlas of Protein Sequence and Structure*, The National Biomedical Research Foundation, Washington, D. C.

Dente, L., Cesarini, G., and Cortese, R., 1983, pEMBL: A new family of single-stranded plasmids, *Nucleic Acids Res.* **11**:1645–1657.

Dibb, N. J., Brown, D. M., Karn, J., Moerman D. G., Bolten, S. L., and Waterston, R. H., 1985, Sequence analysis of mutations that affect the synthesis, assembly and enzymatic activity of the unc-54 myosin heavy chain of *Caenorhabditis elegans*, *J. Mol. Biol.* **183**:543–551.

Dibb, N. J., Maruyama, I., and Karn, J., 1987, The *Caenorhabditis elegans* myosin heavy chain gene family, *J. Mol. Biol.*, submitted for publication.

Doolittle, R. F., Goldbaum, D. M., and Doolittle, L. R., 1978, Designation of sequences involved in the "coiled coil" interdomainal connections in fibrinogen: Construction of an atomic scale model, *J. Mol. Biol.* **120**:311–316.

Dugaiczyk, A., Woo, S. L. C., Lai, E. C., Mace, M. L., McReynolds, L., and O'Malley, B. W., 1978, The natural ovalbumin gene contains seven intervening sequences, *Nature (Lond.)* **274**:328–333.

Elliot, A., and Offer, G., 1978, Shape and flexibility of the myosin molecule, *J. Mol. Biol.* **123:**505–519.

Ellis, R. E., Sulston, J. E., and Coulson, A. R., 1986 The rDNA of *Caenorhabditis elegans:* Sequence and structure. *Nucleic Acids Res.* **14:**2345–2364.

Elzinga, M., and Collins J. H., 1977, Amino acid sequence of a myosin fragment that contains SH-1, SH-2, and N-methylhistidine, *Proc. Natl. Acad. Sci. USA* **74:**4281–4284.

Elzinga, M., and Trus, B. L., 1980, Sequence and proposed structure of a 17,000 dalton fragment of myosin, in: *Methods in Peptide and Protein Sequence Analysis* (C. Birr, ed.), pp. 213–224, Elsevier/North-Holland, Amsterdam.

Emmons, S. W., Yesner, L., Kuan, K.-S., and Katzenberg, D., 1983, Evidence for a transposon in *Caenorhabditis elegans, Cell* **32:**55–65.

Epstein, H. F., and Miller, D. M. III, 1983, Different locations of two myosin isoforms paramyosin, and core filaments within nematode thick filaments, *J. Cell Biol.* **97:**2642.

Epstein, H. F., Waterston, R. H., and Brenner, S., 1974, A mutant affecting the heavy chain of myosin in *Caenorhabditis elegans, J. Mol. Biol.* **90:**291–300.

Epstein, H. F., Miller, D. M. III, Gossett, L. A., and Hecht, R. M., 1982a, Immunological studies of myosin isoforms in nematode embryos, in: *Muscle Development* (M. Pearson and H. Epstein eds.), pp. 7–14, Cold Spring Harbor Laboratory, Cold Spring Harbor, New York.

Epstein, H. F., Berman, S. A., and Miller, D. M. III, 1982b, Myosin synthesis and assembly in nematode body wall muscle, in: *Muscle Development* (M. Pearson and H. Epstein, eds.), pp. 419–427, Cold Spring Harbor Laboratory, Cold Spring Harbor, New York.

Fabian, F., and Muhlrad, A., 1968, Effect of trinitrophenylation on myosin ATPase, *Biochem. Biophys. Acta* **162:**596–603.

Falkenthal, S., Parker, V., and Davidson, N., 1985, Developmental variations in the splicing pattern of transcripts from the *Drosophila* gene encoding myosin alkali light chain result in different carboxy-terminal amino acid sequences, *Proc. Natl. Acad. Sci. USA* **82:**449–453.

Files, J. G., and Hirsh, D., 1981, Ribosomal DNA of *Caenorhabditis elegans, J. Mol. Biol.* **149:**223–240.

Files, J. G., Carr, S., and Hirsh, D., 1983, Actin gene family in *Caenorhabditis elegans, J. Mol Biol.* **164:**355–375.

Flicker, P. F., Wallimann, T., and Vibert, P., 1983, Electron microscopy of Scallop myosin: Location of regulatory light chains, *J. Mol. Biol.* **169:**723–741.

Flicker, P. F., Peltz, G., Sheetz, M. P., Parhaw, P. and Spudich, J. A., 1985, Site specific inhibition of myosin-mediated motility *in vitro* by monoclonal antibodies. *J. Cell Biol.* **100:**1024–1030.

Frank, G., and Weeds, A., 1974, The amino acid sequence of the alkali light chains of rabbit skeletal–muscle myosin, *Eur. J. Biochem.* **44:**317–334.

Fraser, R. D. B., and MacRae, T. P., 1973, *Conformation in Fibrous Proteins,* Academic Press, New York.

Garcea, R. L., Schachat, F., and Epstein, H. F., 1978, Coordinate synthesis of two myosins in wild-type and mutant nematode muscle during larval development, *Cell* **15:**421–428.

Gilbert, W., 1978, Why genes in pieces?, *Nature (Lond.)* **271:**501.

Gossett, L. M., and Hecht, R. M., 1980, A squash technique demonstrating nuclear cleavage of the nematode, *Caenorhabditis elegans, J. Histochem. Cytochem.* **28:**507–510.

Greene, L. E., 1984, Stoichiometry of actin-S-1 cross-linked complex, *J. Biol. Chem.* **259:**7363–7366.

Hardwicke, P. M. D., and Szent-Györgyi, A. G., 1985, Proximity of regulatory light chains in scallop myosin, *J. Mol. Biol.* **183:**203–211.

Hardwicke, P. M. D., Wallimann, T., and Szent-Györgyi, A. G., 1982, Regulatory and essential light chain interactions in scallop myosin I. Protection of essential light chain thiol groups by regulatory light chains, *J. Mol. Biol.* **156**:141–152.

Hardwicke, P. M. D., Wallimann, T., and Szent-Györgyi, A. G., 1983, Light-chain movement and regulation in scallop myosin, *Nature (Lond.)* **301**:478–482.

Harris, H. E., and Epstein, H. F., 1977, Myosin and Paramyosin of *Caenorhabditis elegans*: Biochemical and structural properties of wild-type and mutant proteins, *Cell* **10**:421–428.

Hawthorne, D. G., and Leupold, U., 1974, Suppressor mutations in yeast, *Curr. Topics Microbiol. Immunol.* **64**:1–47.

Heaphy, S., and Treager, R., 1984, Stoichiometry of covalent acting-subfragment-1 complexes formed on reaction with a zero-length cross-linking compound, *Biochemistry* **23**:2211–2214.

Heidecker, G., Messing, J., and Gronenborn, B., 1980, A versatile primer for DNA sequencing in the M13mp2 cloning system, *Gene* **10**:69–73.

Henry, G. D., Winstanley, M. A., Dalgarno, D. C., Scott, G. M. M., Levine, B. A., and Trayer, I. P., 1985, Characterization of the actin binding site on the alkali light chain of myosin, *Biochim. Biophys. Acta* **830**:233–243.

Hiratsuka, T., 1986, Role of the 50-kilodalton tryptic peptide of myosin subfragment-1 as a communicating apparatus between the adenosinetriphosphate and actin binding sites, *Biochemistry* **25**:2101–2109.

Hirsh, D., Files, J. G., and Carr, S. H., 1982, Isolation and genetic mapping of the actin genes of *Caenorhabditis elegans*, in: *Muscle Development* (M. Pearson and H. Epstein, eds.), pp. 77–86, Cold Spring Harbor Laboratory, Cold Spring Harbor, New York.

Hozumi, T., and Muhlrad, A., 1981, Reactive lysyl of myosin subfragment-1: Location on the 27K fragment and labeling properties, *Biochemistry* **20**:2945–2950.

Hulmes, D. H., Miller, A., Parry, A. D., Piez, K. A., and Woodhead-Galloway, J., 1973, Analysis of the primary structure of collagen for the origins of molecular packing, *J. Mol. Biol.* **79**:137–148.

Huxley, H. E., 1969, The mechanism of muscular contraction, *Science (Wash.)* **164**:1356–1366.

Huxley, H. E., and Brown, W., 1967, The low-angle X-ray diagram of vertebrate striated muscle and its behaviour during contraction and rigor, *J. Mol. Biol.* **30**:383–434.

Hvidt, S., Nestler, F. H. M., Greaser, M. L., and Ferry, J. D., 1982, Flexibility of myosin rod determined from dilute solution viscoelastic measurements, *Biochemistry* **21**:4064–4072.

Kagawa, H., Brenner, S., and Karn, J., 1987, Exon-expression cloning of the paramyosin gene from *Caenorhabditis elegans*, *Mol. Cell Biol.*, submitted for publication.

Karn, J., Brenner, S., Barnett, L., and Cesareni, G., 1980, Novel bacteriophage λ cloning vector, *Proc. Natl. Acad. Sci. USA* **77**:5172–5176.

Karn, J., McLachlan, A. D., and Barnett, L., 1982, *unc*-54 myosin heavy chain gene of *Caenorhabditis elegans*: Genetics, sequence, structure, in: *Muscle Development* (M. Pearson and H. Epstein, eds.), Cold Spring Harbor Laboratory, pp. 129–142, Cold Spring Harbor, New York.

Karn, J., Brenner, S., and Barnett, L., 1983a, Protein structural domains in the *Caenorhabditis elegans unc*-54 myosin heavy chain gene are not separated by introns, *Proc. Natl. Acad. Sci. UA* **80**:4253–4257.

Karn, J., Brenner, S. and Barnett, L., 1983b, New Bacteriophage λ vectors with positive selection for cloned inserts, *Methods Enzymol.* **101**:1–19.

Katoh, T., and Morita, F., 1984, Interaction between myosin and F-actin. Correlation with actin-binding sites on subfragment-1. *J. Biochem. (Jpn.)* **96**:1223–1230.

Katoh, T., Imae, S. and Morita, F., 1984, Binding of F-actin to a region between SH-1 and SH-2 groups of myosin subfragment-1 which may determine the high affinity of acto-subfragment-1 complex at rigor. *J. Biochem (Jpn.)***95**:447–454.

Katoh, T., Katoh, H., and Morita, F., 1985, Actin-binding peptide obtained by the cyanogen bromide cleavage of the 20 kDa fragment of myosin subfragment-1, *J. Biol. Chem.* **260**:6723–6727.

Kavinsky, C. J., Emeda, P. K., Sinha, A. M., Elzinga, M., Tong, S. W., Zak, R., Jakovicic, S., and Rabinowitz, M., 1983, Cloned mRNA sequences for two types of embryonic myosin heavy chains from chick skeletal muscle, *J. Biol. Chem.* **258**:5196–5205.

Kendrick-Jones, J., Szentkiralyi, E. M., and Szent-Györgyi, A. G., 1976, Regulatory light chains in myosins, *J. Mol. Biol.* **104**:747–779.

Kensler, R. W., and Levine, R. J. C., 1982, An electron microscopic and optical diffraction analysis of the structure of similar felson muscle thick filaments, *J. Cell Biol* **92**:443–45.

Kensler, R. W., and Stewart, M., 1983, Frog skeletal muscle thick filaments are three-stranded, *J. Cell. Biol.* **96**:1797–1802.

Kiehart, P. D., and Pollard, T. D., 1984, Stimulation of *Acanthamoeba* actomyosin ATPase activity by myosin-II polymerization, *Nature (Lond.)*, **308**:864–866.

Kramer, J. M., Cox, G. N., and Hirsh, D., 1982, Comparisons of the complete sequences of two collagen genes from *Caenorhabditis elegans*, *Cell* **30**:599–606.

Kubo, S., Tokura, S., and Tonomura, Y., 1960, On the active site of myosin A-adenosine triphosphatase I. Reaction of the enzyme with trinitrobenzenesulfonate, *J. Biol. Chem.* **235**:2835–2839.

Kubo, S., Tokuyama, H., and Tonomura, Y., 1965, On the active site of myosin A-adenosine triphosphatase. V. Partial solution of the chemical structure around the binding site of trinitrobenzenesulfonate, *Biochem. Biophys. Acta* **100**:459–470.

Labbe, J. P., Mornet, D., Roseau, G., and Kasab, R., 1982, Cross-linking of F-actin to skeletal muscle myosin subfragment-1 with bis(imido esters): Further evidence for the interaction of myosin-head heavy chain with an actin dimer, *Biochemistry* **21**:6897–6902.

Landel, C. P., Krause, M., Waterston, R. H., and Hirsh, D., 1984, DNA rearrangements of the actin gene cluster in *Caenorhabditis elegans* accompany reversion of three muscle mutants, *J. Mol. Biol.* **180**:497–513.

Leinwand, L. A., Fournier, R. E. K., Nadal-Ginard, B., and Shows, T. B., 1983, Multigene family for sarcomeric myosin heavy chain in mouse and human DNA: localization on a single chromosome, *Science (Wash.)* **221**:766–768.

Levine, R. J. C., Kensler, R. W., Reedy, M. C., Hofman, W., and King, H. A., 1983, Structure and paramyosin content of tarantula thick filaments, *J. Cell. Biol,* **97**:186–195.

Lewin, R., 1982, On the origin of introns, *Science (Wash.)* **217**:921–922.

Lewis, J. A., Wu, C.-H., Berg, H., and Levine, H. H., 1980a, The genetics of levamisole resistance in the nematode *Caenorhabditis elegans*, *Genetics* **95**:905–928.

Lewis, J. A., Wu, C.-H., Levine, J. H., and Berg, H., 1980b, Levamisole resistant mutants of the nematode *Caenorhabditis elegans* appear to lack pharmacological acetylcholine receptors, *Neuroscience* **5**:967–989.

Lowey, S., Slayter, H. S., Weeds, A., Baker, H., 1969, Substructure of the myosin molecule I. Subfragments of myosin by enzymatic degradation, *J. Mol. Biol.* **42**:1–29.

Lu, R. C., 1980, Identification of a region susceptible to proteolysis in myosin subfragment-2, *Proc. Natl. Acad. Sci. USA* **77**:2010–2013.

Lu, R. C., and Wong A., 1982, The primary structure of the susceptible region of long S-2, *Biophys. J.* **37**:52a.

MacKenzie, J. M., Jr., and Epstein, H. F., 1980, Paramyosin is necessary for determination of nematode thick filament length in vivo, *Cell* **22**:747–765.

MacKenzie, J. M., Jr., and Epstein, H. F., 1981, Electron microscopy of nematode thick filaments, *J. Ultrastruc. Res.* **76**:277–285.

MacKenzie, J. M., Jr., Garcea, R. L., Zengel, J. M., and Epstein, H. F., 1978a, Muscle development in *Caenorhabditis elegans;* mutants exhibiting retarded sarcomere construction, *Cell* **15**:751–762.

MacKenzie, J. M., Jr., Schachat, F., and Epstein, H. F., 1978b, Immunocytochemical localization of two myosins within the same muscle cells in *Caenorhabditis elegans, Cell* **15**:413–419.

MacLeod, A. R., Waterston, R. H., and Brenner, S., 1977a, An internal deletion mutant of a myosin heavy chain in *Caenorhabditis elegans, Proc. Natl. Acad. Sci. USA* **74**:5336–5340.

MacLeod, A. R., Waterston, R. H., Fishpool, R. M., and Brenner, S., 1977b, Identification of the structural gene for a myosin heavy chain in *Caenorhabditis elegans, J. Mol. Biol.* **114**:133–140.

MacLeod, A. R., Karn, J., Waterston, R. H., and Brenner, S., 1979, The unc-54 myosin heavy chain gene of *Caenorhabditis elegans:* A model system for the study of genetic suppression in higher eukaryotes, in: *Nonsense Mutations and tRNA Suppressors* (J. E. Celis and J. D. Smith, eds.), pp. 301–311, Academic Press, New York.

MacLeod, A. R., Karn, J., and Brenner, S., 1981, Molecular analysis of the unc-54 myosin heavy chain gene of *Caenorhabditis elegans, Nature* (*Lond.*) **291**:386–390.

Mahdavi, V., Chambers, A. P., and Nadal-Ginard, B., 1984, Cardiac α- and β-myosin heavy chain genes are organized in tandem, *Proc. Natl. Acad. Sci. USA* **81**:2626–2630.

Marston, F. A. D., Lowe, P. A., Doel, M. T., Schoemaker, J M., White, S., and Argal, S., 1984, *Biotechnology* **2**:800–804.

McLachlan, A. D., and Karn, J., 1982, Periodic charge distributions in the myosin rod amino acid sequence match cross-bridge spacings in muscle, *Nature* (*Lond.*) **299**:226–231.

McLachlan, A. D., and Karn, J., 1983, Periodic features in the amino acid sequence of nematode myosin rod, *J. Mol. Biol.* **164**:605–626.

McLachlan, A. D., and Stewart, M., 1976, The 14-fold periodicity in α-tropomyosin and the interaction with actin, *J. Mol. Biol.* **103**:271–298.

Mendelson, R. A., Morales, M. F., and Botts, J., 1973, Segmental flexibility of the S-1 moiety of myosin, *Biochemistry* **12**:2250–2255.

Messing, J., Crea, B., and Seeburg, P. H., 1981, A system for shotgun DNA sequencing, *Nucleic Acids Res.* **9**:309–321.

Miller, A., and Tregear, R. T., 1972, Structure of insect fibrillus flight muscle in the presence and absence of ATP, *J. Mol. Biol.* **70**:85–104.

Miller, D. M. III, and Maruyama, I., 1986, Myosin heavy chain amplification as a suppressor mutation in *Caenorhabditis elegans,* in: *Molecular Biology of Muscle Development, UCLA Symposia on Molecular and Cellular Biology, New Series,* Vol. 29 (C. Emerson, D. A. Fishman, B. Nadal-Ginard, and M. A. Q. Siddiqui, eds.), pp. 124–130, Alan R. Liss, New york.

Miller, D. M. III, MacKenzie, J. M., Bolton, L. H., and Epstein, H. F., 1981, Monoclonal antibodies to nematode myosin heavy chain isoenzymes, *J. Cell. Biol.* **91**:20023a.

Miller, D. M. III, Ortíz, I., Berlíner, G. C., and Epstein, H. F., 1983, Differential localization of two myosins within nematode thick filaments, *Cell* **34**:477–490.

Miller, D. M. III, Stockdale, F. E., and Karn, J., 1986, Immunological identification of the genes encoding the four myosin heavy chain isoforms of *Caenorhabditis elegans, Proc. Natl. Acad. Sci. USA* **83**:2305–2309.

Mitchell, E. J., Jakes, R., and Kendrick-Jones, J., 1986, Localization of light chain and actin binding sites on myosin, *Eur. J. Biochem.* **161**:25–35.

Moerman, D. G., and Baillie, D. L., 1979, Genetic organization in *Caenorhabditis elegans:* Fine structure analysis of the unc-22 gene, *Genetics* **91**:95–103.

Moerman, D. G., Plurad, S., Waterston, R. H., and Baillie, D. L., 1982, Mutations in the unc-54 myosin heavy chain gene of *Caenorhabditis elegans* that alter contractility but not muscle structure, *Cell* **29**:773–781.

Moerman, D. G., Benin, G. M., and Waterston, R. H., 1986, Molecular cloning of the muscle gene unc-22 in *Caenorhabditis elegans* by Tc1 transposon-tagging, *Proc. Natl. Acad. Sci. USA* **83**:2579–2583.

Mornet, D., Pantel, P., Audemard, E., and Kassab, R., 1979, The limited tryptic cleavage of chymotryptic S-1: An approach to the characterization of the actin site in myosin heads, *Biochem. Biophys. Res. Commun.* **89**:925–932.

Mornet, D., Bertrand, R., Pantel, P., Audemard, E., and Kassab, R., 1981*a*, Structure of the actin–myosin interface, *Nature (Lond.)* **292**:301–306.

Mornet, D., Bertrand, R., Pantel, P., Audemard, E., and Kassab, R., 1981*b*, Proteolytic approach to structure and function of actin recognition site in myosin heads, *Biochemistry* **20**:2110–2120.

Mornet, D., Pantel, P., Andemard, C., Derancourt, J., and Kassab, R., 1985, Molecular movements promoted by metal nucleotides in the heavy-chain regions of myosin heads from skeletal muscle, *J. Mol. Biol.* **183**:479–489.

Morse, D. E., Mosteller, R. D., and Yanofsky, C, 1969, Dynamics of synthesis, translation and degredation of *trp* operon messenger RNA in *E. coli, Cold Spring Harbor Symp. Quant. Biol.* **34**:725–741.

Mount, S. M., 1982, A catalogue of splice junction sequences, *Nucleic Acids Res.* **10**:459–472.

Nagai, K., Perutz, M. F., and Poyart, C., 1985, Oxygen binding properties of human hemoglobin mutants synthesized in *Escherichia coli, Proc. Natl. Acad. Sci. USA* **82**:7252–7255.

Nakamaye, K. L., Wells, J. A., Bridenbaugh, R. L., Okamoto, Y., and Young, R. G., 1985, 2-[(4-azido-2-nitrophenyl)amino]ethyl triphosphate, a novel chromophoric and photoaffinity analogue of ATP. Synthesis, characterization and interactions with myosin subfragment-1, *Biochemistry* **24**:5226–5235.

Niederman, R., and Peters, L. K., 1982, Native bare zone assemblage nucleates myosin filament assembly, *J. Mol. Biol.* **161**:505–517.

Offer, G. C., Moos, C., and Starr, R., 1973, A new protein of the thick filaments. Extraction, purification, and characterization, *J. Mol. Biol.* **74**:653–676.

Okamoto, Y., and Yount, R. G., 1983 Identification of an active site peptide of myosin after photoaffinity labeling, *Biophys. J.* **41**:298a.

Otsuka, A., 1987, *Sup-3* suppression affects muscle structure and myosin heavy chain accumulation in *Caenorhabditis elegans, J. Mol. Biol.* submitted for publication.

Pai, E. G., Sachsenheimer, W., Schirmer, R. H., and Schulz, G. E., 1977, Substrate positions and induced-fit in crystalline adenylate kinase, *J. Mol. Biol.* **114**:37–45.

Parry, D. A. D., 1978, Fibrinogen: A preliminary analysis of the amino acid sequences of the portions of the α, β and 4- chains postulated to form the interdomainal link between globular regions of the molecule, *J. Mol. Biol.* **120**:545–551.

Parry, D. A. D., 1981, Structure of rabbit skeletal myosin analysis of the amino acid sequences of two fragments from the rod region, *J. Mol. Biol.* **153**:459–464.

Parry, D. A. D., Crewther, W. G., Fraser, R. D. B., and MacRae, T. P., 1977, Structure of α-keratin: Structural implications of the amino acid sequences of the type I and type II chain segments, *J. Mol. Biol.* **113**:449–454.

Reinach, F. C., Nagai, K., and Kendrick-Jones, J., 1986, Site directed mutagenesis of the regulatory light chain Ca^{++}/Mg^{++} binding site and its role in hybrid myosins. *Nature (Lond.)* 322:80–83.

Reisler, E., Smith, C., and Seegan, G., 1980, Myosin minifilaments, *J. Mol. Biol.* 143: 129–145.

Riddle, D. L., and Brenner, S., 1978, Indirect suppression in *Caenorhabditis elegans*, *Genetics* 89:299–314.

Rozek, C. E., and Davidson, N., 1983, *Drosophila* has one myosin heavy chain gene with three developmentally regulated transcripts, *Cell* 32:23–34.

Rüther, U., and Müller-Hill, B., 1983, Easy identification of cDNA clones, *EMBO J.* 2:1791–1794.

Rüther, U., Koenen, M., Sippel, A. E., and Muller-Hill, B., 1982, Exon cloning: Immunoenzymatic identification of exons of the chicken lysozyme gene, *Proc. Natl. Acad. Sci. USA* 79:6852–6855.

Sanger, F., Nicklen, S., and Coulson, A. R., 1977, DNA sequencing with chain terminating inhibitors, *Proc. Natl. Acad. Sci. USA* 74:5463–5467.

Sanger, F., Coulson, A. R., Barrell, B. G., Smith, A. J. H., and Roe, B. A., 1980, Cloning in single-stranded bacteriophage as an aid to rapid DNA sequencing, *J. Mol. Biol.* 143:161–178.

Sanger, F., Coulson, A. R., Hong, G. F., Hill, D. F., and Petersen, G. B., 1982, Nucleotide sequence of bacteriophage λ DNA, *J. Mol. Biol.* 162:729–773.

Schachat, F. H., Harris, H. E., and Epstein, H. F., 1977, Two homogeneous myosins in body-wall muscle of *Caenorhabditis elegans*, *Cell* 10:721–728.

Schachat, F. H., Garcea, R. L., and Epstein, H. F., 1978, Myosins exist as homodimers of heavy chains: Demonstration with specific antibody purified by nematode mutant myosin affinity chromatography, *Cell* 15:405–411.

Sellers, J. R., and Harvey, E. V., 1984, Localization of a light-chain binding site on smooth muscle myosin revealed by light chain overlay of sodium dodecyl sulfate-polyacrylamide electrophoretic gels, *J. Biol. Chem.* 259:14203–14207.

Sheetz, M. P., and Spudich, J. A., 1983, Movement of myosin-coated fluorescent beads on actin cables *in vitro*, *Nature (Lond.)* 303:31–35.

Sivaramakrishnan, M. and Burke, M., 1982, The free heavy chain of vertebrate skeletal myosin sub fragments shows full enzymatic activity, *J. Biol. Chem.* 257:1102–1105.

Sjöström, M., and Squire, J. M., 1977, Fine structure of the A-band in cryo-sections. The structure of the A-band of human skeletal muscle fibers from ultra-thin cryo sections negatively stained, *J. Mol. Biol.* 109:49–68.

Spudich, J. A., Kron, S. T., and Sheetz, M. P., 1985, Movement of myosin coated beads on orientated filaments reconstituted from purified actin. *Nature (Lond.)* 315:584–586.

Squire, J. M., 1981, *The Structural Basis of Molecular Contraction*, Plenum Press, New York.

Staden, R., 1984, Automation of the computer handling of gel reading data produced by the shotgun method of DNA sequencing, *Nucleic Acids Res.* 12:521–538.

Staden, R., and McLachlan, A. D., 1982, Codon preference and its use in identifying protein coding regions in long DNA sequences, *Nucleic Acids Res.* 10:141–156.

Starr, R., and Offer, G., 1973, Polarity of the myosin molecule, *J. Mol. Biol.* 81:17–31.

Starr, R., and Offer, G., 1983, H-Protein and X-Protein. Two new components of the thick filaments of vertebrate skeletal muscle, *J. Mol. Biol.* 170:675–698.

Strehler, E. E., Strehler-Page, M.-A., Perriard, J.-C., Periamsy, M., and Nadal-Ginard, B., 1986, Complete nucleotide and encoded amino acid sequence of a mammalian myosin heavy chain gene: Evidence against intron-dependent evolution of the rod, *J. Mol. Biol.* 190:291–319.

Sulston, J. E., and Horvitz, H. R., 1977, Postembryonic cell lineages of the nematode *Caenorhabditis elegans, Dev. Biol.* **56:**100–156.

Sulston, J. E., Schierenberg, E., White, J. G., and Thomson, J. N., 1983, The embryonic cell lineage of the nematode *Caenorahabditis elegans, Dev. Biol.* **100:**64–119.

Sutoh, K., 1983, Mapping of actin binding sites on the heavy chain of myosin subfragment-1, *Biochemistry* **22:**1579–1585.

Sutoh, K., Yamomoto, K., and Wakabayashi, T., 1984, Electron microscopic visualization of the SH-1 thiol of myosin by the use of an avidin–biotin system, *J. Mol. Biol.* **178:**323–339.

Sutoh, K., Yamomoto, K., and Wakabayashi, T., 1986, Electron microscopic visualization of the ATPase site of myosin by photoaffinity labeling with a biotinylated photoreactive ADP analogue, *Proc. Natl. Acad. Sci. USA* **83:**212–216.

Szent-Györgyi, A. G., Cohen, C., and Kendrick-Jones, J., 1971, Paramyosin and the filaments of Molluscan catch muscles. II. Native filaments: Isolation and characterization, *J. Mol. Biol.* **56:**239–258.

Szentkiralyi, E. M., 1984, Tryptic digestion of scallop S-1: Evidence for a complex between the two light chains and a heavy chain peptide. *J. Muscle Res. Cell Motil.* **5:**147–164.

Takahashi, K., 1978, Topography of the myosin molecule as visualized by an improved negative staining method, *J. Biochem. (Jpn.)* **83:**905–908.

Takashi, R., Mulrad, A., and Botts, J., 1982, Spatial relationship between a fast-reacting thiol and a reactive lysine residue of myosin subfragment-1, *Biochemistry* **21:**5661 –5668.

Thomas, D. D., Seidel, J. C., Hyde, J. S., and Gergely, J., 1975, Motion of subfragment-1 in myosin and its supramolecular complexes: Saturation transfer electron paramagnetic resonance, *Proc. Natl. Acad. Sci. USA* **72:**1729–1733.

Thomas, D. D., Ishiwata, S., Seidel, J. C., and Gergely, J., 1980, Submillisecond rotational dynamics of spin-labeled myosin heads in myofibrils, *Biophys. J.* **32:**873–889.

Vibert, P., and Craig, R., 1982, Three dimensional reconstruction of thin filaments decorated with a Ca^{2+}-regulated myosin. *J. Mol. Biol.* **157:**299–319.

Vibert, P. and Craig, R., 1983, Electron microscopy and image analysis of myosin filaments from scallop studied muscle, *J. Mol. Biol.* **165:**303–320.

Vibert, P. D., Cohen, C., Hardwicke, P. M. D. and Szent-Györgyi, A. G., 1985, Electron microscopy of cross-linked scallop myosin, *J. Mol. Biol.* **183:**283–286.

Vibert, P. D., Szentkiralyi, E., Hardwicke, P., Szent-Györgyi, A. G., and Cohen, C., 1986, Structural models for the regulatory switch of myosin, *Biophys. J.* **49:**131–133.

Wagner, P. D., and Giniger, E., 1981, Hydrolysis of ATP and reversible binding to F-actin by myosin heavy chains free of all light chains, *Nature (Lond.)* **292:**560–562.

Walker, J. E., Saraste, M., Runswick, M. J., and Gay, N. J., 1982, Distantly related sequences in the α- and β-subunits of ATP synthease, myosin, kinases and other ATP-requiring enzymes and a common nucleotide binding fold, *EMBO J.* **1;**945–951.

Waller, G. G., and Lowey, S., 1985, Myosin subunit interactions. Localization of the alkali light chains, *J. Biol. Chem.* **260:**14368–14373.

Waterston, R. H., 1981, A second informational suppressor, *sup-7 X*, in *Caenorhabditis elegans, Genetics* **97:**307–325.

Waterston, R. H., and Brenner, S., 1978, A suppressor mutation in the nematode acting on specific alleles of many genes, *Nature (Lond.)* **275:**715–719.

Waterston, R. H., Fishpool, R. M., and Brenner, S., 1977, Mutants affecting paramyosin in *Caenorhabditis elegans, J. Mol. Biol.* **117:**825–842.

Waterston, R. H., Thomson, J. N., and Brenner, S., 1980, Mutant with altered muscle structure in *Caenorhabditis elegans, Dev. Biol.* **77:**271–302.

Waterston, R. H., Smith, K. C., and Moerman, D. G., 1982a, A genetic fine structure analysis of the myosin heavy chain gene unc-54 *Caenorhabditis elegans*, *J. Mol. Biol.* **158**:1–15.

Waterston, R. H., Bolton, S., Sive, H. L., and Moerman, D. G., 1982b, Mutationally altered myosins in *Caenorhabditis elegans*, in: *Muscle Development* (M. Pearson and H. E. Epstein, eds.), pp. 119–129, Cold Spring Harbor Laboratory, Cold Spring Harbor, New York.

Waterston, R. H., Hirsh, D., and Lane, T. R., 1984, Dominant mutations affecting muscle structure in *Caenorhabditis elegans* that map near the actin gene cluster, *J. Mol. Biol.* **180**:473–496.

Weeds, A. G., and Lowey, S., 1971, Substructure of the myosin molecule. II. The light chains of myosin, *J. Mol. Biol.* **61**:701–725.

Weeds, A. G., and Pope, B., 1977, Studies on the chymotryptic digestion of myosin effects of divalent cations on proteolytic susceptibility, *J. Biol. Chem.* **255**:1598–1602.

Wells, J., and Young, R. G., 1979, Active site trapping of nucleotides by crosslinking two sulfhydryls in myosin subfragment-1, *Proc. Natl. Acad. Sci. USA* **76**:4966–4970.

Wells, J., and Young, R. G., 1980, Magnesium nucleotide is stoichiometrically trapped at the active site of myosins and its active proteolytic fragments by thiol cross-linking reagents, *J. Biol. Chem.* **255**:1598–1602.

Williams, D. c., Van Frank, R. M., Muth, W. L., and Burnett, J. P., 1982, Cytoplasmic inclusion bodies in *E. coli* producing biosynthetic human insulin proteins, *Science* (Wash.) **215**:687–689.

Wills, N., Gesteland, R. F., Karn, J., Barnett, L., Bolton, S., and Waterston, R. H., 1983, The genes sup-7 X and sup-5 III of *Caenorhabditis elegans* suppress amber nonsense mutations via altered transfer RNA, *Cell* **33**:575–583.

Winkelmann, D. A., Lowey, S., and Press, J. R., 1983, Monoclonal antibodies localize changes on myosin heavy chain isozymes during avian myogenesis, *Cell* **34**:295–306.

Winkelmann, D. A., Almeda, S., Vibert, P., and Cohen, C., 1984, A new myosin fragment: Visualization of the regulatory domain, *Nature* (Lond.) **307**:758–760.

Wozney, J., Hanahan, D., Morimoto, R., Boedtker, H., and Doty, P., 1981, Fine structural analysis of the chicken α-2-collagen gene, *Proc. Natl. Acad. Sci. USA* **78**:712–716.

Wray, J. S., 1979, Structure of the backbone in myosin filaments of muscle, *Nature* (Lond.) **277**:37–40.

Zengel, J. M., and Epstein, H. F., 1980a, Mutants altering coordinate synthesis of specific myosins during nematode muscle development, *Proc. Natl. Acad. Sci. USA* **77**:852–856.

Zengel, J. M., and Epstein, H. F., 1980b, Identification of genetic elements associated with muscle structure in the nematode *Caenorhabditis elegans*, *Cell Motil.* **1**:73–97.

Zengel, J. M., and Epstein, H. F., 1980c, Muscle development in *Caenorhabditis elegans*: A molecular genetic approach, in: *Nematodes as Biological Models* (B. Zuckerman, ed.), pp. 73–126, Academic Press, New York.

Zweig, S. E., 1981, The muscle specificity and structure of two closely related fast-twitch white myosin heavy chain isozymes, *J. Biol. Chem.* **256**:11847–11853.

5

Small Cardioactive Peptides A and B
Chemical Messengers in the *Aplysia* Nervous System

ANNE C. MAHON and RICHARD H. SCHELLER

1. INTRODUCTION

Neurobiologists are conducting investigations ultimately aimed at understanding the cellular and molecular basis of animal behavior. These studies make use of a variety of techniques and approaches, many of which are defining the precise nature of the interactions between a neuron and its target. Of pivotal importance is the characterization of molecules that function as extracellular chemical messengers in the central nervous system (CNS). These messengers have a variety of chemical structures and can act in a direct fashion or across long distances. For example, at the neuromuscular junction, acetylcholine (ACh) acts directly over short distances and is rapidly degraded by acetylcholinesterase (AChE). Furthermore, high-affinity reuptake systems remove many neurotransmitters from the synaptic cleft (Hall, 1973; Cooper *et al.*, 1978). By contrast, a variety of other substances, such as biologically active peptides, may act either synaptically or at targets quite distant from their

ANNE C. MAHON and RICHARD H. SCHELLER • Department of Biological Sciences, Stanford University, Stanford, California, 94305.

sites of release. These longer-range effects often have slower onsets and longer durations than interactions mediated at a classic synapse (Krieger, 1983). More recent studies have demonstrated that single neurons may elicit multiple actions mediated by more than one chemical messenger (Lundberg and Hokfelt, 1983).

The specificity of intercellular communication mediated by the more classic transmitters is largely determined by spatial constraints of the synapses involved. Conversely, the activities of diffusible substances such as peptide hormones depend mainly on the distribution of high-affinity and very specific receptors. Animal behavior is largely governed by the set of interactions between neurons determined by these spatial and biochemical constraints.

This chapter discusses our studies directed at characterizing the structure and behavioral roles of biologically active peptides in the nervous system. The numerical simplicity and accessibility of the *Aplysia californica* CNS permits a correlation of the activity of individual nerve cells with specific behaviors (Kandel, 1976). The *Aplysia* CNS consists of about 20,000 neurons distributed among four symmetrically paired ganglia in the head and a single asymmetrical abdominal ganglion. The buccal ganglion is situated on the buccal musculature and mediates a number of activities related to feeding behavior (Cohen *et al.*, 1978). The pedal, pleural, and cerebral ganglia are joined by connectives that encircle the esophagus and are involved in the control of a variety of behaviors, including locomotion. The 2000 neurons of the abdominal ganglion regulate many visceral functions as well as defensive and reproductive behaviors, such as gill withdrawal and egg laying (Kandel, 1976).

Perhaps the most outstanding characteristic of the *Aplysia* CNS is the remarkable size (up to 1 mm) of many neurons. It is possible to identify these large neurons reproducibly on the basis of several criteria, including size, morphology, color, and endogenous electrical activity (Frazier *et al.*, 1967). This ability has greatly facilitated physiological studies, which over the past two decades have defined an impressive array of neural circuits. The identification of sensory, motor, and interneurons and studies of their interactions have made it possible to assign neural correlates to a number of simple behaviors (Kandel, 1976).

Considerable progress has been made in the biochemical and molecular characterization of many *Aplysia* neurons. This can be attributed in part to the fact that the neurons identified contain large amounts of DNA, RNA, and protein that can be isolated and analyzed. The large neurons in *Aplysia* are highly polyploid, which is the result of up to 16 rounds of DNA replication in the absence of cell division (Coggeshall *et al.*, 1971; Lasek and Dower, 1971). Thus, the neurons contain as much as 130,000 times the haploid content of DNA. *In vitro* translation of the mRNA isolated from identified neurons and *in vivo* labeling of the proteins present in these neurons has demonstrated that the cells contain correspondingly large amounts of RNA and protein (Nambu *et al.*, 1983). Analysis of the proteins from individual neurons following *in situ* labeling with radiolabeled amino acids reveals that a large portion (up to 50%) of the biosynthetic capacity of the cell is dedicated to the synthesis of a neuropeptide precursor protein unique to that cell or to a small subset of cells in the CNS (Aswad, 1978; Berry, 1976; Loh and Gainer, 1975). This fact, in addition to the size and accessibility of the neurons, has made it possible to isolate the genes encoding a number of neuropeptide precursors using differential screening procedures (Scheller *et al.*, 1983; Nambu *et al.*, 1983; Taussig *et al.*, 1984; Mahon *et al.*, 1985*a,b*; Schaefer *et al.*, 1985). This discussion is restricted to two related molluskan neuropeptides: the small cardioactive peptides A and B (SCP$_A$ and SCP$_B$).

The small cardioactive peptides (SCPs) were initially identified in the terrestrial pulmonate, *Helix aspersa,* using an isolated snail heart assay. In this assay, nervous system extracts were found to increase the amplitude and rate of the heartbeat in an analogous fashion to serotonin (Lloyd, 1978; Lloyd *et al.*, 1980 *a–c*). Two low-molecular-weight protease-sensitive peaks of bioactivity were resolved using reverse-phase high-pressure liquid chromatography (HPLC) and designated SCP$_A$ and SCP$_B$ (Fig. 1). The SCPs were subsequently identified in other mollusks, including *Aplysia californica* and *Aplysia brasiliana*. The isolation and purification of peptide extracts from the nervous systems of 2500 *Aplysia brasiliana* led to the elucidation of the amino acid sequence of SCP$_B$ by fast atom-bombardment mass spectrometry (Morris *et al.*, 1982).

2. THE DISTRIBUTION OF SCP-IMMUNOREACTIVE NEURONS IN THE CNS

The distribution of SCP-expressing cells and processes was determined using immunohistochemistry. In these studies, synthetic SCP$_B$

SCP$_A$ Ala Arg Pro Gly **Tyr Leu Ala Phe Pro Arg Met–amide**

SCP$_B$ Met Asn **Tyr Leu Ala Phe Pro Arg Met–amide**

Figure 1. Amino acid sequences of the small cardioactive peptides A and B. The carboxyl terminal 7-amino acids (bold type) are shared between the two peptides, and both sequences are amidated.

was coupled to serum albumin with carbodiimide, and rabbit antibodies were raised to the conjugate. Immunohistochemical staining of the nervous system was done either on cryostat sections of adult tissue or on whole-mount preparations of late-juvenile animals (Fig. 2) (Mahon *et al.*, 1985*a*; Lloyd *et al.*, 1985*a*).

Anti-SCP$_B$ immunoreactive neurons are most prevalent in the buccal ganglion. This finding is consistent with biochemical and bioassay results indicating that this ganglion contains approximately 10 times more SCP$_B$ than any other ganglion (Lloyd *et al.*, 1985*a*). Three groups of immunoreactive cell bodies were observed in the buccal ganglion: (1) the large neurons B1 and B2, and three to five smaller cells in this area of the ganglion; (2) a characteristic cluster of 20–30 smaller neurons on the dorsal surface, usually about 10–20 μm in diameter; and (3) several large neurons on the ventral surface, including three of the four neurons B3, B6, B9, and/or B10. The buccal ganglion innervates the gut and the buccal mass, a muscular organ that moves the mouth during feeding. Both immunohistochemical experiments and biochemical assays demonstrate that the SCPs are present in neurons that innervate both the gut and the buccal mass. The axons from neurons B1 and B2 exit the buccal ganglion via the esophageal nerve and innervate the esophagus. SCP-immunoreactive varicosities and processes are found in the accessory radula closer muscle of the buccal mass and in the nerve to this muscle (Lloyd *et al.*, 1984).

In addition to the buccal ganglion, all other ganglia exhibit anti-SCP staining cells as well as intensely staining processes and varicosities in the neuropile. The large amount of immunoreactivity in the buccal ganglion and the physiological activities of the peptides (discussed in Section 7) suggest that the SCPs are involved in feeding behavior. However, the widespread distribution of the SCPs makes it likely that the peptides are involved in other behavioral processes as well (Mahon *et al.*, 1985*a*; Lloyd *et al.*, 1985*a*).

Figure 2. Schematic diagram of the buccal ganglion and immunohistochemical localization of SCP$_B$. (A) Identified cells in the buccal ganglion of *Aplysia californica*. Cells that show SCP$_B$ immunoreactivity are shaded. The large neurons B1 and B2 send processes containing SCPs to the esophagus. (B–H) Cells and processes in the *Aplysia* nervous system that stain with antiserum directed against SCP$_B$: (B) Buccal neurons B1 and B2. (C) B1 or B2 and several smaller cells. (D) A cluster of small buccal neurons. (E–G) Processes and cells in the cerebral ganglion. (H) Processes in the abdominal ganglion. Size bar: 50 μm in B–E, G, and H, or 10 μm in F.

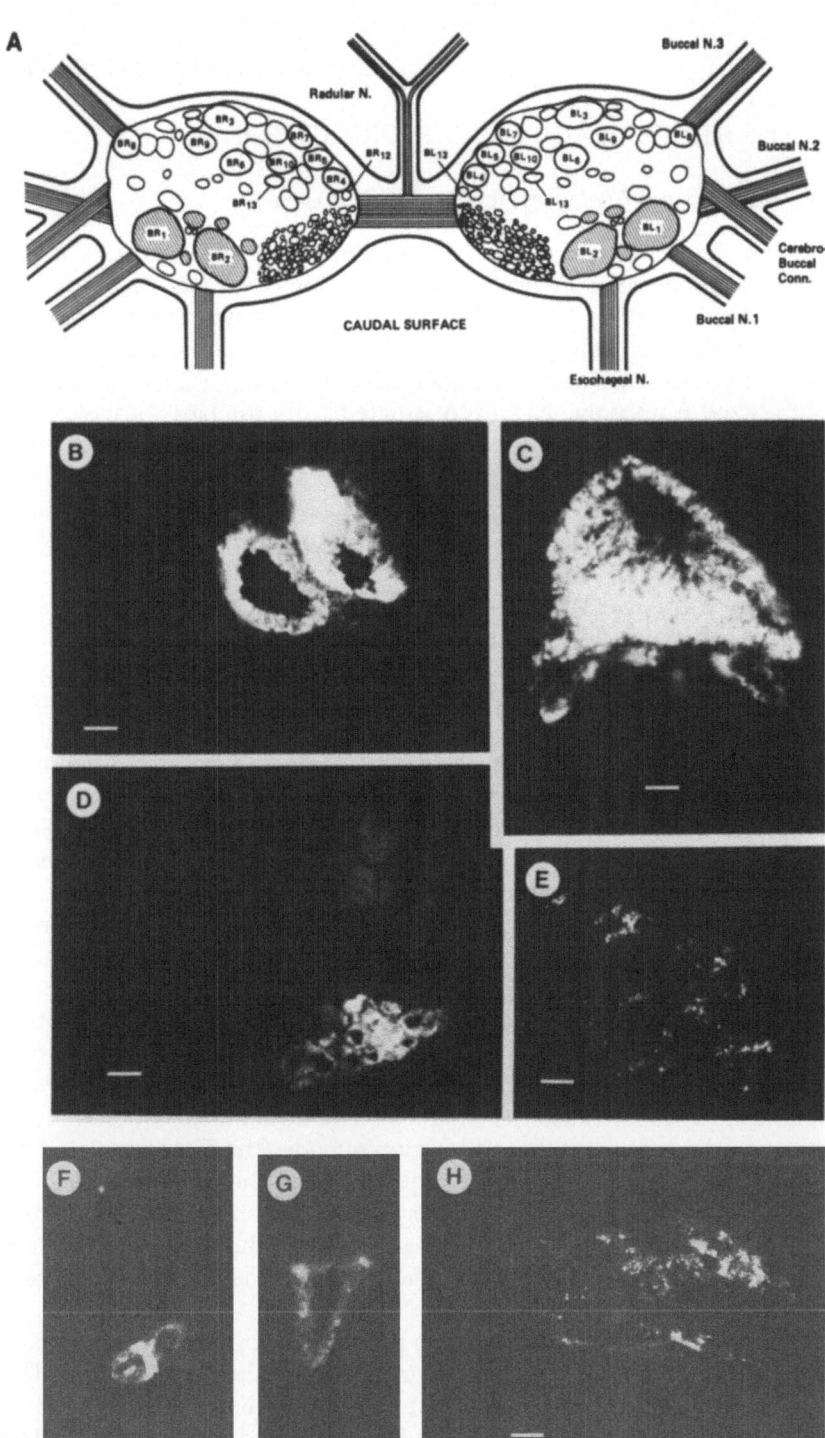

3. THE SCP GENE AND PRECURSOR PROTEIN

3.1. Cloning and Characterization of the SCP Gene

The studies described in Section 2 provided the necessary information for molecular genetic experiments aimed at isolating a cDNA clone encoding SCP_B. A cDNA library was constructed in the cloning vector λgt10 using poly (A)$^+$ RNA from buccal ganglia, where the SCPs are most prevalent. This library was screened using radiolabeled cDNA synthesized from poly (A)$^+$ RNA isolated from the large anti-SCP immunoreactive buccal neurons B1 and B2, and from the bag cells, which do not stain with the antisera. Clones were selected which hybridized to B1 and B2 cDNA but not to the bag cell cDNA. DNA sequence analysis confirmed that these clones encode the precursor for both the SCP_A and SCP_B.

A representative SCP cDNA clone was used to screen a phage lambda genomic library resulting in the isolation of genomic fragments encoding the peptides (Fig. 3) (Mahon et al., 1985a). Two sets of overlapping clones were obtained: one that encodes the 5' region of the mRNA and another that encodes the remainder of the message. We have not yet linked these sets of clones even after extensive screening of several genomic libraries. Although further studies are necessary to characterize the gene completely, we are able to make some preliminary conclusions. First, there appears to be a single gene encoding the SCP peptides; however, since fewer than 50 nucleotides are required to encode the peptides, we cannot completely rule out the possibility of other SCP-encoding genes. Second, while the precursor protein is only 136 amino acids, the gene is split into several exons and spans at least 15,000 nucleotides of genomic DNA. Large primary transcripts are associated with other neuropeptide genes we have characterized, including those that encode the egg-laying hormone as well as the L11 and R14 peptides (Mahon et al., 1985b; Nambu et al., 1983; Taussig et al., 1985). The nucleotide sequence of the first axon demonstrates that this region encodes the 5'-untranslated region of the message, the signal sequence, and the first amino acid of the SCP_B peptide, splitting the codon of the second amino acid. A substantial potential exists for recombination within the large introns of the SCP gene, a mechanism that may lead to the formation of new peptide products (Gilbert, 1978).

Analysis of the cDNA clone has provided information on the structure of the SCP mRNA and precursor protein (Mahon et al., 1985a). The SCP-encoding cDNA clone is about 1400 nucleotides and contains 37 consecutive adenine residues that define the 3' end of the message. The

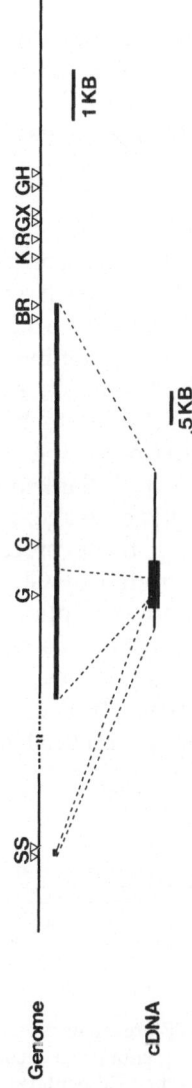

Figure 3. The gene encoding the SCP peptides. The top line represents the restriction enzyme map for the region of the genome containing the SCP gene. Restriction sites were determined by single, double, and partial digests on two sets of nonoverlapping clones. The broken line indicates the region that has not been cloned. The cDNA clone (bottom line) has a 136-amino acid open reading frame (filled box). Three ³²P-labeled restriction fragments were generated from the cDNA clone and were hybridized to the genomic clone. The areas of hybridization on the genomic clone are indicated below the genomic DNA (bars). The direction of transcription is from right to left. S, Sal I; G, Bgl II; B, Bam HI, R, Eco RI; H, Hind III; K, Kpn I; X, Xho I.

initiator methionine is 112 nucleotides downstream from the 5' end of the cDNA clone and is followed by a 136-amino acid open reading frame. Similar to the transcripts for the R14 and L11 peptide precursor proteins, the 3' untranslated region represents a major portion of the SCP transcript, comprising more than 800 nucleotides in the SCP message. The 3'-untranslated region contains three AXUAAA sequences that may serve as recognition signals for the polyadenosine tail addition. One of these sequences is 21 nucleotides from the poly (A) track and therefore may have been the signal used in the message we have cloned.

3.2. The SCP Precursor Protein

A schematic diagram of the SCP precursor protein inferred from the nucleotide sequence of the cDNA clone is shown in Fig. 4. The initiator methionine is followed by a glutamic acid residue at position 2 and an arginine at position 7. The next 15 amino acids do not contain any charged residues, and 11 are hydrophobic, suggesting this region functions as the signal peptide (Blobel and Dobberstein, 1975). The 9-amino acid sequence of SCP_B begins following the leader peptide at residue 25. The carboxyl terminal of SCP_B is flanked by Gly-Arg residues, which signal amidation and proteolytic cleavage, respectively (Loh and Gainer, 1983). Immediately following these processing signals is the sequence of SCP_A, which is also bounded by Gly-Arg. The remaining 78 amino acids do not contain any known peptides; however, the presence of six cysteine residues suggests that this region of the molecule may exist in a highly crosslinked configuration.

In the first neuropeptide precursors studied, the biologically active peptides were located in the carboxyl terminal part of the molecule

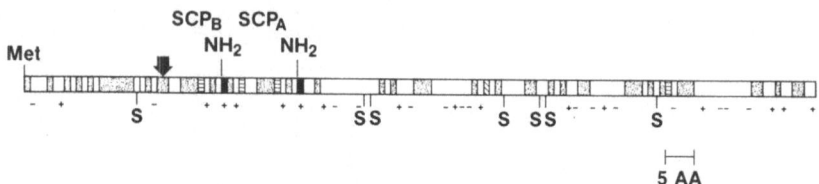

Figure 4. Schematic representation of the SCP precursor protein. The coding region was identified as the only open reading frame long enough to encode a protein of the appropriate molecular weight, which also encodes the SCP peptides. Positions of hydrophobic ( ), proline (\\\), histidine (≡), charged (+ or −), and cysteine residues (S) are indicated. The arrow indicates the position of the signal sequence cleavage, the lines above the sequence represent internal proteolytic cleavage sites, and an NH₂ above the line indicates positions of carboxyl-terminal amidation.

(Herbert *et al.*, 1981). However, as the present study illustrates, there appear to be no set rules to peptide precursor organization. The SCP precursor may use the signal protease cleavage site to generate the amino terminal of the active SCP$_B$ moiety. By contrast, the egg-laying hormone, which is located in the carboxyl-terminal region of its precursor, requires cleavage at two pairs of basic residues to free the active peptide from the precursor (Scheller *et al.*, 1983).

The presence of the both SCP$_A$ and SCP$_B$ on the same precursor is consistent with biochemical and bioassay experiments that detect equimolar amounts of the peptides in single identified cells. These results suggest that SCP$_{A,B}$ may be co-released (Lloyd *et al.*, 1985a). The release of more than one biologically active peptide expands the potential array of responses that a single neuron may evoke. It is not clear whether the presence of the two closely related peptides on the SCP precursor simply affords a molar increase in the amount of active peptide per precursor or whether the peptides have distinct functions.

4. SCP GENE EXPRESSION

The expression of the SCP gene was investigated by analyzing mRNA from the nervous system and by characterizing cDNA clones homologous to the RNA from these tissues. The poly (A)$^+$ RNA from all the major ganglia was size-fractionated on agarose gels, transferred to nitrocellulose and hybridized to a ^{32}P-labeled SCP cDNA clone (Fig. 5). A single transcript of approximately 1900 nucleotides is detected in all the ganglia. cDNA clones have been analyzed from libraries representing the entire *Aplysia* CNS. These clones are of approximately the same size and appear to represent a single RNA species. Thus, the SCP gene is transcribed and processed into a single major RNA species in the adult CNS (Mahon *et al.*, 1985a). From these data, we infer that there is a single SCP precursor protein. The presence of an mRNA in each of the ganglia, along with the histochemical experiments described earlier, demonstrates that the gene is expressed in a dispersed set of neurons.

5. SUBCELLULAR LOCALIZATION

We have studied the ultrastructure of neurons B1 and B2 as well as the subcellular location of the SCPs using electron microscopy (Fig. 6). The B1 and B2 somata contain a large number of dense-core vesicles. Histochemical studies demonstrate that the peptides are localized to

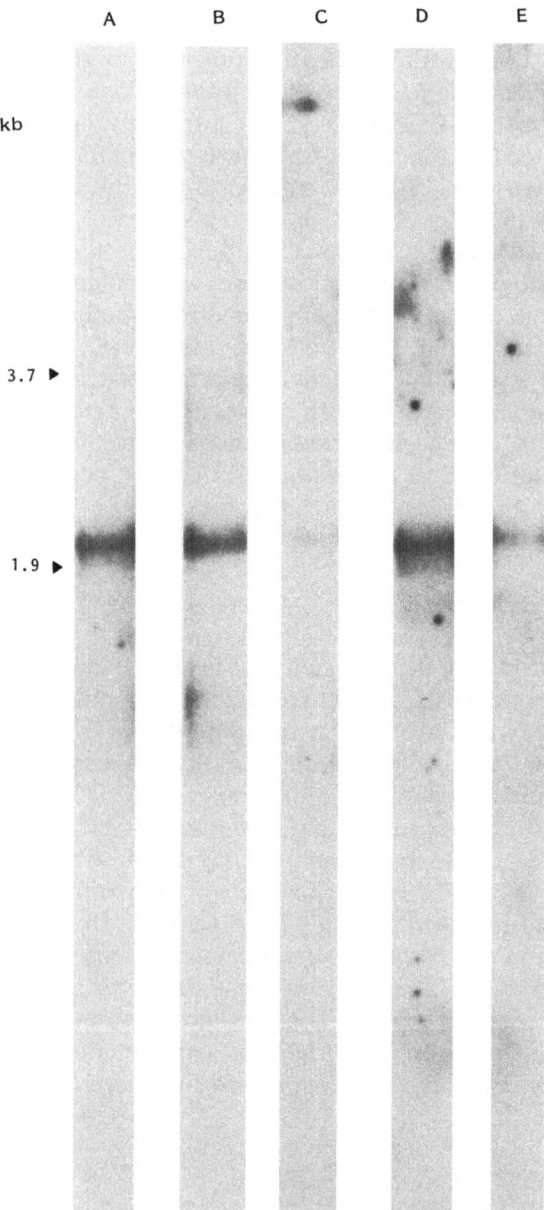

Figure 5. RNA blot hybridization analysis. Poly (A)⁺ RNA was isolated from the major ganglia and fractionated according to size on agarose gels containing formaldehyde. (A) abdominal ganglion, (B) buccal ganglion, (C) cerebral ganglion, (D) pedal ganglion, and (E) pleural ganglion.

these granules. Measurements of the granule size revealed a bimodal distribution with major peaks centered between 80–90 nm and 100–110 nm. We do not know the function of the multiple dense-core vesicle size classes. Other workers have noted that various physical constraints restrict the size of vesicles formed from dissociated plasma membrane. Different vesicle sizes may result in varied quantal sizes if indeed these organelles are also found at release sites (Sossin *et al.*, 1985).

6. COEXISTENCE OF MULTIPLE TRANSMITTERS

6.1. SCP$_A$ and SCP$_B$

The structure of the SCP precursor suggests that both peptides are synthesized in equimolar ratios. To test this prediction, protein extracts from isolated B1 and B2 neurons were separated by reverse-phase HPLC. The fractions were assayed for their ability to increase the frequency and amplitude of the snail heartbeat. Two approximately equal peaks of activity were observed with retention times corresponding to those for SCP$_A$ and SCP$_B$ (Fig. 7) (Lloyd *et al.*, 1985a).

6.2. Acetlycholine in Neuron B2

The experiments described above provide evidence that the B1 and B2 neurons use at least two peptide transmitters. The possibility that other chemical messengers are used by these cells was also investigated (Lloyd *et al.*, 1985a). In these studies radiolabeled choline was injected into neuronal cell bodies, and the radiolabeled compounds present in the cells after 1-hr incubation were analyzed (Fig. 8). In neuron B2 approximately one-half the radiolabeled choline was converted into ACh, while in B1 no significant amounts of ACh were detected. These data suggest that the enzyme choline acetyltransferase, a marker protein for cholinergic cells, is present in neuron B2 but not in B1. Although preliminary, these findings suggest that neurons B1 and B2 differ in that B2 may use ACh as a cotransmitter with the SCPs (Lloyd *et al.*, 1985a).

7. PHYSIOLOGICAL ACTIVITIES

The precise roles of the SCPs in *Aplysia* have not been defined; however, recent studies have attributed modulatory activities to the

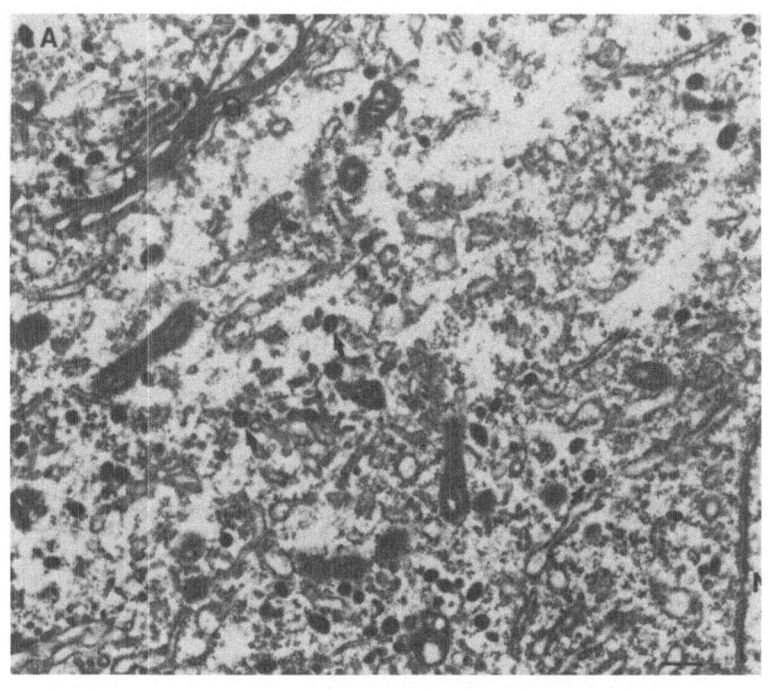

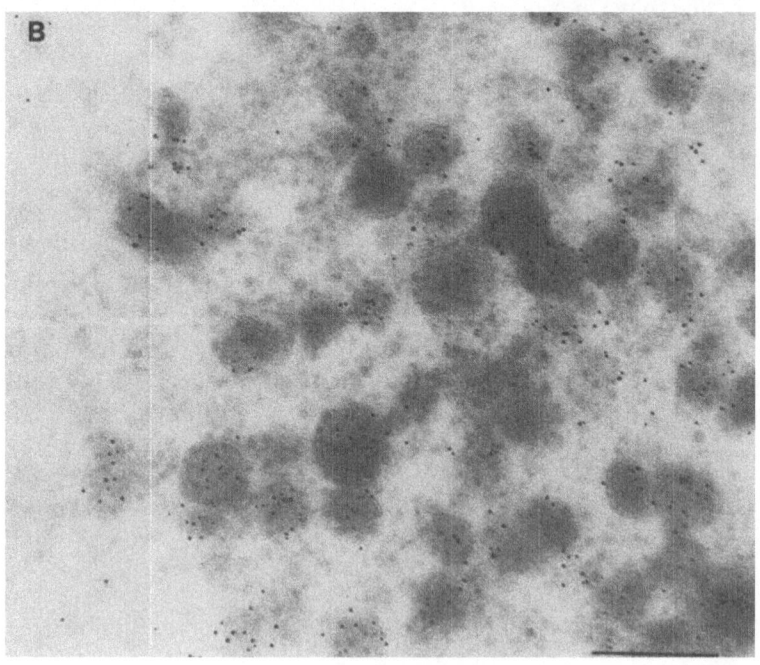

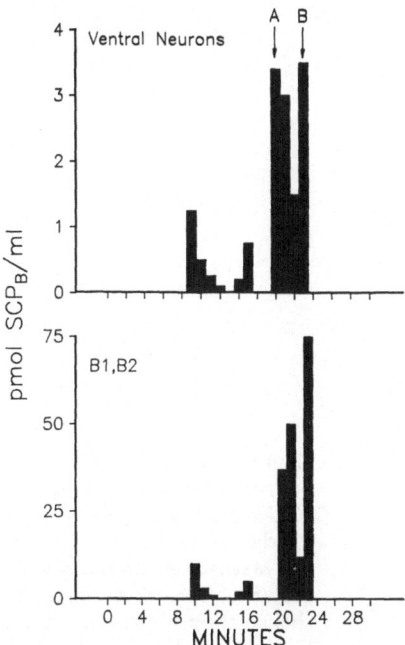

Figure 7. Analysis of the SCPs in identified buccal neurons. Identified neurons were dissected and pooled from 25 animals; the protein extract was chromographed by reverse-phase HPLC without carrier tissue. Fractions were dried and assayed for cardioactivity on the isolated snail heart. The activity is compared with standard solutions of synthetic SCP_B. The arrows designate the retention times for purified native SCP_A (A) and for synthetic SCP_B (B). (Top panel) Large ventral neurons (B3, B6, B9, and B10). (Bottom panel) B1, B2.

peptides in both the CNS and the periphery. The SCPs have excitatory actions on the *Aplysia* heart, increasing both the amplitude and the rate of beating in a manner similar to that observed in the land snail. The SCPs mimic serotonin in this effect, although the peptides are at least 10-fold more potent (Lloyd, 1980c). The effects of both serotonin and SCP_B appear to be mediated by an increase in cyclic adenosine monophosphate (cAMP) levels, primarily in the atrioventricular (AV) valves, which are thought to act as pacemaker regions. Short exposure of the AV valves to either compound generates an approximately 100-fold increase in the levels of cAMP (Lloyd *et al.*, 1985b). The time course of AV valve contractions paralleled the SCP_B-induced levels of cAMP, having a rapid onset that decreased to control levels within about 2 min. Other areas of the heart, such as the atrium and ventricle, showed only weak responses to either serotonin or SCP_B. The threshold for SCP activity

---

Figure 6. The dense-core granules in buccal ganglion neurons. (A) Osmium-stained electron micrograph of a B1 or B2 neuron. N, nucleus; G, Golgi; arrows point to examples of dense-core granules. Bar: 300 nm. (B) Cryoultramicrotome section of a B1 or B2 neuron reacted with an antibody against SCP_B followed by protein A with bound colloidal gold. The gold particles represent SCP_B immunoreactivity and are localized to the dense-core granules. Bar: 300 nm.

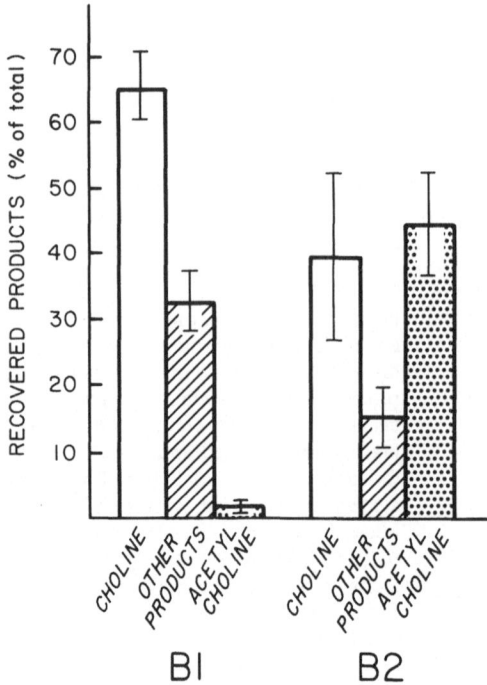

Figure 8. Assay of choline acetyltransferase activity in B1 and B2. Neurons were injected with [^{3}H]choline, incubated for 1 hr and the radiolabeled products were analyzed. Open bars, unreacted choline; stippled bars, acetylcholine; striped bars, unidentified products.

on the heart is 6×10^{-12}M. This concentration is lower than the sensitivity of the assay used to detect the SCPs in the hemolymph ($\geq 10^{-11}$), and no direct SCP innervation of the heart has been observed. Therefore, the role of the SCPs in regulating cardiac output remains unclear (Lloyd et al., 1985b).

The SCPs have also been identified as potential facilitatory transmitters during sensitization of the gill- and siphon-withdrawal reflex (Abrams et al., 1984). Sensitization is attributed to an increase in the duration of action potentials in sensory neurons that synapse on gill and siphon motor neurons. The increased duration of the action potentials results from the closure of unique K$^+$ channels, the serotonin-sensitive I$_S$ channels, due to cAMP-mediated protein phosphorylation in the sensory neurons, producing an increase in Ca^{2+} influx during the action potential that leads to increased transmitter release. These cellular activities result in a more vigorous withdrawal of the gill and siphon (Kandel and Schwartz, 1982). The gill and siphon reflex also can be modified using associative learning paradigms that use the same cellular processes involved in sensitization (Kandel et al., 1983).

When extracts of abdominal ganglia were fractionated and tested for their ability to produce presynaptic facilitation, three compounds were found to mediate the response: serotonin, the SCPs, and an unidentified compound, probably related to tryptophan. The onset of the SCP response is faster and the potency about 10-fold greater than that of serotonin. While the specific role of the SCPs in heterosynaptic facilitation is not yet defined, their physiological activities suggest another action parallel to that of serotonin, possibly mediating a simple form of learning (Abrams *et al.*, 1984).

The high concentration of the SCPs in the buccal ganglion suggests that the peptides play a prominent role in feeding behavior. The physiological activities of the SCPs in two aspects of this behavior have been investigated. First, the large SCP-containing buccal cells, B1 and B2, innervate the esophagus and gut. The SCPs enhance the peristaltic contractions in these tissues, resulting in a more efficient translocation of food to the digestive glands. Second, SCP_B has been shown to enhance cholinergic motor neuron-induced contractions of the accessory radula closer (ARC) muscle (Fig. 9). These actions are very similiar to the effects

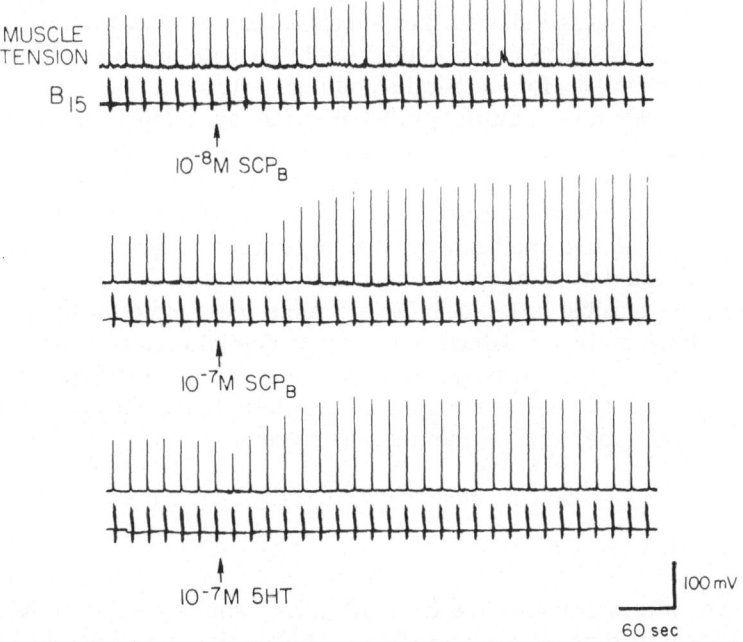

Figure 9. Enhancement of ARC muscle contractions by SCP_B and serotonin. Muscle contraction was elicited by stimulating the cholinergic motor neuron B15. In each of the three sets of traces the top is a record of mechanical activity in the muscle and the bottom records electrical activity in neuron B15. The time that SCP_B or serotonin was added to the bath to give the indicated final concentration is indicated by the arrow.

generated by the serotonergic metacerebral cell (MCC), again suggesting parallel actions for serotonin and the SCPs (Lloyd *et al.*, 1984). Both serotonin and SCP_B increase the levels of cAMP in the ARC in a dose-dependent manner. Food-induced arousal produces MCC-mediated potentiation of the muscle contraction (Kupfermann, 1974). These combined observations suggest that SCP peptides may play an important role in eliciting and maintaining the aroused state generated by food or appetite. Collectively, the physiological data demonstrate that the SCPs have a variety of activities in both the central nervous system and the periphery. Clearly, as may be the case with all chemical messengers, no unique role can be ascribed to the small cardioactive peptides (Lloyd *et al.*, 1984).

8. CONCLUSIONS

Whereas the term neuropeptide is an adequate description of a structural class of secreted chemical messengers, the functional diversity of the group necessitates further subclassification as hormones, transmitters, and modulators. This functional variability increases the diversity of mechanisms for chemical signaling available to the nervous system. For example, neuropeptides provide an additional means of communication beyond the synapse. The relative contribution of extra-synaptic and synaptic chemical communication to the total number of interactions between neurons is unknown. This chapter reviews current information on the small cardioactive peptides in *Aplysia*. These studies were facilitated by the large neuron size in *Aplysia*. This characteristic also makes feasible future cellular and molecular studies of the peptide biosynthetic pathway, which has many potential regulatory sites from the control of gene transcription to secretion of the messenger. Further investigations of this system should also provide insight into the functional roles of peptides in the nervous system.

ACKNOWLEDGMENTS

The authors would like to thank John Nambu and Mark Kirk for the critical reading of this manuscript. This work is supported by grants to RHS from the National Institutes of Health (NIH) and the National Science Foundation (NSF). ACM is supported by an NIH postdoctoral fellowship. RHS is a McKnight Foundation Scholar, an Alfred P. Sloan Fellow, and a Klingenstein Fellow in the Neurosciences.

REFERENCES

Abrams, T. W., Castellucci, V. F., Camardo, J. S., Kandel, E. R., and Lloyd, P. E., 1984 Two endogenous neuropeptides modulate the gill and siphon withdrawal reflex in *Aplysia* by presynaptic facilitation involving cAMP dependent closure of a serotonin-sensitive potassium channel, *Proc. Natl. Acad. Sci. U.S.A.* **81:**7956–7960.

Aswad, D., 1978, Biosynthesis and processing of presumed neurosecretory proteins in single identified neurons of *Aplysia californica*, *J. Neurobiol.* **9:**267–284.

Berry, R. W., 1976, Processing of low molecular weight proteins by identified neurons of *Aplysia*, *J. Neurochem.* **26:**229–231.

Blobel, G., and Dobberstein, B., 1975, Transfer of proteins across membranes. I. Presence of proteolytically processed andd unprocessed nascent immunoglobulin light chains and membrane bound ribosomes of murine myeloma, *J. Cell Biol.* **67:**835–852.

Coggeshall, R. E., Yaksta, B. A., and Swartz, F. J., 1971, A cytometric analysis of the DNA in the nucleus of the giant cell R2 in *Aplysia*, *Chromosoma* **32:**205–212.

Cohen, J. L., Weiss, K. R., and Kupfermann, I., 1978, Motor control of buccal muscles in *Aplysia*, *J. Neurophysiol.* **41:**157–180.

Cooper, J. R., Bloom, F. E., and Roth, R. H., 1978, *The Biochemical Basis of Neuropharmacology*, 3rd ed, Oxford University Press, New York.

Frazier, W. T., Kandel, E. R., Kupfermann, I., Waziri, R., and Coggeshall, R. E., 1967, Morphological and functional properties of identified neurons in the abdominal ganglion of *Aplysia californica*, *J Neurophysiol.* **30:**1288–1315.

Gilbert, W., 1978, Why genes in pieces? *Nature (Lond.)* **271:**501.

Hall, Z. W., 1973, Multiple forms of acetylcholinesterase and their distribution in endplate and non-endplate regions of rat diaphragme muscle, *J. Neurobiol.* **4:**343.

Herbert, E., Brunberg, N., Lissitsky, J. C., Civelli, O., and Uhler, M., 1981, Pro-opiomelanocortin: A model for the regulation of expression of neuropeptides in pituitary and brain. *Neurosci. Commun.* **1:**16–27.

Kandel, E. R., 1976, *The Cellular Basis of Behavior*, W. A. Freeman, San Francisco.

Kandel, E. R., and Schwartz, J. H., 1982, Molecular biology of learning: Modulation of transmitter release, *Science* **218:**433–443.

Kandel, E. R., Abrams, T., Bernier, L., Carew, T. J., Hawkins, R. D., and Schwartz, J. H., 1983, Classical conditioning and sensitization share aspects of the same molecular cascade in *Aplysia Cold Spring Harbor Symp. Quant. Biol.* **48:**821–830.

Krieger, D. T., 1983, Brain peptides: What, where and why? *Science* **222:**975–985.

Kupfermann, I., 1974, Feeding hehavior in *Aplysia*: A simple system for the study of motivation, *Behav. Biol.* **10:**1–26.

Lasek, R. J., and Dower, W. J., 1971, *Aplysia californica*: Analysis of the nuclear DNA in individual nuclei of giant neurons, *Science* **172:**278–280.

Lloyd, P. E., 1978, Neurohumoral control of cardiac activity in the snail, *Helix aspersa, J. Comp. Physiol.* **128:**277–283.

Lloyd, P. E., 1980*a*, Biochemical and pharmacological analysis of endogenous cardioactive peptides in the snail, *Helix aspersa, J. Comp. Physiol.* **139:**265–270.

Lloyd, P. E., 1980*b*, Modulation of neuromuscular activity by 5-hydroxytryptamine and endogenous peptides in the snail, *Helix aspersa, J. Comp. Physiol.* **139:**333–339.

Lloyd, P. E., 1980*c*, Mechanisms of action of 5-hydroxytryptamine and endogenous peptides on a neuromuscular preparation in the snail, *Helix sapersa, J. Comp. Physiol.* **139:**341–347.

Lloyd, P. E., Kupfermann, I., and Weiss, K. R., 1984, Evidence for parallel actions of a molluscan neuropeptide and serotonin in mediating arousal in *Aplysia*, *Proc. Natl. Acad. Sci. USA* **81:**2934–2937.

Lloyd, P. E., Mahon, A. C., Kupfermann, I., Cohen, J. L., Scheller, R. H., and Weiss, K. R., 1985a, Biochemical and immunocytological localization of molluscan small cardioactive peptides (SCPs) in the nervous system of *Aplysia*, *J. Neurosci.* **5:**1851–1861.

Lloyd, P. E., Kupfermann, I., and Weiss, K. R., 1985b, Two endogenous neuropeptides (the SCPs) produce a cAMP-mediated stimulation of cardiac activity in *Aplysia*, *J. Comp. Physiol.* (in press).

Loh, Y. P., and Gainer, H., 1975, Low molecular weight specific proteins in identified molluscan neurons. 1. Synthesis and storage, *Brain Res.* **92:**181–192.

Loh, Y. P., and Gainer, H., 1983, Biosynthesis and processing of neuropeptides, in: *Brian Peptides* (D. T. Kreiger, M. J. Brownstein, J. B. Martin, eds.), pp. 79–116, Wiley, New York.

Mahon, A. C., Lloyd, P. E., Weiss, K. R., Kupfermann, I., and Scheller, R. H., 1985a, The small cardioactive peptides A and B of *Aplysia* are derived from a common precursor molecule, *Proc. Nat. Acad. Sci. USA* **82:**3925–3929.

Mahon, A. C., Nambu, J. R., Taussig, R., Shyamala, M., Roach, A., and Scheller, R. H., 1985b, Structure and expression of the egg-laying hormone gene family, *J. Neurosci.* (in press).

Morris, H. W., Panico, M., Karplus, A., Lloyd, P. E., and Riniker, B., 1982, Elucidation by FAB-MS of the structure of a new candidate peptide from *Aplysia*, *Nature (Lond.)* **300:**643–645.

Nambu, J. R., Taussig, R., Mahon, A. C., and Scheller, R. H., 1983, Gene isolation with cDNA probes from identified *Aplysia* neurons: Neuropeptide modulators of cardiovascular physiology. *Cell* **35:**47–56.

Scheller, R. H., Jackson, J. F., McAllister, L. B., Rothman, B. S., Mayeri, E., and Axel, R., 1983, A single gene encodes multiple neuropeptides mediating a stereotyped behavior, *Cell* **35:**7–22.

Sossin, W., Kreiner, T., and Scheller, R. H., 1986, Dense core vesicles: Multiple populations in *Aplysia* neurosecretory cells, in: Fast and Slow Chemical Signalling in the Nervous System (L. L. Iverson and E. Goodman, eds.) pp. 260–278, Oxford University Press.

Taussig, R., Kaldany, R. R., and Scheller, R. H., 1984, A cDNA clone encoding neuropeptides isolated from *Aplysia* neuron L11, *Proc. Natl. Acad. Sci USA* **81:**4988–4992.

Taussig, R., Kaldany, R. R., Rothbard, J. B., Schoolnik, G., and Scheller, R. H., 1985, Expression of the L11 neuropeptide gene in the *Aplysia* central nervous system, *J. Comp. Neurol.* **238:**53–64.

6

Molecular Biology Approach to the Expression and Properties of Mammalian Cholinesterases

HERMONA SOREQ, DINA ZEVIN-SONKIN,
ORA GOLDBERG, and CATHERINE PRODY

1. INTRODUCTION: EXPRESSION OF CHOLINESTERASES AS A RESEARCH SUBJECT—SCIENTIFIC SIGNIFICANCE, ADVANTAGES, AND DIFFICULTIES

At the cholinergic synapse, the enzyme acetylcholinesterase terminates the electrophysiological response to the neurotransmitter acetylcholine (ACh) by degrading it very rapidly. Acetylcholine is a principal, and the most studied, neurotransmitter in all the higher eukaryotic organisms. The cholinergic acetylcholine receptor and the hydrolyzing enzyme, AChE, are thus the two major elements in a regulatory system that controls the response to ACh in numerous cell types, tissues, and organisms.

Several properties make the regulation of the expression of AChE a particularly fascinating research subject. This applies to the properties

HERMONA SOREQ, DINA ZEVIN-SONKIN, and CATHERINE PRODY • Department of Neurobiology, The Weizmann Institute of Science, Rehovot 76100, Israel.   ORA GOLDBERG • Department of Organic Chemistry, The Weizmann Institute of Science, Rehovot 76100, Israel.

of the enzyme itself, as well as to the mechanism(s) controlling its expression at different steps along the pathway of gene expression.

1.1. Detection of Enzymatic Activity

The enzymatic activity of AChE is well characterized, and minute quantities of the active enzyme can be detected with exceptionally high sensitivity. The turnover number of AChE from various genetically remote species ranges around 10^4 molecules of neurotransmitter hydrolyzed per second for each catalytic site (Silver, 1974; Vigny *et al.*, 1978). This implies that with a sensitive and long-term stable assay, one can reliably detect as little as 10^7 catalytic sites, or a few picograms, for an average molecular weight of about 80,000 (Sorenson *et al.*, 1982).

Biochemical characterizations of AChE have so far been carried out only in tissues enriched in this enzymatic activity, using several methods for the quantitative monitoring. A widely used spectrophotometric assay is that of Ellman *et al.* (Ellman *et al.*, 1961) based on the enzymatic hydrolysis of acetylthiocholine (AThCh) to yield thiocholine (ThCh). The latter produces the yellow anion of 5-thio-2-nitrobenzoic acid, when reacted with 5,5'-dithiobis(2-nitrobenzoic acid). A more sensitive technique is based on the hydrolysis of ACh labeled in its acetate moiety by ^{3}H. The [^{3}H]acetate is extracted into the scintillation fluid and counted (Johnson and Russell, 1975). The lower limit of sensitivity for this method has been reported to be in the range of micrograms neuronal tissue per sample. More recently, we developed a method with a wide linear range of accuracy and a limit of detection 100-fold lower than that of the radiometric method (Parvari *et al.*, 1983). The assay is based on reacting ThCh produced by the enzyme from AThCh with the fluorogenic maleimide N-(4-(7-diethylamino-4-methylcoumarin-3-yl)phenyl)maleimide (Sippel, 1981).

Minute amounts of AChE can be histochemically detected by a method developed by Koelle and Friedenwald (1949) and modified by Karnovsky and Roots (1964). In this assay, ThCh is liberated from AThCh by cholinesterase (ChE) into a solution containing copper ferricyanide and citrate ions.

ThCh reduces ferricyanide to ferrocyanide, which combines with Cu^{2+} ions to form the insoluble copper ferrocyanide as a finely granular brown precipitate. Thus, a color contrast is produced directly at the site of enzymatic activity. This method was used by Couteaux and Taxi (1952) to visualize AChE at end plates, by Wilson and Walker (1974) to demonstrate AChE reappearing around the nucleus after inhibition of the existing AChE, and by Smilowitz (1980) to locate AChE in the Golgi membranes. We currently use this technique to detect immunoreacted cholinesterase (ChE) precipitates in crossed immunoelectrophoretic analyses of oocyte translation products (Dziegielewska *et al.*, 1986). Once sufficiently long DNA sequences of ChE are inserted into the appropriate expression vectors, the histochemical detection of active ChE will be highly useful for transfection experiments of ChE-deficient cells in culture (see details in following sections).

1.2. Polymorphism of Cholinesterases

Mammalian ChEs exhibit extensive polymorphism (Silver, 1974; Massoulie and Bon, 1982) at several levels. They are classified by their substrate specificity as acetylcholinesterase (acetylcholine hydrolase, EC 3.1.1.7, AChE) or pseudocholinesterase (acylcholine acylhydrolase, EC 3.1.1.8, ΨChE). The two enzymes also differ in their susceptibility to various inhibitors (Austin and Berry, 1953).

AChE occurs in multiple molecular forms, which exhibit different sedimentation coefficients on sucrose gradients (Massoulie *et al.*, 1984) and can be further distinguished according to their hydrodynamic interactions with nonionic detergents (Rosenberry and Scoggin, 1984; Steiger *et al.*, 1984; Futerman *et al.*, 1983). The latter property may be explained by the presence of integral hydrophobic domains (probably a C'-terminal peptide (Rosenberry and Scoggin, 1984; Steiger *et al.*, 1984) or by tight interactions with phospholipid moieties (i.e., phosphatidylinositol) (Futerman *et al.*, 1983). Secreted, cytoplasmic and membrane-associated pools of AChE are found in the mammalian nervous tissue (Massoulie and Bon, 1982). Such a divergent subcellular segregation generally implies different polypeptide sequences in the nascent molecule. However, pharmacological studies and enzyme kinetic analyses indicated that the different forms of AChE possess similar catalytic properties (Vigny *et al.*, 1978), thus suggesting that different AChE forms may share common domains, including the ACh binding site and, possibly, a peptidergic site (Chubb, 1984). Homologies among different forms of AChE from several tissues and species, revealed by various antibodies, may indeed indicate the existence of such common domains (Fambrough *et al.*, 1982; Brimijoin *et al.*, 1983; Marsh *et al.*, 1984). However, each of the AChE forms may also contain distinct polypeptide regions, responsible for the subcellular segregation of these various enzyme forms, for their different amphipathic properties (Grassi *et al.*, 1982) and for their different modes of assembly into multisubunit protein molecules. This hypothesis is sup-

Table I. Levels of ChE Polymorphism[a–c]

1. Substrate specificity	Substrate	Inhibitor
Acetylcholinesterase (AChE)	Acetylcholine	BW284C51
Butyrylcholinesterase (ψChE)	Butyrylcholine	Iso-OMPA
2. Hydrophobicity	Interaction with nonionic detergents	Properties
Hydrophylic	No effect	Globular, soluble
Amphipathic	Change in sedimentation	Bound to PI *or* contains hydrophobic peptide
3. Molecular forms	Number of subunits	"Collagen" tail
Symmetric	1–4	None
Asymmetric	1–3 tetramers	1–3 chains
4. Subcellular localization	Putative membrane anchorage	Expected signal peptide
Cytoplasmic	None	No
Membrane integral protein	Integral hydrophobic peptide	Yes
Extracellular matrix-associated	Collagen tail *or* PI interaction	Yes
Secreted	None	Yes

[a] Mammalian cholinesterases (ChEs) exhibit extensive polymorphism at several levels.
[b] Such divergent segregation in properties and subcellular localization generally implies different polypeptide sequences in the nascent molecule. However, enzyme kinetics analyses, pharmacological data, and cross-reactivity of anti-ChE antibodies with ChEs from different tissues and species suggest that various ChEs also possess common domains.
[c] It is not known whether the different forms of ChEs are produced from distinct genes or whether their biosynthesis is regulated by post-transcriptional and/or post-translational processing events.

ported by reports on antibodies that are capable of differentiating between low salt-soluble and detergent-soluble forms of AChE from *Torpedo* electric organ (Doctor *et al.*, 1983).

Some of the different levels at which the polymorphism of cholinesterases is manifested are summarized in Table I.

1.3. Tissue and Cell-Type Specificity

Cholinesterases are particularly abundant in nerve and muscle but can also be found in many other tissue and cell types, such as the adrenal medulla or in megakaryocytes (Massoulie and Bon, 1982). Transiently high levels of ChEs have also been reported in a number of embryonic tissues (Drews, 1975), where their appearance has been correlated with cell migration and tissue reorganization. One example is the mesenchymal layer in the developing head of the chick embryo (Layer, 1983). In addition, considerable levels of ChEs have been detected in various neoplastic tissues, such as ovarian carcinomas (Drews, 1975) and brain tumors of glial and mesenchymal origin (Ord and Thompson, 1952; Razon *et al.*, 1984). Cholinesterases from different tissue and cell types differ in molecular forms, in their hydrophobicity, and in glycosylation pattern (Razon *et al.*, 1984; Zakut *et al.*, 1985; Meflah *et al.*, 1984).

1.4. Putative Biological Role(s)

It has generally been accepted for a number of decades that in muscle and brain, AChE breaks down the neurotransmitter ACh. The role of cholinesterases in tissue sources other than brain or muscle remains completely unknown. Accumulating evidence now suggests, however, that in many tissues, including brain and muscle, a substantial portion of ChEs is probably involved in other, unrelated processes (recently reviewed by Chubb (1984), Greenfield (1984), and Balasubramanian (1984). The evidence is based on the following findings:

1. High levels of AChE are not necessarily accompanied by high levels of the neurotransmitter synthesizing enzyme, cholineacetyltransferase, of ACh (Silver, 1974; Greenfield, 1984), or of ACh-binding sites (Gurwitz *et al.*, 1984; Egozi *et al.*, 1986).
2. Certain AChE-containing nerve cells do not contain choline acetyltransferase and do not respond to ACh, but do, however, interact with other neurotransmitters (Greenfield, 1984; Graybiel and Ragsdale, 1982).
3. AChE exists not only as a membrane protein, as expected of a neurotransmitter hydrolyzing enzyme, but also in cytoplasmic and secreted buffer-soluble forms (Massoulie and Bon, 1982; Zakut *et al.*, 1985; Soreq *et al.*, 1982).
4. Release of AChE into the cerebrospinal fluid (CSF) is not parallel to ACh release and is not affected by stimulation and/or blockage of ACh receptors (Greenfield, 1984).
5. Several activities of AChE were shown to be unaffected by blocking the cholinergic site with selective inhibitors.

Thus, the peptidergic activity of purified *Torpedo* AChE (Chubb, 1984) and mammalian ψChE (Lockridge, 1982) on substance P and both Met and Leu enkephalins (but not on endorphins) is also manifested by the organophosphate-inhibited enzyme. In the amacrine cells in the retina, application of AChE increases the immunological response to both substance P and enkephalin (Chubb, 1984). An effect on the motor behavior of rats was observed after injection of purified AChE into the substantia nigra, and without any relationship to substance P or enkephalin degradation (Greenfield, 1984), and application of AChE to neuronal cells in culture alters the dendritic Ca^{2+} currents (Llinas *et al.*, 1984), thus modulating neuronal activity.

1.5. *Genetic Evidence for Allelic Polymorphism*

Genetic evidence suggests that in *Drosophila*, the *Ace* locus is the only one controlling the expression of AChE (Hall, 1982). In nematodes, at least two separate mutations are involved in the expression of the two different species of cholinesterase (Johnson *et al.*, 1981; Johnson and Russell, 1983) and the expression of the enzyme is regulated in a tissue-specific manner (Herman and Kari, 1985). In humans, ψChE appears to be encoded by one gene with at least five different alleles (Silver, 1974). However, different genetic linkage studies assign this gene to chromosome 16 (Lovrien *et al.*, 1978), 1 (Chautard-Freie-Maia, 1977), or 3 (Sparkes *et al.*, 1984). The various electrophoretic migration patterns of AChE from human erythrocytes are compatible with the existence of several alleles for a single gene coding for "true" AChE (Coates and Simpson, 1972). The amino acid composition of human AChE (Rosenberry and Scoggin, 1984) differs considerably from that of human ψChE (Lockridge, 1984). Moreover, genetic defects in human ψChE are not accompanied by parallel defects in AChE (Silver, 1974). This implies that ψChE and AChE in humans are encoded by at least two distinct DNA sequences. Since AChE and ψChE in humans resemble each other in so many properties, however, they might be at least partially homologous to each other. Thus, we may expect to find cholinesterase-coding sequences in at least two interrelated human DNA fragments. Both enzymes are serine esterases and display biochemical peptidergic activities (Chubb, 1984). It is therefore also possible that the genes coding for AChE and ψChE share common domains with genes coding for other esterases, peptidases, or both.

1.6. *Molecular Approach to Cholinesterases*

In spite of the apparent importance of AChE as a research subject and the extremely high sensitivity with which its activity can be detected, very little is known about its biological role(s) and mode of regulation.

Even in a clearly cholinergic site such as the neuromuscular junction, at which AChE does hydrolyze ACh, it is not clear how its activity is controlled. The finding that muscle denervation results in the disappearance of the tailed, asymmetrical forms of AChE from the neuromuscular junction (Massoulie and Bon, 1982), suggests the existence of a close relationship between the neuronal signal, the binding of ACh to its receptor and the way in which the expression of AChE is regulated. It is completely unknown, however, which level of gene expression is affected by the neuronal signal so as to alter the expression of AChE and what is the detailed molecular mechanism that leads to such alterations.

One reason for the lack of knowledge on this subject is certainly the extreme sparcity of the enzyme AChE, coupled with its polymorphism. Even in a relatively rich source such as brain tissue, AChE comprises only about 0.001% of the total protein content. Thus, in spite of the effort invested for many years by numerous research groups, the amino acid sequence of this enzyme is not yet known, the three-dimensional structure of its catalytic site has not been revealed, and there is no way to immunoprecipitate mammalian AChEs as distinct polypeptide bands out of crude mixtures—although antibodies that precipitate the enzymatic activity have been developed in several laboratories.

The techniques of genetic engineering, and in particular of gene cloning, now undergoing a rapid expansion, offer the possibility of replicating the mammalian ChE gene(s) in various types of DNA vectors. DNA sequencing, which is by now technically much easier than polypeptide sequencing, could provide the amino acid sequence of mammalian AChE. Using the correct vectors and expression system, gene expression could be achieved. This would eventually lead to production of sizable amounts of catalytically active AChE. The purified enzyme could be employed in biochemical studies on structure–function relationships of the enzyme with substrates and putative inhibitors and reactivators. The availability of the enzyme might also permit crystallization, eventually leading to elucidation of the three-dimensional structure of human AChE and supplying good quality probes with which at least part of the scientific problems raised above could be solved. In addition, the molecular cloning approach will yield the opportunity for a thorough study of several subjects of major importance:

1. Regulation of cholinergic elements and their interrelationship in denervated and reinervated muscle
2. The role and mode of expression of ChEs in developing neuronal systems (such as the role of ChEs in the process leading

from the neural tube of *Xenopus* embryos to the mature nervous system)

3. Putative relationships between the clinical expression of neuromuscular disorders and defects in the expression of ChEs—*not* necessarily in the cholinergic site (see details in following sections).

1.7. General Research Strategy: Simultaneous Experiments Approaching Various Levels of Gene Expression

There is no direct information as to whether ChEs of different substrate specificites and sedimentation properties are produced by different genes or result from either post-transcriptional or post-translational processing, or both. A possible precursor–product relationship has been suggested to exist between ψChE and AChE (Koelle *et al.*, 1977). However, this view has been refuted by the lack of cross-reactivity between antibodies to AChE and ψChE and by the existence of parallel sets of multiple forms of both enzymes. Accordingly, it has been proposed that the monomeric and dimeric forms of ψChE serve as precursors to the more complex forms (Wilson and Walker, 1974; Rieger *et al.*, 1976). The possible metabolic relationships among different cholinesterase forms have been investigated by analysis of the recovery of ChE activity after irreversible inhibition of preexisting enzyme by diisopropylfluorophosphate (DFP), of the effects of transcription inhibitors, or of the pattern of appearance of the molecular forms in developing cells *in vivo* and in culture (Wilson and Walker, 1974; Rieger *et al.*, 1976; Rotundo, 1984). One should bear in mind, however, that the nonselective DFP and transcription inhibitors may affect various biosynthetic processes in the cells and that it is difficult to distinguish between enzyme interconversion and sequential appearance and breakdown of cholinesterase activities. Some of these difficulties could be overcome by direct measurements of cholinesterase synthesis as directed by mRNA in a heterologous translation system, while other problems would be solved once good quality ChE DNA probes are available.

The scarcity of ChEs naturally causes difficulties also when approaching the subject by molecular biology techniques. Since very little is known on the structure and amino acid sequence of these proteins and on the mode by which their production is regulated, the expression of cholinesterases must be studied simultaneously at three levels along the pathway of gene expression—the levels of DNA, of mRNA, and of the active protein. The study of each of these levels yields a distinct body of information that will eventually be combined to obtain a clearer picture of how ChEs are being regulated.

2. EXPRESSION OF CHOLINESTERASE mRNAS IN MICROINJECTED *XENOPUS* OOCYTES

Xenopus oocytes offer an attractive experimental approach to study the different processes that start with translation of mRNA and lead to the formation and compartmentalization of a mature protein (Gurdon *et al.*, 1971). During the past decade, *Xenopus* oocytes have been used extensively as an efficient, long-lasting translation system for a variety of microinjected mRNAs, correctly performing translation (Marbaix and Huez, 1980), processing (Ghysdael *et al.*, 1977), and various post-translational modifications (Lane, 1983; Soreq, 1985). In a number of cases, the injected mRNAs have been translated into biologically active enzymes, sequestered within the injected oocytes into their natural localization (Labarca and Paigen, 1977) or secreted into their incubation medium (Soreq *et al.*, 1982; Miskin and Soreq, 1981). Injection of mRNA from different cholinesterase-expressing sources into these oocytes produces cholinesterase activities (Soreq *et al.*, 1982, 1984). Figure 1 demonstrates the biosynthesis of secretory, cytoplasmic, and membrane-bound ChEs in oocytes injected with poly(A)$^+$ RNA from mouse brain, muscle, and liver. It can be seen that brain mRNA is most potent in directing the production of secreted ChEs, while liver mRNA mostly directs the synthesis of the cytoplasmic enzyme and muscle mRNA produces membrane-integral cholinesterases. Determination of ChE mRNA levels has proved time and concentration dependent (Soreq *et al.*, 1982), and the enzyme produced from brain poly(A)$^+$ RNA was biochemically defined as "true" mammalian AChE (Soreq *et al.*, 1982). The concentration of ChE mRNA out of the total poly(A)$^+$ RNA was calculated to be about 1×10^{-5} of the total mRNA (Soreq *et al.*, 1982). This indicates that the level of ChE mRNA is one of the limiting factors determining the level of the enzyme protein in the brain.

Translation Evidence for Multiple mRNA Species Inducing Cholinesterase Synthesis in Microinjected Oocytes

The levels, substrate specificities, and molecular forms of ChEs were studied in detail in discrete regions of fetal human brain (Zakut *et al.*, 1985), mesenchyme-originated meningiomas, and gliogenous glioblastomas, as a prerequisite for the characterization of ChE mRNAs in these

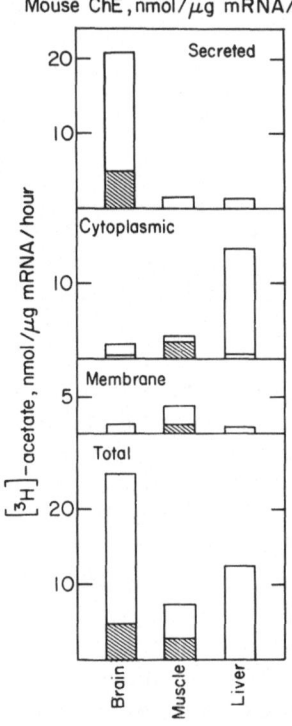

Figure 1. Biosynthesis of ChEs in *Xenopus* oocytes injected with mRNA from various mouse tissues. Mature *Xenopus* oocytes were microinjected with 50 ng each of poly(A)$^+$ RNA from 10-day-old mouse brain, muscle, or liver. Oocytes were incubated for 24 hr in the presence of protease inhibitors. Incubation medium, containing secreted proteins, was separated. Following homogenization, cytoplasmic fractions were prepared by centrifugation and membrane-bound proteins were extracted with detergent and salt. ChE activity, in nanomoles [^{3}H]acetate released/μg mRNA/hr, was determined in each fraction by hydrolysis of [^{3}H]acetylcholine in the presence of 10^{-5} M iso-OMPA, which selectively inhibits butyrylcholinesterase activity (□), or of 10^{-5} M BW284C51, which specifically blocks AChE activity (▧). ChE activity in fractions of oocytes injected with incubation medium (without protease inhibitors) served as controls and were subtracted from each value. Total activities represent sums of secreted, cytoplasmic, and membrane activities.

tissue sources. In each of these tissues, we found ChE activities with distinct characteristic properties in terms of substrate specificities and migration patterns (Razon *et al.*, 1984; Zakut *et al.*, 1985). These observed biochemical differences do not, however, furnish a clear-cut conclusion as to whether different forms of ChE are produced from distinct mRNA species. For example, a possible explanation for the differences in ChE properties between the examined tissue sources could be that ChE properties are modified post-translationally by protein(s) translated from other, cell-type specific, mRNA(s).

Another implication of these findings is that in human brain ChE expression is not limited to nerve cells. Thus, developing gliogenous and meningeal cells may produce high levels of this enzymatic activity. In addition, in several non-neuronal cell lines, AChE has been implicated in controlling cellular proliferation and differentiation. For example, it has been demonstrated that in high-density cultures of WRL-10A mouse fibroblasts, AChE synthesis is greatly increased when growth is arrested and the enzyme is secreted quantitatively into the medium (Massoulie and Bon, 1982). The stimulation of megakaryocytopoiesis in culture by ACh derivatives (Burstein *et al.*, 1980) and the demonstration that megakaryocytes release AChE (Paulus *et al.*, 1981) suggest that interplay of ACh and AChE biosynthesis may be related to growth-control mechanisms operating at the transcriptional level.

Sucrose-gradient fractionation of DMSO-denatured mRNA from primary gliomas, meningiomas, and embryonic brain revealed three size classes of ChE-inducing mRNAs, sedimenting at ~32 S, 20 S, and 9 S; the relative and absolute amounts of these different classes of ChE-inducing mRNAs varied between the three tissue sources examined (Soreq *et al.*, 1984). To determine which types of ChEs are produced in oocytes, the activity of the enzymes was measured in the presence of selective inhibitors. Both AChE and ψChE multimeric ChE activities were found in the mRNA-injected oocytes. Moreover, human brain mRNAs inducing "true" and pseudo-ChE activities had different size distribution, indicating that different mRNAs might be translated into various types of ChEs. These findings imply that the heterogeneity of ChEs in the human nervous system is not limited to the post-translational level but extends to the level of mRNA (Soreq *et al.*, 1984).

The biochemical characterization of ChEs in primary brain tumors also suggested the presence of a mammalian ChE variant that is susceptible to both the AChE-specific inhibitor, 1,5-bis(4-allyldimethylammonium phenyl)pentan-3-one dibromide (BW284C51), and the ψChE-specific inhibitor, tetraisopropyl pyrophosphoramide (iso-OMPA) (Razon *et al.*, 1984). This finding agreed with the parallel result obtained in ChE measurements in mRNA-microinjected oocytes, in the presence and absence of inhibitors (Soreq *et al.*, 1984).

3. IDENTIFICATION OF *DROSOPHILA* DNA FRAGMENT THAT HYBRIDIZES WITH CHOLINESTERASE mRNA

The genetics of the nervous system in *Drosophila melanogaster* is well characterized, and various mutants in which the expression of specific neurotransmitter hydrolyzing enzymes is defective are available (Hall, 1982). The *Ace* locus, regulating the activity of AChE, was first found to be located on the third chromosome at the 87DE region by segmental aneuploid analysis (Hall and Kankel, 1976). The specific activity of the *Drosophila* enzyme has been reported to depend linearly on the dosage of this particular region in haploid, euploid, and triploid flies. Four allelic lethal mutations were isolated, all of which have approximately half-normal activity in the hetrozygous state, suggesting that there is only one structural locus for AChE in *Drosophila* and that it is contained in the above region, next to the *rosy* locus. The vicinity of the *Ace* locus was further delimited by two deletion breakpoints: Df(3R)ry[1301] and Df(3R)Kar[SZ11] (Hilliker *et al.*, 1980). These breakpoints were localized on the DNA sequence of a 315-kb segment, which extends from region 87D 5–7 to 87E 5–6 on the third chromosome and was cloned and mapped by chromosomal walking (Bender *et al.*, 1983; Spierer *et al.*, 1983). The RNA transcripts that hybridize with this 315-kb segment were mapped by RNA–DNA hybridization (Hall *et al.*, 1983). Twenty discrete poly-adenylated RNA species transcribed from nonrepetitive DNA hybridized with DNA from this region at various developmental stages. Four different transcripts hybridized with DNA from a 50-kb fragment internal within the 87E 1,2 region, which was previously associated with a single complementation group (Hall *et al.*, 1983). This large segment was further divided, and the fragments obtained were tested by hybridization with labeled RNA isolated from ecdysone-treated *Drosophila* cells in culture, in which formation of AChE is induced by 50-fold by ecdysone (Cherbas *et al.*, 1977). A 10.5-kb fragment resulting from digestion of the 50-kb DNA fragment with the restriction enzyme SalI (which we designated DroS segment) was the only fragment hybridizing specifically to RNA that became abundant in AChE-induced cells (P. Spierer, unpublished observation, as quoted) (Soreq *et al.*, 1985). The DroS fragment was given to us by Dr. P. Spierer. (DroS is a 10.5-kb fragment of genomic DNA from the vicinity of the *Ace* locus in *Drosophila melanogaster*, which was indicated to be directly correlated with the biosynthesis of AChE.)

3.1. *Assignment of Drosophila Transcripts that Hybridize with DroS*

The DroS segment of the *Drosophila Ace* locus, shown to be directly correlated with AChE biosynthesis, was cloned into the pBR322 plasmid and then cut with EcoRI restriction enzyme. The segments obtained, of length 950, 1650, 2000, 2300, and 3000 base pairs (bp), were separated by agarose gel electrophoresis and recovered by electroelution. Each of the eluted fragments was ^{32}P-labeled by nick translation and employed individually as a probe for hybridization with poly(A)[+] mRNA from *Drosophila* larvae, fractionated by agarose gel electrophoresis and transferred to nitrocellulose filters. This blotting analysis demonstrated that the various DNA fragments of the DroS segment hybridized with several different poly(A)[+] RNA species (Soreq *et al.*, 1985).

In order to examine which of the mRNA species that hybridized with DroS could be translated into active AChE, poly(A)[+] RNA from *Drosophila* larvae was fractionated according to size by sucrose-gradient centrifugation. The separated RNA fractions were microinjected into

Xenopus oocytes and the enzymatically active AChE (Soreq *et al.*, 1982, 1984) was assayed radiometrically (Johnson and Russell, 1975). The RNA fraction sedimenting at about ≥28 S was enriched in mRNA capable of inducing AChE activity in oocytes (Soreq *et al.*, 1985). This fraction included two RNA species that hybridized with the DroS DNA segment, a major species of 4.5-kb length and a minor species of 5.2-kb length. Both RNA species hybridized with the central part of the DroS segment, which includes the 2.0-kb EcoRI fragment designated DroSR. Therefore, a comparison of the RNA blot hybridization and microinjection of size-fractionated RNA indicated that the DroSR fragment might encode AChE-inducing mRNA in *Drosophila* larvae.

3.2. Hybrid Selection of Cholinesterase-Inducing mRNA by DroSR

The DroSR DNA fragment was isolated, bound to a nitrocellulose filter, and hybridized with poly(A)$^+$ mRNA from *Drosophila* larvae and pupae. The hybridized mRNA fractions were eluted from the filters (Ricciardi *et al.*, 1979) and injected into oocytes, to assay their capacity to induce AChE activity. The hybridization-selection procedure resulted, in both cases, in a considerable enrichment of AChE-inducing mRNA, supporting the conclusion that the DroSR fragment hybridizes specifically with AChE-inducing mRNA from *Drosophila* larvae and pupae. We therefore chose the DroSR fragment as a probe of the *Drosophila* AChE gene.

3.3. RNA Blot Hybridization Reveals Homology between DroSR and mRNA from Human Cholinesterase-Expressing Tissues

Recently, 69 of a panel of 146 monoclonal antibodies (mAb) obtained against proteins from *Drosophila melanogaster* nervous system (Fujita *et al.*, 1982) showed cross-reactivity with one or more proteins from the human central nervous system (CNS) (Miller and Benzer, 1983). This remarkably high degree of cross-reactivity between fly and human brain proteins implies a conservation that appears to be particularly conspicuous in elements of the cholinergic system. Thus, muscarinic cholinergic receptors isolated from human and *Drosophila* heads show similar sizes and a common isoelectric point, and both display cross-reactivity to monoclonal antibodies raised against mouse muscarinic receptors (Venter *et al.*, 1984). This indicates structural similarities between human and *Drosophila* receptors, suggesting that the primary structure of muscarinic cholinergic receptors has been highly conserved over a substantial period of evolution.

Pharmacological studies and enzyme kinetic analyses suggest that all the molecular species of AChE, in different tissues and species, have identical or nearly identical enzymatic sites (Vigny *et al.*, 1978). Monoclonal antibodies against human erythrocyte AChE could cross-react with AChE at the neuromuscular junction (Fambrough *et al.*, 1982). It has also been found that AChEs from various species contain immunologically homologous domains (Fambrough *et al.*, 1982; Brimijoin *et al.*, 1983; Marsh *et al.*, 1984). The major molecular form of *Drosophila* AChE (Dudai, 1977; Zingde *et al.*, 1983) is a tetramer, similar to that found in the human brain (Sorenson *et al.*, 1982; Zakut *et al.*, 1985; Muller *et al.*, 1985). It is thus reasonable to assume that there is some homology between human and *Drosophila* AChE. This putative homology could exist at the protein level, and could also extend to the mRNA and the DNA sequence, possibly allowing the isolation of the human AChE gene using a *Drosophila* probe.

To test this hypothesis, we examined the ability of a very highly ^{32}P-labeled DroSR probe to hybridize with human poly(A)$^+$ RNA. A positive signal was obtained with an mRNA species of ≥ 7 kb in length, present in blotted samples of RNA from postnatal and embryonic brain and AChE-positive meningioma and glioma. No hybridization was observed with RNA from AChE-deficient HEp carcinoma, where both the specific activity of the enzyme and the concentration of AChE-inducing mRNA are 10-fold lower than in the brain (Razon *et al.*, 1984; Soreq *et al.*, 1984). The four other fragments obtained from the DroS segment by EcoRI restriction showed no hybridization to brain poly(A)$^+$ mRNA, suggesting that the interaction of DroSR with human brain mRNA is highly selective.

4. ISOLATION AND PARTIAL CHARACTERIZATION OF HUMAN DNA FRAGMENTS HOMOLOGOUS TO DroSR

Our reasoning for the use of the DroSR DNA fragment as a probe was based on the following arguments:

1. The DroS fragment is located in the vicinity of the genetic *Ace* locus in *Drosophila*, regulating the expression of AChE. This is based on the chromosomal walking from region 87D 5–7 to 87E 5–6 on the third chromosome in *Drosophila* (Spierer *et al.*, 1983).
2. The DroS fragment hybridized specifically to RNA that became abundant in AChE-induced cells, whereas other fragments of the *Ace* locus did not (unpublished observation, as quoted in Soreq *et al.*, 1985).
3. The DroSR DNA fragment could hybrid-select *Drosophila* RNA molecules that induced AChE biosynthesis in microinjected *Xenopus* oocytes (Soreq *et al.*, 1985; Avni, 1985).

4. The DroSR fragment cross-hybridized with mRNA from human brain, but not with mRNA from the AChE-deficient HEp carcinoma (Soreq *et al.*, 1985; Avni, 1985).

4.1. Isolation and Characterization of a Human Genomic Fragment: Huache1

The DroSR fragment was used as a probe to isolate homologous sequences from a human genomic DNA library cloned in Charon 4a λ phages. The expected average size of insert DNA in the library is 20 kb; the human haploid genome includes a total of about 4×10^6 kb (Manniatis *et al.*, 1982). In order to ensure full representation of the whole human genome in the screened phages, we plated three genome equivalents (about 7×10^5 phages). Twenty-eight 150-mm-diameter petri dishes were plated with 25,000 plaque-forming units (PFU) per dish of library phages. The phages were transferred to duplicate nitrocellulose filters (Benton and Davies, 1977), and hybridized under low stringency conditions (Shilo and Weinberg, 1981) with the DroSR DNA probe ^{32}P-labeled by nick translation. The hybridization signals were rather weak, probably due to a high degree of mismatch between the *Drosophila* and the human DNA sequences. However, six phages displayed reproducible positive hybridization in duplicates, under these low stringency conditions. One of these, a phage containing a 13.5-kb human genomic DNA fragment, designated Huache1, was amplified; its DNA was purified and further characterized by restriction analysis. A 2.6-kb fragment of the Huache1 fragment, produced by EcoRI digestion and designated Huache1R, was found to hybridize with DroSR (Soreq *et al.*, 1985). All further analyses were therefore focused on this Huache1R fragment.

4.2. Hybridization of Huache1R DNA with Poly(A)$^+$ RNA Species from Fetal Human Brain

Poly(A)$^+$ RNA from human fetal brain and human epidermoid (HEp-3) carcinoma was fractionated by agarose gel electrophoresis, transferred to nitrocellulose filters, and hybridized with the Huache1R DNA probe. The blot revealed an intense diffuse band of about 7 kb in length in poly(A)$^+$ RNA from fetal human brain and a very faint, more slowly migrating but sharp band in mRNA from HEp carcinoma tissue (Soreq *et al.*, 1985). A similar intense diffuse $\geq$7-kb band was also observed in mRNA from AChE-containing human leukemic cells, indicating a possible size heterogeneity of the RNA complementary to the Huache1R fragment. DMSO-denatured poly(A)$^+$ mRNA from fetal brain was also size-fractionated by sucrose-gradient centrifugation and microinjected into *Xenopus* oocytes to test for induction of AChE synthesis. The mRNA that induces AChE activity in oocytes was most enriched in the fraction sedimenting faster than 28S ribosomal RNA, as would be expected of an mRNA 7 kb in length. A similar length of 7 kb, although with lower activities, was also found for the major species of AChE-inducing mRNA from primary gliomas and meningiomas (Soreq *et al.*, 1984).

4.3.   *Hybrid Selection of Acetylcholinesterase-Inducing mRNA with Huache1R DNA*

The Huache1R DNA fragment was tested for its ability to hybrid-select mRNAs capable of inducing AChE in oocytes. Samples of 20 μg poly(A)$^+$ RNA from various regions of human brain were hybridized with either Huache1R DNA or control DNA from Charon 4a phages, bound to nitrocellulose. Hybridized RNA was eluted and microinjected into oocytes, in parallel with unfractionated poly(A)$^+$ mRNA. Within each set of samples, translationally active AChE-inducing mRNA was reproducibly retained on the nitrocellulose filters (Soreq *et al.*, 1985).

Cholinesterase activity in the mammalian brain is mostly (>90%) of the "true" AChE type, which can be distinguished by its susceptibility to the selective AChE inhibitor BW284C51, and by its resistance to the organophosphorus poison iso-OMPA (Soreq *et al.*, 1982; 1984). In oocytes injected with mRNA from human parietal cortex, ChE activity was composed of a major fraction resistant to iso-OMPA and a minor part resistant to BW284C51. The enzyme produced in oocytes injected with hybrid-selected mRNA from this source was partially sensitive to iso-OMPA and was completely inhibited by BW284C51. By contrast, the ChE activity in control oocytes could be almost completely blocked by both inhibitors, implying it was of amphibian origin. It can thus be concluded that unfractionated mRNA from human parietal cortex induced in oocytes both "true" AChE and (although to a much lower extent) "pseudo", BW284C51-resistant ChE. Finally, these results suggest that the Huache1R fragment hybridized to mRNA, which induced the synthesis of true brain AChE, entirely resistant to iso-OMPA and completely susceptible to BW284C51.

4.4.   *Immunoprecipitation of Acetylcholinesterase Polypeptides from Ooyctes Injected with Hybrid-Selected mRNA*

The immunoprecipitation approach was applied in order to further corroborate the fact that the Huache1R hybridized to mRNA capable of inducing the formation of human AChE in *Xenopus* oocytes and to exclude the possibility that the injected mRNA induced the formation of amphibian AChE. Immunoprecipitation was carried out with AE-2 monoclonal antibodies against human erythrocyte AChE (Fambrough *et al.*, 1982). Irrelevant monoclonal antibodies were used for control. Precipitates were separated by polyacrylamide gel electrophoresis (PAGE) and analyzed by autoradiography.

Immunoprecipitation of products from oocytes injected with unfractionated brain mRNA but not with mRNA from the AChE-deficient HEp carcinoma revealed a 85,000-M_r protein. Oocytes injected with hybrid-selected mRNA produced a 85,000-M_r protein, which was selectively precipitated by anti-AChE antibodies, but not by irrelevant antibodies. Moreover, this 85,000-M_r protein was absent both from the total translation products produced in oocytes injected with control mRNA (selected by Charon 4a DNA) and from the proteins precipitated from control oocyte extract by anti-AChE. The size of the precipitate corresponds to the reported size of the AChE subunit from human brain (Sorenson *et al.*, 1982). The results of this independent approach support our conclusion that the Huache1R fragment hybridized selectively to mRNA, which induced the formation of human AChE in *Xenopus* oocytes (Zevin-Sonkin *et al.*, 1985) and rule out the possibility that the induced enzyme was of amphibian origin, since the antibodies employed do not cross-react with frog AChE (Fambrough *et al.*, 1982).

4.5. Preliminary Characterization of Huache1R DNA

Sequence analysis of Huache1R was performed in order to search for putative sequence(s) coding for AChE in this fragment. It should again be pointed out that the amino acid sequence of mammalian AChE is not known at all. However, interspecies cross-homologies with anti-AChE antibodies (Chubb, 1981; Fambrough *et al.*, 1982; Brimijoin *et al.*, 1983; Marsh *et al.*, 1984) indicate that some amino acid sequence homology may exist even between ChE from sources that are genetically remote, such as *Torpedo*, *Drosophila*, and man. In any case, we found only minor homologies at the nucleotide level between Huache1R and DroSR (which was sequenced by Dr. P. Spierer and L. M. C. Hall).

4.6. Isolation of Huache1R-Homologous Genomic DNA Fragments

In order to find out whether additional DNA sequences within the human genome contain domains that are homologous to the Huache 1R fragment, three genome equivalents of the human genomic library were screened using labeled Huache1R DNA as a probe. Twenty-four different phages displayed positive hybridization signals, and twelve of these were amplified for further analysis. DNA samples from these phages were digested with EcoRI; the fragments were separated on a 1% agarose gel, transferred to nitrocellulose filters by blotting, and hybridized with the Huache1R probe. The hybridization revealed different restriction patterns for the isolated phages, suggesting that at least 10 different phages contained domains with sequence homologies to Huache1R.

4.7. Preparation of Huache1R-Homologous Fetal cDNA Clones

Hybridization of Huache1R DNA with mRNA from fetal brain resulted in the formation of RNA–DNA hybrids (Soreq *et al.*, 1985). This suggested that the Huache1R DNA was actively transcribed in the fetal brain into mRNA. We therefore prepared and screened a cDNA library from fetal human brain, using labeled Huache1R DNA as a probe and found 42 positive clones out of 37,000. The insert from one of these clones, designated C3, was used as a probe for hybridization with the EcoRI digestion products of the 12 phages that contain Huache1R-homologous inserts. The C3 probe (of the length of 2600 bp) yielded a positive pattern of hybridization with the restriction fragments, different from that observed with the Huache1R probe (Zevin-Sonkin *et al.*, 1985).

The C3 cDNA clone, and three other positive cDNA clones, 350, 850, and 2550 bp long were studied in further detail. DNA from all four inserts was isolated and labeled. Gel electrophoresis of mRNA from fetal brain and from HEp epidermoid carcinoma, followed by blotting and hybridization with the DNA probes, revealed that all four inserts hybridized with heavy mRNA chains present in the brain. We then searched for homologies between these cDNA chains and the original Huache1 genomic sequence, from which Huache1R was derived (Soreq *et al.*, 1985). Gel electrophoresis of Huache1 DNA cut with EcoR1, followed by blotting and hybridization with these four DNA probes, revealed that all four inserts hybridized with different regions of the Huache1 DNA. This could either indicate that the Huache1 DNA contains nontranscribed intervening sequences or reflects the existence of cross-homologous domains within the Huache1 DNA itself. Mapping of restriction sites within the two longer inserts, designated A7 and C3, showed clear differences in restriction patterns, indicating that they were either derived from two different mRNA species, or from two different regions on a single, high-molecular weight mRNA. Finally, hybridization-selection with mRNA from fetal brain was performed, and revealed that DNA from two of these inserts hybridized with AChE mRNA (Zevin-Sonkin *et al.*, 1985).

4.8. Conclusions

The different Huache1R-homologous phages that we have selected could contain different parts of the AChE gene, with their inserts located in adjacent positions to that of Huache1 in the human genome. Another possibility is that they code for different cholinesterases, or represent parts of the gene(s) coding for other esterases or peptidases, with partial homologies to the AChE gene(s). Parallel assumptions may be made regarding the selected cDNA clones. It is thus plausible that not only the genes coding for AChE, but also other genes could interact with both Huache1R and AChE mRNA. Accordingly, in order to focus on the particular DNA sequences that actually code for the ChE proteins, we therefore decided to search for probes of higher specificity and selectivity than those of DroSR and Huache1R. Such properties could be offered by synthetic short oligonucleotides, prepared according to amino acid sequence(s) recently found unique to cholinesterases (see details in following sections).

5. PREPARATION OF SYNTHETIC OLIGONUCLEOTIDE PROBES ACCORDING TO THE CONSENSUS SEQUENCE AT THE ORGANOPHOSPHATE-BINDING SITE

Synthetic oligodeoxyribonucleotides have been shown to hybridize specifically to complementary DNA sequences (Wallace et al., 1981). Under appropriate hybridization conditions, only perfectly base-paired oligonucleotide xDNA duplexes will form whereas duplexes containing mismatched base pairs will not be stable. The high degree of hybridization specificity has led to the development of a general method for using synthetic oligonucleotides as specific probes to identify cloned DNAs coding for proteins of interest. This technique has been applied to the successful isolation of several cDNA clones, for example—those coding for human β_2-microglobulin (Suggs et al., 1981), for murine transplantation antigen (Reyes et al., 1981) and for human apolipoprotein CII (Myklebost et al., 1984). We have recently used this approach to screen for ChE genomic and cDNA clones.

The amino acid sequence of AChE is unknown. It has recently been found, however, that organophosphorus-binding sites of both Torpedo AChE tetramers (Macphee-Quigley et al., 1985) and human "pseudo"-ChE tetramers (Lockridge, 1984) include a common sequence of six amino acids identical to a hexapeptide previously found in electric eel AChE (Schaffer et al., 1973). This primary structure was determined by peptide

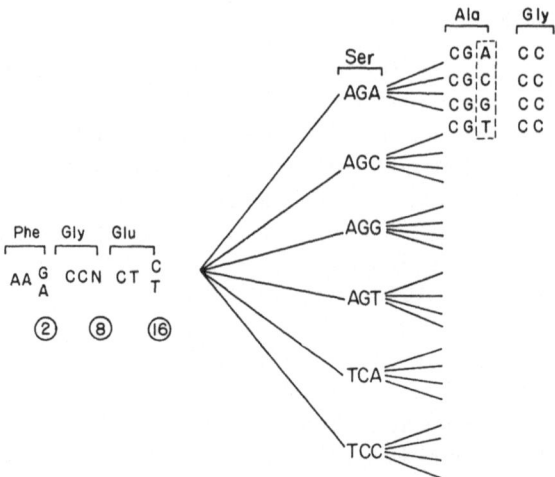

Test with Poly (A)$^+$ RNA from brain vs. HEp

Figure 2. OPSYN probes. A consensus hexapeptide sequence has recently been found to be included in the organophosphate (OP) binding site of cholinesterases from various tissues and species. Synthetic mixtures of oligodeoxyribonucleotides (OPSYN), all of which are complementary to sequences that could code for the consensus peptide, were designed and prepared. Twenty-four mixtures of such oligodeoxynucleotides, each containing 16 different 17-mer chains, were prepared by solid-phase synthesis using phosphoramidite chemistry and purified by polyacrylamide gel electrophoresis followed by Sephadex G-25 chromatography. For hybridization experiments, separate OPSYN mixtures were end-labeled with [^{32}P]-ATP by polynucleotide kinase.

sequencing of proteolytic degradation products of the purified ChEs, accompanied by identification of the organophosphate-binding serine residue using [^{3}H]diisopropylfluorophosphate (DFP). On the basis of these findings, it is reasonable to expect that some sequence homology, at least at the level of amino acid sequence, also exists between ChE from sources that are genetically remote such as *Drosophila* and human.

5.1. Strategy for Preparation and Selection of the Correct Mixture of Synthetic Oligonucleotides

The consensus hexapeptide sequence from the organophosphate-binding site of ChEs (designated OPSYN) is Phe-Gly-Glu-Ser-Ala-Gly. Because of the ambiguity in the genetic code (Chen *et al.*, 1982), this peptide could be encoded by one of 384 different oligonucleotides (after deletion of the third nucleotide in the terminal glycine codon) as presented in Figure 2. Lacking sufficient information concerning codon preferences in human genes, and considering that the genes coding for

ChEs may not behave as species-specific genes (since they represent a case of proteins that have been conserved through evolution), we decided to prepare the entire collection of the OPSYN sequences. It would have been rather impractical, however, to use a mixture of 384 different sequences, out of which only one 17-nucleotide-long oligomer would have had the right sequence. In such a case, the specific activity of the correct probe would inevitably be very low, thus leading to considerable chances for false-positive hybridization signals. To overcome this difficulty, we prepared 24 oligonucleotide mixtures, each comprising 16 different 17-mers (see Figure 2). Each of these mixtures included only one of the sequences coding for serine and alanine but permitted all the amibiguities for phenylalanine, the central glycine, and glutamic acid.

The oligonucleotide designs used ensured that at least nine successive nucleotides of the correct OPSYN sequence were present in all the analyzed mixtures and that the correct sequence represented one-sixteenth of the total labeled OPSYN probe included in each reaction mixture. However, we used hybridization conditions having a minimal length necessary to stabilize the DNA–DNA hybrids of 11–13 nucleotides (Wallace et al., 1981). Therefore, only mixtures containing the correct codon for serine (and, preferably, also for alanine) could hybridize. To select the most suitable mixture for screening purposes, we prepared identical Northern RNA blots from poly(A)$^+$ RNA from brain and from epidermoid carcinoma. Each of the various OPSYN mixtures was labeled separately and tested by hybridization to one of these RNA blots. The mixture that displayed the highest hybridization signal with brain RNA, but not with HEp RNA, was selected for further use.

5.2. Probing of Selected DNA Fragments with the Synthetic Oligonucleotide

We tested the ability of several DNA fragments to hybridize with the selected OPSYN mixture. These included the human genomic clone Huache1, all the phage clones that displayed cross-homology to it, the human cDNA clones prepared in pBR322 from fetal brain mRNA, the DroS clone from the vicinity of the Ace locus, and several other DNA fragments from the Ace locus. The experiment was first carried out with blotted bacterial colonies (cDNA clones) or phage plaques (genomic clones). In this experiment the Huache1 clone, but none of the phages that hybridized to it, and about 40% out of the 42 pBR322 colonies gave positive hybridization signals. We then prepared DNA from positively hybridizing colonies, digested it with EcoRI, and separated the fragments by agarose gel electrophoresis. The separated DNA fragments were blotted onto a nitrocellulose filter that was then hybridized with the OPSYN mixture and labeled with ^{32}P by polynucleotide kinase. None of the human DNA fragments came out positive by this analysis, indicating that the positive hybridization with the blotted colonies was false. The DroSR fragment and two other EcoRI-cut fragments from DroS gave a positive signal, whereas several other cloned DNA fragments from the *Ace* locus did not hybridize. The hybridization signal obtained for the DroS fragment was clearly specific, although not strong. We therefore assume that OPSYN-like sequences are located along this fragment. These sequences could possibly code for organophosphate-binding sites in *Drosophila* cholinesterase. However, the *Drosophila* enzyme does not display the same interaction with organophosphates as the human one (Dudai, 1977; Zingde *et al.*, 1983), suggesting that they may include a similar, but not an identical, OPSYN sequence. The DroS fragment should therefore be tested for its ability to hybridize with the other OPSYN mixtures.

The above-mentioned results indicate that the DroSR fragment may include the region coding for the organophosphate-binding site in *Drosophila* AChE and that the Huache1 DNA fragment, although homologous to DroSR, does not include such a region. This fragment is therefore more likely to represent another part of a human AChE gene.

5.3. Screening for OPSYN-Containing cDNA Sequences from cDNA Libraries in λgt Vectors

In order to obtain ultimate proof of the identity of a gene coding for a protein of a known function (but unknown sequence), it should be necessary to produce this protein from cloned DNA sequence(s) within the bacterial host. This goal became feasible with the recent development of expression vector libraries.

λ vectors for cDNA cloning offer two major advantages over the plasmid vectors of the previous generation, such as pBR322:

1. They can be used (particularly the λgt10 vector) to prepare very large cDNA libraries, containing 10^5–10^7 recombinants. This property is an important prerequisite for the isolation of cDNA clones of rare mRNAs.
2. The λgt11 vector is an expression vector, capable of producing a polypeptide specified by the DNA insert fragment. cDNA sequences cloned in λgt11 can therefore be tested with antibody probes to examine whether they encode the antigen against which the antibodies are directed (Huynh *et al.*, 1985).

These advantages answer the two main difficulties in identifying cDNA clones for AChE. We therefore decided to screen for such colonies using the λgt vectors.

Since AChE is a ubiquitous protein, appearing in most tissues, we first assumed that ChE mRNA chains would be present in λgt cDNA libraries from most tissues and cell types, albeit in very low concentrations. We therefore screened all the libraries from human origin that were made available to us. λgt10 libraries derived poly(A)$^+$ RNA from human placenta, from SV80 human fibroblasts, and from human lymphocytes were probed with the ^{32}P-labeled OPSYN mixture; 7×10^5 phages from each library were screened, to ensure complete representation of the entire cDNA repertoire. However, there were no positives

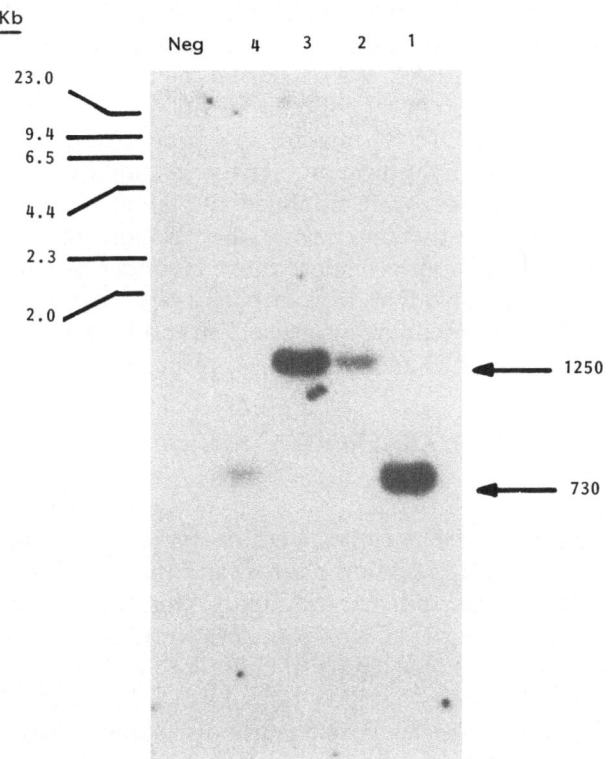

Figure 3. Hybridization of Neuroche inserts with the synthetic OPSYN oligonucleotide probe. A cDNA library prepared from mouse neuroblastoma poly(A)$^+$ RNA and packaged in λgt11 phages (gratefully received from C. Gorridis) was screened with ^{32}P-labeled OPSYN oligodeoxynucleotide mixture; 4 out of 200,000 phage plaques gave positive hybridization signals. DNA was purified from the four positive Neuroche phages (1–4) as well as from a single negative phage for control (Neg). cDNA inserts were restricted with EcoRI, separated by agarose gel electrophoresis and blotted onto nitrocellulose. The filter was hybridized with [^{32}P]-OPSYN, washed and exposed for autoradiography. Hind III-digested λ-DNA served for molecular weight markers.

in both screens. This implies either that both SV80 fibroblasts and human placenta do not produce cholinesterase at all (and that the ChE activities measured in these sources were derived entirely from the contribution of blood or added serum) or that the level of ChE mRNAs is, in both cases, too low to permit its representation in the cDNA library prepared from these sources (even though it was packaged in the highly efficient λgt10 vector). We therefore decided to screen only cDNA libraries from cell or tissue sources known to be particularly enriched with AChE.

Mouse neuroblastoma cells are well known as a rich source for "true" AChE, about 10-fold richer than human neuroblastoma (Kimhi *et al.*, 1980; Lazar and Vigny, 1980; Soreq *et al.*, 1983). AChE from human and mouse tissues was found to be immunologically cross-reactive with several antibodies (Fambrough *et al.*, 1982; Brinijoin *et al.*, 1983; Marsh *et al.*, 1984; Grassi *et al.*, 1982). We screened a cDNA library from poly(A)$^+$ RNA from mouse neuroblastoma cells in the λgt11 expression vector, using the ^{32}P-labeled OPSYN mixture as a probe. Four out of 2×10^5 phages gave positive hybridization signals, in duplicates and in three successive screens. These were amplified and their DNA was prepared, cut with the EcoR1 restriction enzyme, electrophoresed on an agarose gel, and blotted onto a nitrocellulose filter. Hybridization with the OPSYN probe then revealed that the four positives were, in fact, two pairs of identical OPSYN-containing inserts, of 730 and 1250 bp in length (Fig. 3) designated Neuroche3 and Neuroche4, respectively.

6. PRELIMINARY CHARACTERIZATION OF NEUROCHE cDNA CLONES

Both Neuroche cDNA clones were analyzed by enzymatic restriction, inserted into M13 Messing phages, and their DNA sequences examined by the Sanger dideoxy technique. This analysis revealed that 560 nucleotides in the two cDNA sequences were identical, whereas the 5' regions were not, suggesting that these two cDNA chains were reverse-transcribed from two largely homologous, but not identical, species of ChE mRNA. The sequence analysis also showed that both inserts code for putative polypeptides in which the organophosphate-binding peptide encoded by OPSYN is included (Fig. 4).

6.1. Blot Hybridizations with Restricted Genomic DNA (Human and Mouse) and with mRNA

One of the methodologies used to estimate how many genes code for a particular protein consists of the hybridization of genomic DNA with the corresponding cDNA probe, following enzymatic restriction,

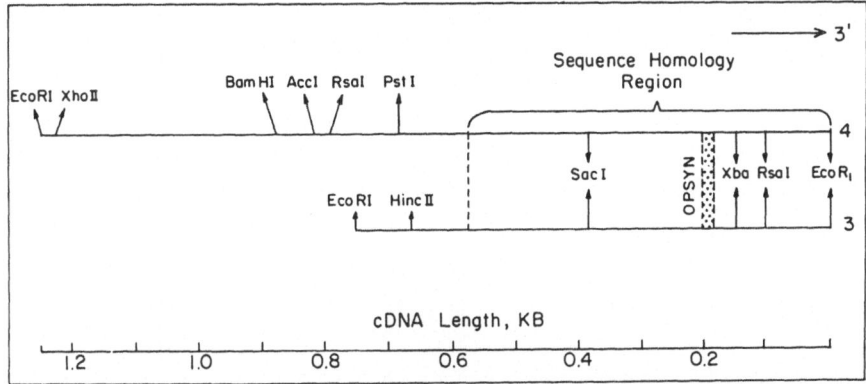

Figure 4. Characterization of the Neuroche inserts. Neuroche cDNA inserts 3 and 4 were characterized by enzymatic restriction and by nucleotide sequencing in M13 single-stranded phages using the Sanger dideoxy technique. The 3'-terminal part of Neuroche3 and Neuroche4, 580 nucleotides in length, was found identical in its nucleotide sequence in both inserts and to include the OPSYN sequence. In contrast, the 5'-terminal parts of these cDNAs were found to differ in enzymatic restriction sites and sequence. This implies that the Neuroche3 and Neuroche4 inserts are either derived from two distinct mRNA species or that they were reverse-transcribed from two partially homologous regions on one very long mRNA chain.

agarose gel electrophoresis, and blotting onto a nitrocellulose filter. The Neuroche4 probe, labeled with ^{32}P by nick translation, was subjected to this procedure and was found to hybridize with very few DNA fragments of mouse genomic DNA, of the total length of about 40 kb. Although Neuroche4 does not cover the entire length of ChE mRNA (see following sections for details), this result implies that Neuroche4 does not hybridize with a multigene family. A parallel hybridization experiment demonstrated a considerable homology between Neuroche4 and human genomic DNA, as could have been expected for ChEs from cross-immunological analyses. Agarose gel electorphoresis of poly(A)$^+$ RNA from the brain, muscle, and liver of 10-day-old BALB/c mice, followed by blotting to nitrocellulose and hybridization with ^{32}P-labeled Neuroche probes, was also carried out. This experiment showed that both Neuroche3 and Neuroche4 hybridize with mRNA of the size of about 27S and that although this mRNA is expressed in all three tissues, it is present in brain and muscle in larger amounts than in the liver (Fig. 5). A parallel microinjection experiment, followed by ChE bioassay, confirmed that translationally active ChE mRNA is indeed present in all tissues. However, mouse brain mRNA directed in the oocytes mainly the biosynthesis of secretory ChE, whereas liver mRNA only produced cytoplasmic ChE and muscle mRNA was most potent in directing the production of membrane-associated ChE (see Fig. 1). This finding suggests that the mRNAs

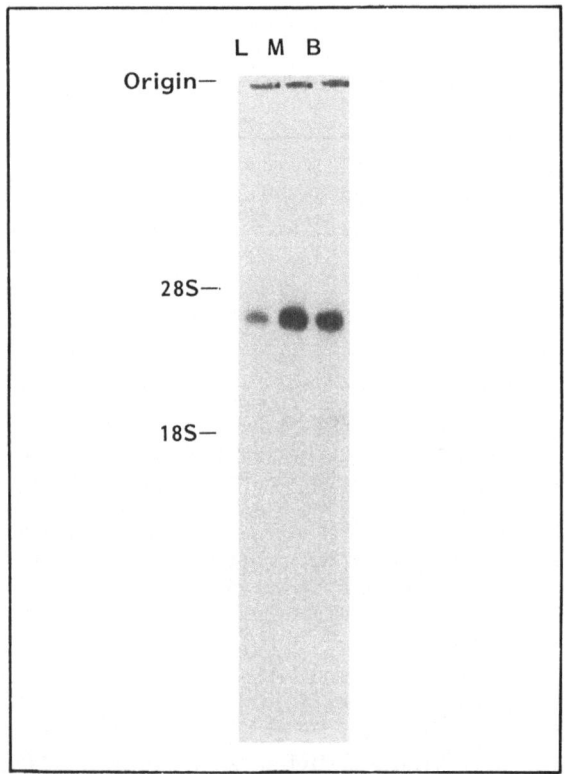

Figure 5. Hybridization of ^{32}P-labeled Neuroche4 DNA with blotted poly(A)± RNA from mouse tissues. Poly(A)[+] RNA was extracted from 10-day-old mouse brain (B), muscle (M), or liver (L), separated by agarose gel electrophoresis, and blotted onto nitrocellulose. The filter was hybridized with ^{32}P-labeled Neuroche4 DNA, washed, and exposed for autoradiography. A major band of RNA of about 27 S and a minor one of about 20 S could be seen in all three tissue sources, with muscle being most enriched with Neuroche4-hybridizable mRNA and liver being relatively deficient in it. Hybridization with [^{32}P]-Neuroche3 gave essentially similar results (not shown).

directing the production of these different ChEs, in various tissues, are largely homologous and of a similar size. It also puts forward the possibility that post-translational processing, such as glycosylation, could be involved in directing the various ChEs to their different subcellular localizations. Finally, it shows that the Neuroche4 probe, of 1250 nucleotides in length, corresponds to only about 25% of the length of ChE mRNA. It is therefore expected that this cDNA probe, inserted into the λgt11 expression vector, could be translated to yield only up to 25% of the subunit size of ChE, or about 15,000 M_r.

6.2. Identification of Acetylcholinesterase-Immunoactive Fusion Protein by Crossed Immunoelectrophoresis and Immunoprecipitation

Since mouse neuroblastoma cells express "true" AChE (Kinki *et al.*, 1980; Lazar and Vigny, 1980; Soreq *et al.*, 1983), it may be concluded that the Neuroche probes were reverse-transcribed from AChE mRNA. However, one would like to be especially cautious when a gene is pursued that codes for a protein of an unknown amino acid sequence. Thus, we were particularly interested to find out whether the Neuroche probes,

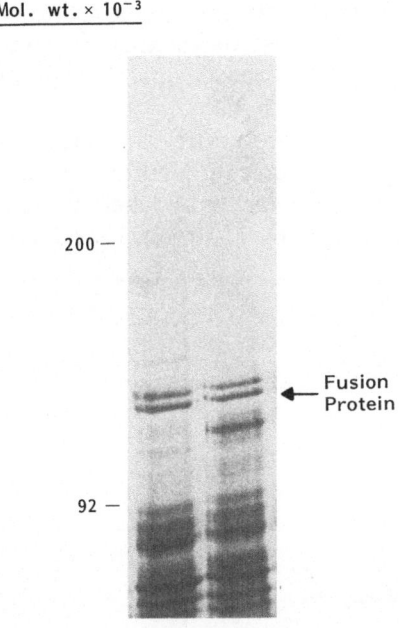

Figure 6. Induction of synthesis of a fusion protein in bacteria infected with Neuroche4 clones and induced with isopropyl thio-β-D-galactoside (IPTG). The Neuroche4 insert, of 1250 nucleotides in length, was inserted into the 3′-terminal part of the β-galactosidase coding sequence in the λgt11 expression vector. *Lon* protease-deficient lysogenic y1089 bacteria were infected with Neuroche4 phages, heat-treated to prevent the expression of the *hfla* 150 lysogeny mutation, and induced with IPTG to block the *lac* repressor for β-galactosidase synthesis (Huynh *et al.*, 1985). [^{35}S]methionine was added to the cultures together with IPTG, and they were further incubated for 60 min. Bacterial extracts were then separated and analyzed by gradient polyacrylamide gel electrophoresis followed by autoradiography. A considerable induction was observed in the synthesis of a high-molecular-weight [^{35}S]methionine-labeled fusion protein (marked by an arrow), demonstrating that the Neuroche4 DNA contained an open reading frame and that it was inserted in the correct position for protein synthesis. (−) Noninduced; (+) induced. Molecular-weight markers are noted in the attached scale.

inserted into the λgt11 expression vector, direct the production of parts of an AChE-like protein in their host bacteria.

Within the λgt11 expression vector, the cloned insert is ligated at the 3'-terminal part of the gene coding for β-galactosidase. When the cloned sequence is inserted in the correct reading frame, the host bacteria produce a fusion protein, composed of the N-terminal part of β-galactosidase (114,500 M_r) and of the polypeptide product directed by the inserted cDNA. This fusion protein would be subjected to the repression mechanisms controlling the production of bacterial β-galactosidase. Thus, isopropyl-thio-β-D-galactoside (IPTG), a competitive substrate analogue for β-galactosidase, would induce the constitutive production of this fusion protein. When protease-deficient strains of bacteria are employed as hosts to the λgt11 phages, ample quantities of the fusion protein may be produced (Huynh *et al.*, 1985).

When this induction procedure was carried out with bacteria infected with λgt11 phages carrying Neuroche inserts, the production of fusion proteins could be detected by labeling the synthesized proteins with [^{35}S]methionine (Fig. 6). Neuroche3 produced a fusion protein of

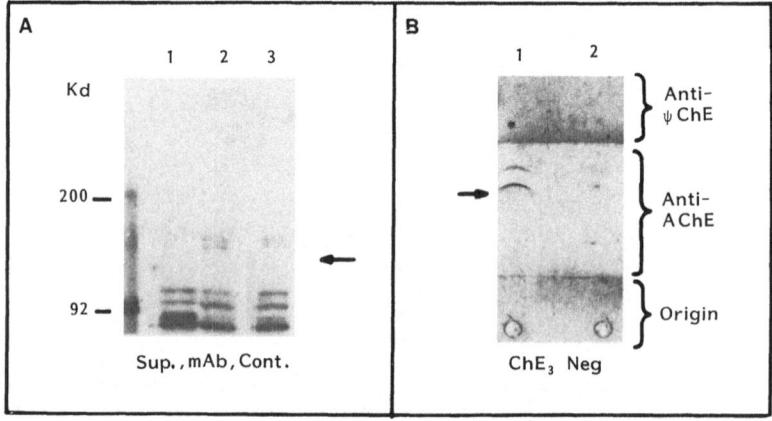

Figure 7. Neuroche 3 insert directing the synthesis of a fusion protein immunoreactive with anti-AChE antibodies. Bacteria were infected with λgt11 phages carrying the Neuroche3 cDNA insert (ChE₃) or a negative cDNA insert for control (Neg). The synthesis of fusion proteins was induced with IPTG. (A) Immunoprecipitation was carried out using complexes of *Staphylococcus aureus* protein A coupled with goat anti-mouse IgG and with AE-2 monoclonal antibodies against human AChE (mAb) (Fambrough *et al.*, 1982) or with an irrelevant monoclonal antibody for control (Cont.). Fusion protein (marked by an arrow) was removed from the supernatant (Sup.) by the specific mAb complexes, but not by the control ones. (B) Immunoelectrophoresis was carried out using rabbit anti-AChE serum and commercial anti-human ψChE serum in the intermediate and top gels, respectively. Precipitation arcs were formed with Neuroche3 proteins (ChE₃) but not with proteins from control bacteria (Neg) in the gel containing anti-AChE antibodies and not in the gel with anti-ψChE antibodies. Staining was with Coomassie Brilliant Blue.

about 120,000 and Neuroche4 of about 140,000M_r. Crossed immunoe-lectrophoresis of bacterial proteins against anti-AChE rabbit antiserum, followed by staining with Coomassie Brilliant Blue, was then carried out (Fig. 7). Clear precipitation arcs were produced with protein extracts from bacteria carrying the two Neuroche probes, but not from bacteria carrying an irrelevant insert (negative in hybridization with the OPSYN probe).

Finally, immunoprecipitation of the ^{35}S-labeled bacterial proteins with monoclonal mouse antibodies directed against human AChE (AE-2)(Fambrough et al., 1982) has proved that it was the fusion protein that precipitated with the anti-AChE antibodies (Fig. 7).

7. SUMMARY AND CONCLUSIONS

During the past 3 years, we have run several screening experiments in the search for ChE genes. These were carried out using various li-braries and different DNA probes and led to the following conclusions:

1. The human AChE gene shows some homology to both the *Ace* region from *Drosophila* and to several DNA sequences from the human genome. The latter are probably parts of genes coding for other proteins with partial domains homologous to regions in the AChE gene.
2. Human AChE and ψChE are probably encoded by single-copy genes, or even by one gene for the two proteins, since the se-quence coding for the organophosphate-binding site is not re-peated many times in the human genome.
3. AChE mRNA is represented (in a number of copies allowing reliable selection of cDNA clones) only in tissue sources or cell types that are particularly enriched with AChE.
4. The genes coding for ChEs in the mouse and the human genome show considerable sequence homology.

ACKNOWLEDGMENTS

We are grateful to Drs. C. Gorridis, J. Massoulie, B. P. Doctor, Lucinda Hall, Pierre Spierer, Haim Zakut, Avi Matzkel, Eduardo Schejter, K. Djiegielewska, E. Zeelon, V. Roter, and C. Gorridis, and to Mr. Adi Avni, and Ms. Rivka Zisling, for their contributions to this work. This research was supported by the U.S. Army Medical Research and De-velopment Command (contract DAMD 17-85C-5025, to H.S.) and by the Hermann and Lilly Schilling Foundation for Medical Research (to H.S.).

REFERENCES

Austin, L., and Berry, W. K., 1953, Two selective inhibitors of cholinesterase, *Biochem. J.* **54**:695–700.

Avni, A., 1985, Isolation and partial characterization of a human acetylcholinesterase gene identified by homology to the *Drosophila* gene, M. Sc. thesis, The Weizmann Institute of Science.

Balasubramanian, A. S., 1984, Have cholinesterases more than one function, *TINS* **7**:467–468.

Bender, W., Spierer, P., and Hogness, D. S., 1983, Chromosomal walking and jumping to isolate DNA from the Ace and rosy loci and bithorax complex in *Drosophila melanogaster*, *J. Mol. Biol.* **168**:17–33.

Benton, W. D., and Davies, R. W., 1977, Screening Lambda gt recombinant clones by hybridization to single plaque in situ, *Science* **196**:180–184.

Brimijoin, S., Mintz, K. P., and Alley, M. C., 1983, Production and characterization of separate monoclonal antibodies to human acetylcholinesterase and butyrylcholinesterase, *Mol. Pharmacol.* **24**:513–520.

Burstein, S. A., Adamson, J. W., and Walker, L. A., 1980, Megakaryocytopoiesis in culture: Modulation by cholinergic mechanisms, *J. Cell Physiol.* **103**:201–208.

Chautard-Freie-Maia, E. A., 1977, Probable assignment of the serum cholinesterase (E_1) and Transferrin (Tf) Loci to chromosome 1 in man, *Human Hered.* **27**:134–142.

Chen, H. R., Dayhoff, M. O., Barker, W. C., Hunt, L. T., Yeh, L.-S., George, D. G., and Orcutt, B. C., 1982, Nucleic acid sequence database IV, *DNA* **1**:365–374.

Cherbas, P., Cherbas, L., and Williams, C. M., 1977, Induction of Acetylcholinesterase Activity by beta-Ecdyson in a Drosophila cell line, *Science* **197**:275–277.

Chubb, I. W., 1984, Acetylcholinesterase—Multiple functions, in: *Cholinesterases—Fundamental and Applied Aspects* (M. Brzin, T. Kiauta, and E. A. Barnard, eds.), pp. 345–359, Walter de Gruyter, Berlin.

Coates, P. M., and Simpson, N. E., 1972, Genetic variation in human erythrocyte acetylcholinesterase, *Science* **175**:1466–1477.

Couteaux, R., and Taxi, J., 1952, Recherches histochimiques sur la distribution des activités cholinesterasiques an niveau de la synapse myoneurale, *Arch. Anat. Microsc. Morphol. Exp.* **41**:352–392.

Doctor, B. P., Camp, S., Gentry, M. K., Taylor, S. S., and Taylor, P., 1983, Antigenic and structural differences in the catalytic subunits of the molecular forms of acetylcholinesterase, *Proc. Natl. Acad. Sci. USA* **80**:5767–5771.

Drews, U., 1975, Cholinesterase in embryonic development, *Prog. Histochem. Cytochem.* **7**:1–52.

Dziegielewska, K. M., Saunders, N. R., Schejter, E. J., Zakut, H., Zevin-Sonkin, D., Zisling, R., and Soreq, H., 1986, Synthesis of plasma proteins in fetal, adult and neoplastic human brain tissue, *Dev. Biol.* **115**:93–104.

Dudai, Y., 1977, Molecular states of acetylcholinesterase from *Drosophila melanogaster*, *Dros. Inform. Serv.* **52**:65.

Egozi, Y., Sokolovsky, M., Matzkel, A., Schejter, E., Blatt, I., Zakut, H., and Soreq, H., 1986, Divergent regulation of muscarinic binding sites and acetylcholinesterase in discrete regions of the developing human fetal brain. *Cell Molec. Neurobiol.* **6**:55–70.

Ellman, G. L., Courtney, D. K., Andres, V., and Featherstone, R. M., 1961, A new and rapid colorimetric determination of acetylcholinesterase activity, *Biochem. Pharmacol.* **7**:88–95.

Fambrough, D. M., Engel, A. G., and Rosenberry, T. L., 1982, Acetylcholinesterase of human erythrocytes and neuromuscular junctions: Homologies revealed by monoclonal antibodies, *Proc. Natl. Acad. Sci. USA* **79**:1078–1082.

Fujita, S. C., Zipursky, S. I., Benzer, S., Ferrus, A., and Shotwell, S. L., 1982, Monoclonal antibodies against the *Drosophila* nervous system, *Proc. Natl. Acad. Sci. USA* **79:**7929–7933.

Futerman, A. H., Low, M. G., and Silman, I., 1983, A hydrophobic dimer of acetylcholinesterase from Torpedo californica electric organ is solubilized by phosphatidylinositol-specific phospholipase C, *Neurosci. Lett.* **40:**85–89.

Ghysdael, J., Hubert, H., Travnicek, M., Bolognesi, D. P., Burny, A., Cleuter, Y., Kettman, G. F. R., Marbaix, G., Portelle, D., and Chantrenne, H., 1977, Frog oocytes synthesize and completely process the precursor polypeptide to virion structural proteins after microinjection of avian myeloblastosis virus RNA, *Proc. Natl. Acad. Sci. USA* **74:**3230–3234.

Grassi, J., Vigny, M., and Massoulie, J., 1982, Molecular forms of acetylcholinesterase in bovine caudate nucleus and superior cervical ganglion: Solubility properties and hydrophobic character, *J. Neurochem.* **38:**457–469.

Graybiel, A. M., and Ragsdale, C. W., 1982, Pseudocholinesterase staining in the primary visual pathway of the macaque monkey, *Nature (Lond.)* **299:**439–441.

Greenfield, S., 1984, Acetylcholinesterase may have novel functions in the brain, *Trends Neurosci.* **7:**364–368.

Gurdon, J. B., Lane, C. D., Woodland, H. R., and Marbaix, G., 1971, Use of frog eggs and oocytes for the study of messenger RNA and its translation in living cells, *Nature (Lond.)* **233:**177–182.

Gurwitz, D., Razon, N., Sokolovsky, M., and Soreq, H., 1984, Expression of muscarinic receptors in primary brain tumors, *Dev. Brain Res.* **14:**61–70.

Hall, J. C., 1982, Genetics of the nervous system in Drosophila, *Q. Rev. Biophys.* **15:**3–479.

Hall, J. C., and Kankel, D. R., 1976, Genetics of acetylcholiesterase in *Drosophila melanogaster, Genetics* **83:**517–535.

Hall, L. M. C., Mason, P. J., and Spierer, P., 1983, Transcripts genes and bands in 315 Kb of Drosophila DNA, *J. Mol. Biol.* **169:**83–96.

Herman, R. K., and Kari, C. K., 1985, Muscle-specific expression of a gene affecting acetylcholinesterase in the nematode caenorhabditis elegans, *Cell* **40:**509–514.

Hilliker, A. J., Clark, S. H., Chovnick, A., and Gelbart, W. M., 1980, Cytogenetic analysis of chromosomal region adjacent to the rosy locus in *Drosophila* melanogaster, *Genetics* **95:**95–110.

Huynh, T. V., Young, R. A., and Davis, R. W., 1985, Constructing and screening cDNA libraries in Lambda-gt10 and Lambda-gt11, in: *DNA Cloning Techniques: A Practical Approach* (D. Glover, ed.), pp. 49–78, IRL Press, Oxford.

Johnson, C. D., and Russell, R. L., 1975, A rapid, simple radiometric assay for cholinesterase, suitable for multiple determinations, *Anal. Biochem.* **64:**229–238.

Johnson, C. D., and Russell, R. L., 1983, Multiple molecular forms of acetylcholinesterase in the nematode caenorhabditis elegans, *J. Neurochem.* **41:**30–46.

Johnson, C. D., Duckette, J. G., Culotti, J. G., Herman, R. K., Meneley, P. M., and Russell, R. L., 1981, An acetylcholinesterase-deficient mutant of the nematode caenorhabditis elegans, *Genetics* **97:**261–279.

Karnovsky, M. J., and Roots, L., 1964, A "direct-coloring" thiocholine method for cholinesterase, *J. Histochem. Cytochem.* **12:**219–221.

Kimhi, Y., Mahler, A., and Saya, D., 1980, Acetylcholinesterase in mouse neuroblastoma cells: Intracellular and released enzyme, *J. Neurochem.* **34:**554–559.

Koelle, G. B., and Friedenwald, J. S., 1949, A histochemical method for localizing cholinesterase activity, *Proc. Soc. Exp. Biol. Med.* **70:**617–622.

Koelle, W. A., Smyrl, E. G., Ruch, G. A., Siddons, V. E., and Koelle, G. B., 1977, Effects of protection of butyrylcholinesterase on regeneration of ganglionic acetylcholinesterase, *J. Neurochem.* **28:**307–311.

Labarca, C., and Paigen, K., 1977, mRNA-directed synthesis of catalytically active mouse beta-glucuronidase in *Xenopus* oocytes, *Proc. Natl. Acad. Sci. USA* **74**:4462–4465.

Lane, C. D., 1983, The fate of genes, messengers and proteins introduced into *Xenopus* oocytes, *Current Rev. Dev. Biol.* **18**:89–119.

Layer, P. G., 1983, Comparative localization of acetylcholinesterase and pseudocholinesterase during morphogenesis of the chicken brain, *Proc. Natl. Acad. Sci. USA* **80**:6413–6417.

Lazar, M., and Vigny, M., 1980, Modulation of the distribution of acetylcholinesterase molecular forms in a murine neuroblastoma x sympathetic ganglion cell hybrid line, *J. Neurochem.* **35**:1067–1079.

Llinas, R., Greenfield, S. A., and Jansen, H., 1984, Electrophysiology of para-compacta cells in the *in vitro* substantia nigra—A possible mechanism for dendritic release, *Brain Res.* **294**:127–132.

Lockridge, O., 1982, Substance P hydrolysis by human serum cholinesterase, *J. Neurochem.* **39**:106–110.

Lockridge, O., 1984, Amino acid composition and sequence of human serum cholinesterase: A progress report, in: *Cholinesterase Fundamental and Applied Aspects* (M. Brzin, E. A. Barnard, and D. Sket, eds.), pp. 5–12, Walter de Gruyter, Berlin.

Lovrien, E. W., Magenis, R. E., Rivas, M. L., Lamvik, N., Rowe, S., Wood, J., and Hemmerling, J., 1978, Serum cholinesterase E_2 linkage analysis: Possible evidence for localization to chromosome 16, *Cytogenet. Cell, Genet.* **22**:324–326.

Macphee-Quigley, K., Taylor, P., and Taylor, S. S. 1985, Sequence of selected peptides from *Torpedo californica* acetylcholinesterase, *Fed. Proc.* **44**:1068.

Maniatis, T., Fritsch, E. F., and Sambrook, J., 1982, *Molecular Cloning, A Laboratory Manual*, Cold Spring Harbor Laboratory, Cold Spring Harbor, N.Y.

Marbaix, G., and Huez, G., 1980, Expression of messenger RNA injected into *Xenopus* oocytes, in: *Transfer of Cell Constituents into Eukaryotic Cells* (J. E. Celis, A. Graessmann, and A. Loyter, eds.), pp. 347–381, Plenum Press, New York.

Marsh, D., Grassi, J., Vigny, M., and Massulie, J., 1984, An immunological study of rat acetylcholinesterase: Comparison with acetylcholinesterase from other vertebrates, *J. Neurochem.* **43**:204–213.

Massoulie, J., and Bon, S., 1982, The molecular forms of cholinesterase and acetylcholinesterase in vertebrates, *Annu. Rev. Neurosci.* **5**:57–106.

Massoulie, J., Bon, S., Lazar, M., Grassi, J., Marsh, D., Meflah, K., Toutant, J. P., Vallette, F., and Vigny, M., 1984, The polymorphism of cholinesterases: Classification of molecular forms; interactions of solubilization characteristics; metabolic relationships and regulations, in: *Cholinesterases, Fundamental and Applied Aspects* (M. Brzin, E. A. Barnard, and D. Sket, eds.), pp. 73–97, Walter de Gruyter, Berlin.

Meflah, K., Bernard, S., and Massoulie, J., 1984, Interactions with lectins indicate differences in the carbohydrate composition of the membrane-bound enzymes acetylcholinesterase and 5′-nucleotidase in different cell types, *Biochimie* **66**:59–69.

Miller, C. A., and Benzer, S., 1983, Monoclonal antibody cross-reaction between *Drosophila* and human brain, *Proc. Natl. Acad. Sci. USA* **80**:7641–7645.

Miskin, R., and Soreq, H., 1981, Microinjected *Xenopus* oocytes synthesize active human plasminogen activator, *Nucl. Acid Res.* **9**:3355–3364.

Muller, F., Dumez, Y., and Massoulie, J., 1985, Molecular forms and solubility of acetylcholinesterase during the embryonic development of rat and human brain, *Brain Res.* **331**:295–302.

Myklebost, O., Williamson, B., Markham, A. F., Myklebost, S. R., Rogers, J., Woods, D. E., and Humphries, S. E., 1984, The isolation and characterization of cDNA clones for human apolipoprotein CII, *J. Biol. Chem.* **259**:4401–4404.

Ord, G. M., and Thompson, R. H. S., 1952, Pseudocholinesterase activity in the central nervous system, *Biochem. J.* **51**:245–251.

Parvari, R., Pecht, I., and Soreq, H., 1983, A microfluorometric assay for cholinesterases, suitable for multiple kinetic determinations of picomoles of released thiocoline, *Anal. Biochem.* **133**:450–456.

Paulus, J. P., Maigen, J., and Keyhani, E., 1981, Mouse megakaryocytes secret acetylcholinesterase, *Blood* **58**:1100–1106.

Razon, N., Soreq, H., Roth, E., Bartal, A., and Silman, I., 1984, Characterization of levels and forms of cholinesterases in human primary brain tumors, *Exp. Neurol.* **84**:681–695.

Reyes, A. A., Johnson, M. J., Schold, M., Ito, H., Ike, Y., Morin, C., Itakura, K., and Wallace, R. B., 1981, Identification of an H-2K^b-related molecules by molecular cloning, *Immunogenetics* **14**:383–397.

Ricciardi, R. P., Miller, J. S., and Roberts, B. E., 1979, Purification and mapping of specific mRNA by hybridization selection and cell free translation, *Proc. Natl. Acad. Sci. USA* **76**:4927–4931.

Rieger, F., Faivre-Bauman, A., Benda, P., and Vigny, M., 1976, Molecular forms of acetylcholinesterase: Their de novo synthesis in mouse neuroblastoma cells, *J. Neurochem.* **27**:1059–1063.

Rosenberry, T. L., and Scoggin, D. M., 1984, Human erythrocyte acetylcholinesterase is an amphipathic protein whose short membrane-binding domain is removed by papain digestion, *J. Biol. Chem.* **250**:5643–5652.

Rotundo, R. L., 1984, Asymmetric acetylcholinesterase is assembled in the Golgi apparatus, *Proc. Natl. Acad. Sci. U.S.A.* **81**:479–483.

Schaffer, N. K., Michel, H. O., and Bridges, A. F., 1973, Amino acid sequence in the region of the reactive serine residue of electric eel acetylcholinesterase, *Biochemistry* **12**:2946–2950.

Shilo, B., and Weinberg, R. A., 1981, DNA sequences homologous to vertebrate oncogenes are conserved in *Drosophila melanogaster*, *Proc. Natl. Acad. Sci. USA* **78**:6789–6792.

Silver, A., 1974, *The Biology of Cholinesterases*, North-Holland, Amsterdam.

Sippel, T. O., 1981, Microfluorometric analysis of protein thiol groups with a coumarinylphenyl-maleimide, *J. Histochem. Cytochem.* **29**:1377–1381.

Smilowitz, H., 1980, Routes of intracellular transport of acetylcholine receptor and esterase are distinct, *Cell* **19**:237–244.

Sorenson, K., Gentinetta, R., and Brodbeck, U., 1982, An amphiphile-dependent form of human brain caudate nucleus acetylcholinesterase: Purification and properties, *J. Neurochem.* **39**:1050–1060.

Soreq, H., 1985, The biosynthesis of biologically active proteins in mRNA-injected *Xenopus* oocytes, *CRC Crit. Rev. Biochem.* **18**:199–238.

Soreq, H., Parvari, R., and Silman, I., 1982, Biosynthesis and secretion of active acetylcholinesterase in *Xenopus* oocytes microinjected with mRNA from rat brain and from *Torpedo* electric organ, *Proc. Natl. Acad. Sci. USA* **79**:830–835.

Soreq, H., Miskin, R., Zutra, A., and Littauer, U., 1983, Modulation in the levels and localization of plasminogen activator in differentiating neuroblastoma cells, *Dev. Brain Res.* **7**:257–269.

Soreq, H., Zevin-Sonkin, D., and Razon, N., 1984, Expression of cholinesterase gene(s) in human brain tissues: Translational evidence for multiple mRNA species, *EMBO J.* **3**:1371–1375.

Soreq, H., Zevin-Sonkin, D., Avni, A., Hall, L. M. C., and Spierer, P., 1985, A human acetylcholinesterase gene identified by homology to the *Drosophila* gene, *Proc. Natl. Acad. Sci. USA* **82**:1827–1831.

Sparkes, R. S., Field, L. L., Sparkes, M. C., Crist, M., Spence, M. A., James, K., and Garry, P. J., 1984, Genetic linkage studies of transferrin, pseudocholinesterase and chromosome 1 loci, *Hum. Hered.* **34:**96–100.

Spierer, P., Spierer, A., Bender, W., and Hogness, A. S., 1983, Molecular mapping of genetic and chrommomeric units in *Drosophila melanogaster, J. Mol. Biol.* **168:**35–50.

Steiger, S., Brodbeck, U., Reber, B., and Brunner, J., 1984, Hydrophobic labeling of the membrane binding domain of acetylcholinesterase from Torpedo marmorata, *FEBS Lett.* **168:**231–233.

Suggs, S. V., Wallace, R. B., Hirose, T., Kawashima, E. H., and Itakura, K., 1981, Use of synthetic oligonucleotides as hybridization probes: Isolation of cloned cDNA sequences for human β_2-microglobulin, *Proc. Natl. Acad. Sci. USA* **78:**6613–6618.

Venter, J. C., Eddy, B., Hall, L. M., and Fraser, C. M., 1984, Monoclonal antibodies detect the conservation of muscarinic cholinergic receptor structure from *Drosophila* to human brain and detect possible structure homology with alpha-adrenergic receptors, *Proc. Natl. Acad. Sci. USA* **81:**272–276.

Vigny, M., Gisiger, V., and Massoulie, J., 1978, "Nonspecific" cholinesterase and acetylcholinesterase in rat tissues: Molecular forms, structural and catalytic properties, and significance of the two enzyme systems, *Proc. Natl. Acad. Sci. USA* **75:**2588–2592.

Wallace, R. B., Johnson, M. J., Hirose, T., Miyake, T., Kawashima, E. H., and Itakura, K., 1981, The use of synthetic oligonucleotides as hybridization probes. II. Hybridization of oligonucleotides of mixed sequence to rabbit beta-globin DNA, *Nucl. Acid Res.* **9:**879–894.

Wilson, B. W., and Walker, C. R., 1974, Regulation of newly synthesized acetylcholinesterase in muscle cultures treated with diisopropylfluorophosphate, *Proc. Natl. Acad. Sci. USA* **71:**3194–3198.

Zakut, H., Matzkel, A., Schejter, E., Avni, A., and Soreq. H., 1985, Polymorphism of acetylcholinesterase in discrete regions of the developing human fetal brain, *J. Neurochem.* **45:**382–389.

Zevin-Sonkin, D., Avni, A., Zisling, R., Koch, R., and Soreq, H., 1985, Expression of acetylcholinesterase gene(s) in the human brain: Molecular cloning evidence for cross-homologous sequences, *J. Physiol. (Paris)* **80:**221–228.

Zingde, S., Rodrigues, V., Joshi, S. M., and Krishnan, K. S., 1983, Molecular properties of *Drosophila* acetylcholinesterase, *J. Neurochem.* **41:**1243–1252.

7

Genes and Gene Families Related to Immunoglobulin Genes

GLEN A. EVANS

1. INTRODUCTION

The vertebrate immune response involves two major classes of effector cells: T lymphocytes and B lymphocytes. In recent years, the molecules involved in antigen recognition and cell–cell interactions of T and B lymphocytes have been defined at the molecular level. One of the remarkable results of this analysis has been the realization that many of the cell-surface molecules involved in the immune response demonstrate structural similarities at the protein and gene levels. While the surface molecules of T and B cells are functionally diverse, similarities in their structure have suggested that they may have been derived from a single evolutionary precursor.

This chapter discusses the molecules that are related to immunoglobulins and that can be grouped into what is now known as the immunoglobulin superfamily. A gene family is a group of homologous genes encoding products with similar functions. By contrast, a superfamily is a group of gene families or single genes that may encode unrelated functions but that can be grouped together based on similarities in structure. The immunoglobulin superfamily is named for its first

GLEN A. EVANS • Cancer Biology Laboratory, The Salk Institute, San Diego, California 92138.

described members: the immunoglobulin λ, κ, and heavy-chain gene families. However, the immunoglobulin superfamily is now known to include the α-, β-, and γ-chains of the T-cell antigen receptor, the class I and class II gene families of the major histocompatibility complex, β_2-microglobulin, the cell-surface receptor for polymeric immunoglobulins, the T-cell accessory molecules T8 and T4, and the Thy-1 and OX-2 antigens. The discussion focuses on the similarity of these gene products to immunoglobulins emphasizing the use of a single structural motif, the immunoglobulin homology unit, in various functional contexts.

Why are genes that function in immune reactions of importance to molecular neurobiology? First, the analysis of the immunoglobulin superfamily provides an example of the application of molecular techniques that could be used to define genes of importance in neurological development. The production of antisera and monoclonal antibodies permitted the initial definition and purification of many of the cell surface antigens of T and B cells. Recombinant DNA technology then provided the ability to clone genes that encoded these antigens, rapidly determine their sequence and structural relationships, and make defined structural changes within the genes *in vitro*. DNA-mediated gene transfer and similar gene-transfer techniques permitted cloned genes to be stably introduced into the genome of differentiated cells in culture so that the biological significance of *in vitro* manipulations could be assessed. The combination of gene transfer, genetic manipulation, and *in vitro* biological assays, such as T-cell-mediated cytotoxic assays, permitted the precise correlation of molecular structure with function. More recently, the injection of cloned genes into mouse embryos and the creation of transgenic strains of mice has provided an *in vivo* extension to this gene transfer technology (Le Meur *et al.*, 1985).

Second, not all members of the immunoglobulin superfamily are clearly related to the immune response. For instance, two members of the superfamily, Thy-1 and OX-2, are expressed in the mammalian nervous system. Thy-1 is, in fact, the major cell-surface glycoprotein found in rodent brain (Williams and Gagnon, 1982) and may be functionally important in neural cell interactions. As additional neural antigens are defined through the use of monoclonal antibodies (Zipursky *et al.*, 1984; Mirsky *et al.*, 1982; Schachner *et al.*, 1983) and characterized through molecular cloning, additional members of the immunoglobulin superfamily may be found as nervous system antigens.

Third, the immunoglobulin superfamily provides a paradigm of a single structural unit diversified for various functions related to cellular interactions. Other gene families important to brain function and development, such as those encoding neurotransmitter receptors, ion channels, and adhesion structures, are likely to show similar functional diversification of a single structural unit. Finally, cells carrying primitive immunoglobulinlike genes have been proposed as the primordial sensory cells which, through evolution, could give rise to both immune and nervous systems (Williams, 1985). The investigation of molecular structure may ultimately clarify how complex recognition and information processing systems evolved.

The Immunoglobulin Superfamily

Table I is a compilation of genes or gene families whose detailed molecular structure suggests a relationship to immunoglobulins. A schematic diagram of the structure of each gene product, showing regions of similarity to immunoglobulin V or C domains, is shown in Fig. 1. For each, the structures of the gene or gene product have been determined by genomic or cDNA sequencing. Where the structure of the gene has been analyzed, the intron–exon organization is similar to that of immunoglobulin genes (Fig. 2). With some gene families, in particular the class I and class II molecules of the major histocompatibility complex, gene transfer, *in vitro* recombination, and site-specific mutagenesis has provided a detailed correlation of primary amino acid sequence with biological function. Only in the cases of the immunoglobulins, and to a lesser extent, the class I major histocompatibility antigen, has the three dimensional structure been determined through X-ray crystallography (Amzel and Poljak, 1979; Hood *et al.*, 1983). While Table I includes genes or gene families for which sequence data are currently available, several additional genes may ultimately be included in this superfamily, in particular the T4 surface molecule of T lymphocytes (Littman *et al.*, 1985) and the mast cell/macrophage receptors for immunoglobulin (Williams, 1985) and some cell surface antigens of the mammalian nervous system. It should also be noted that not all surface molecules of lymphocytes have immunoglobulinlike structures. The subunits of the T3 surface molecule of T lymphocytes (Van den Elsen *et al.*, 1985) and the L-CA/T200 antigen (Thomas *et al.*, 1985) have been cloned and sequenced and demonstrate virtually no structural similarity to immunoglobulins.

2. IMMUNOGLOBULIN GENES AND THE IMMUNOGLOBULIN DOMAIN

The major function of the immunoglobulin molecule is to bind specifically to foreign antigen and to effect inactivation or expulsion of that substance from the body. In producing proteins that perform this function, immunoglobulin genes have developed a remarkable ability to gen-

Table I. The Mouse Immunoglobulin Superfamily

Gene/Family	No. of chromosomal genes	Location[a]	Molecular weight (kd)	Subunit structure	Function	Expression
Ig heavy chain	100's V, 8C	12	52	H_2L_2 tetramer	Antigen-binding/removal	B lymphocytes, plasma cells
λ	2V, 4C	16	23	H_2L_2 tetramer	Antigen-binding removal	B lymphocytes, plasma cells
κ	100's V, 1C	6	23	H_2L_2 tetramer	Antigen-binding removal	B lymphocytes, plasma cells
TCR α	10's V, 2C	14	40–45	α–β heterodimer	Antigen-binding/MHC recognition	T lymphocytes
β	10's V, 2C	6	40–45	α–β heterodimer	Antigen-binding/MHC recognition	T lymphocytes
γ	?V, 2C	13	40–45	?	?	Cytotoxic T lymphocytes
Class I MHC						
H-2	2–3	17	45	$H-2/b_2$-m heterodimer	Restrictive recognition (cytotoxic T cells)	All somatic cells
Qa/TLa	25–31	17	35–45	$H-2/b_2$-m heterodimer	?	Thymocytes, lymphocytes liver, testis
β₂-Microglobulin	1	2	12	$H-2/b_2$-m heterodimer	Restrictive recognition (cytotoxic T cells)	All somatic cells
Class II MHC						
α	2	17	30–33	α–β heterodimer	Restrictive recognition (helper T cells)	B lymphocytes/macrophages
β	2–3	17	27–29			
Poly Ig receptor	1	?	80	Single subunit	Transepithelial transport of Ig	Gut/mammary epithelial cells
T4 (L3T4)	1	?	55	Homodimer	?/MHC recognition	Helper T lymphocytes
T8 (Lyt-2)	1	6	38	Homodimer	?/MHC recognition	Cytotoxic T lymphocytes
Thy-1	1	9	25	Single subunit	?	T lymphocytes (mouse only) neural cells/fibroblasts
OX-2	1	?	47	Single subunit	?	thymocytes, neural cells, endothelium, some B cells, follicular dendritic cells

[a] Mouse chromosomal location.

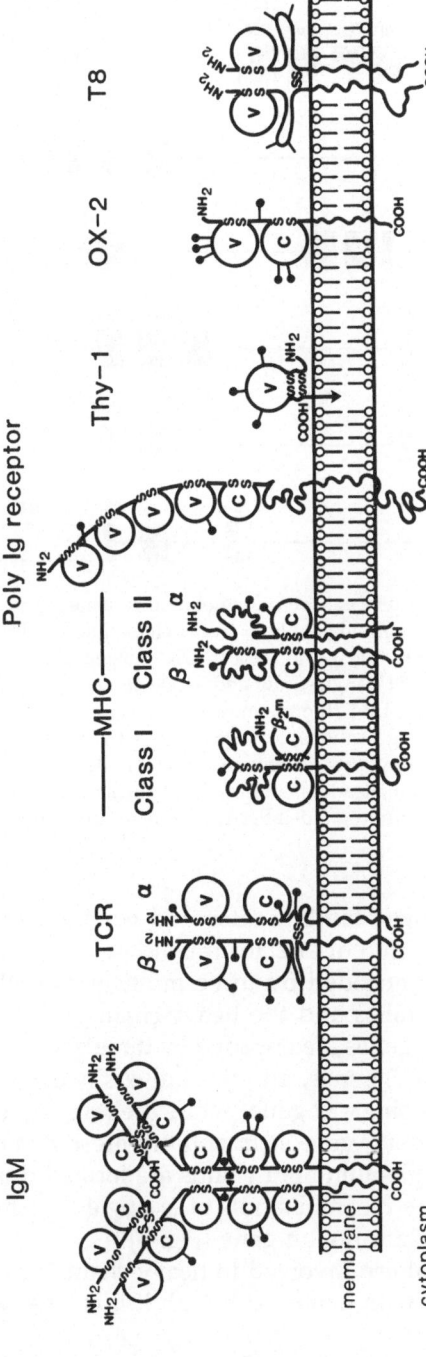

Figure 1. Cell-surface molecules of the immunoglobulin superfamily. The structures of gene products listed in Table I are shown diagrammatically. Circular structures indicate domains with homology to immunoglobulin variable (V) or constant (C) domains with central disulfide bridges (SS). Blackened circles indicate the probable location of N-linked carbohydrate. The locations of O-linked carbohydrate in the T8 molecule are indicated by solid lines. The attachment of the Thy-1 glycoprotein to the cytoplasmic membrane is through a carboxy-terminal lipid, indicated by an arrow.

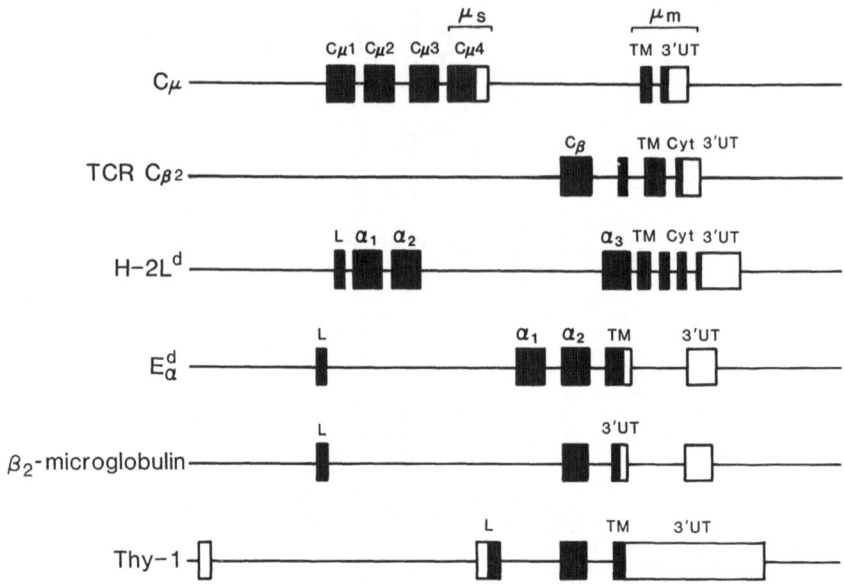

Figure 2. Structural relationships between immunoglobulinlike genes. The intron–exon structure of genes encoding the immunoglobulin μ heavy chain (Honjo, 1983), the T-cell receptor β2 constant region (Davis, 1985), the class I MHC antigen H-2Ld (Evans et al., 1982a), the class II MHC gene Eαd (Mengle-Graw and McDevitt, 1985), β2-microglobulin (Parnes and Seidman, 1982), and Thy-1 (Ingraham et al., 1985) are shown. In the Thy-1 gene, the first exon has two differing locations, corresponding to two promoters (Ingraham and Evans, 1986) (■) Exons encoding protein sequence; (□) encode 5′ or 3′ untranslated mRNA sequences. L, exons encoding leader peptides; TM, transmembrane domains; Cyt, cytoplasmic peptides; 3′UT, 3′ untranslated mRNA. The remaining exons encode immunoglobulinlike domains.

erate diversity in the antigen-binding sites and couple these binding sites with one of several different effector functions.

Immunoglobulins are encoded by three multigene families, the λ and κ light-chain gene families and the heavy-chain gene family (Fig. 3). The λ and κ light-chain genes are encoded by three genetic elements: variable (V) and constant (C) genes, and (J), a diversity element. In the mouse, the κ family has a single C gene, while the λ family has four C genes. Hundreds of different V gene segments are present in the mouse germline. Active immunoglobulin light chains are formed during B-cell differentiation through the recombination of a single V gene element with a J element and constant-region gene (Fig. 3).

The recombination pattern involved in heavy-chain rearrangement is similar to that of light-chain genes, although it involves two recom-

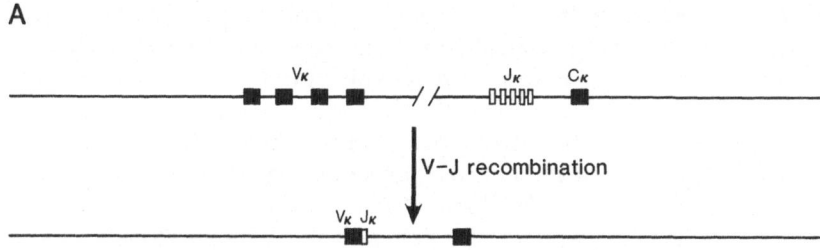

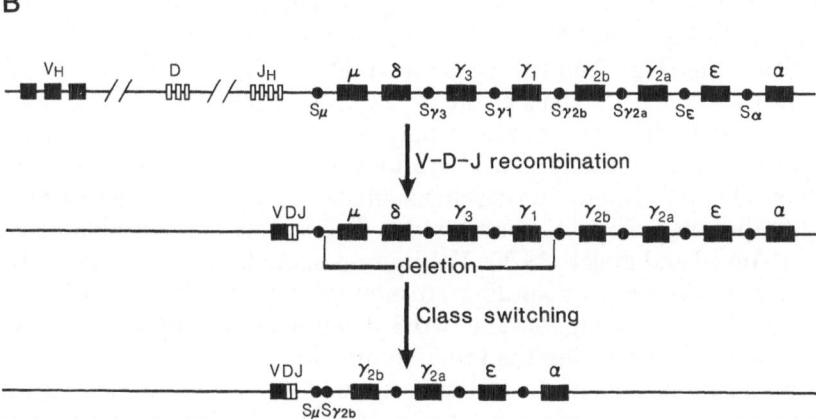

Figure 3. Developmental rearrangements of immunoglobulin genes. (A) The light-chain gene, in this case the mouse κ gene, consists of a single constant (C) region and multiple variable (V) regions. Rearrangement takes place between a single V_k and J_k to generate an active κ gene. (B) The mouse immunoglobulin heavy-chain locus consists of multiple V regions and eight functionally distinct C genes, with different effector functions. Initial recombination takes place between single V, J, and D segments to generate an active heavy-chain gene producing IgM. Later, recombination between the μ switch region and another immunoglobulin switch region replaces the μ constant-region gene with another constant-region gene. Boxes represent protein encoding regions. Circles represent the sites of recombination during class switching.

bination events. The heavy-chain gene family has 8 C genes that are able to specify at least five different effector functions. The μ gene encodes the constant region of the IgM heavy-chain molecule, the γ genes the various isotypes of IgG, and the α and ε genes the IgA and IgE constant regions (Fig. 3). Through a complex series of developmental rearrangements and alterations in RNA splicing a single V region gene is brought together with a D and J segment and coupled to one or more C region genes giving several functional types of antibody. The first recombination event forms an active μ gene. Further recombination events between immunoglobulin "switch" sequences convert the μ gene to production of one of the other types of immunoglobulin. The details of this elegant process have been well described in several recent reviews (Max, 1984; Honjo, 1983; Wall and Kuehl, 1983).

The similarities between the genes of the immunoglobulin super-family and immunoglobulin genes are based on the similarities to a single type of structure: the immunoglobulin homology unit. With the exception of exons encoding several specialized domains, such as leader peptides, transmembrane domains, hinge regions, intracytoplasmic peptides, both light- and heavy-chain genes are composed of structurally similar exons (Figs. 2 and 3). Each exon is about 330 nucleotides in length with RNA splicing signals at the 5' and 3' junctions (Max, 1984). Each immunoglobulin exon encodes a polypeptide of about 110 amino acids in length with centrally placed cystines separated by about 65 amino acids. This polypeptide represents a single domain structure that folds into a discrete, compact subunit with a distinct three-dimensional structure (Amzel and Poljak, 1979). The immunoglobulin fold consists of two β-sheet structures surrounding an internal volume filled with hydrophobic side chains (Fig. 4). The two β sheets are joined by an intrachain disulfide bridge crossing the inner volume in a direction perpendicular to the plane of the sheets. Subunits encoded by exons of the V genes differ slightly from those encoded by C genes by the presence of an extra length of polypeptide forming a two-stranded loop (Fig. 4). The length of this loop varies among different V-region domains.

Nonimmunoglobulin genes that demonstrate similarities to the immunoglobulin gene exon or the immunoglobulin polypeptide domain are considered members of the immunoglobulin superfamily. These genes contain exons encoding immunoglobulinlike homology units of about 110 amino acids in length and with centrally placed cystines. Like immunoglobulins, mRNA splicing takes place between the second and third nucleotide of the junctional codon (Sharp, 1981; Hood et al., 1983). In many cases, amino acid sequence homology has been found between immunoglobulin V or C regions, particularly in areas surrounding the cystine residues. Where DNA sequence homology is absent, the domains still retain the same length and spacing between cystine residues. It is assumed that a similarity in amino acid sequence implies a similarity in three-dimensional structure. Although limited data from X-ray crystallography is available only for the class I MHC antigen, it indicates a twofold axis of symmetry. As an additional indication of similarity in

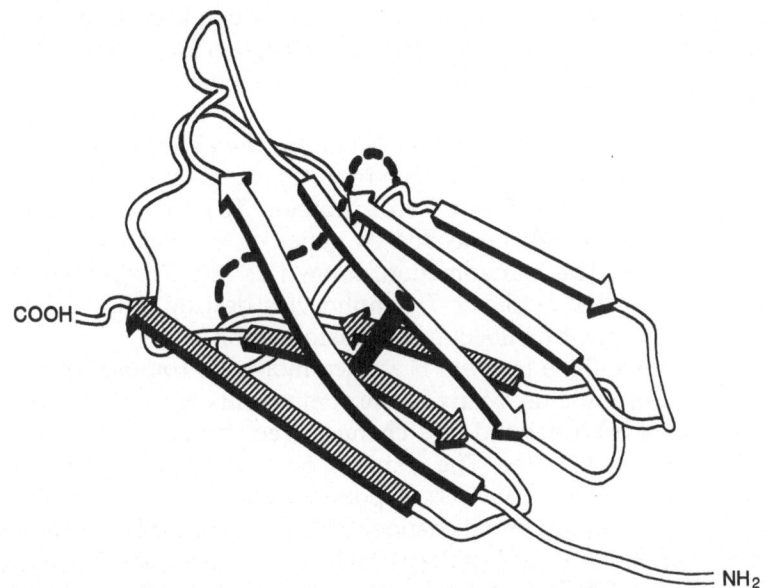

Figure 4. The immunoglobulin domain. Schematic diagram of the α-carbon backbone of the immunoglobulin C_L homology unit containing two planes of parallel β-sheet structure. The positions of the intrachain disulfide bridge is shown by a dark band perpendicular to the planes of the β-sheet structure. Dotted lines indicate the approximate position of an additional loop of polypeptide chain characteristic of V domains. (Adapted from Amzel and Poljak, 1979.)

three-dimensional structure, computer analysis of secondary structure has been carried out. Predictive computer analysis, although only marginally accurate, often demonstrates centrally placed β-sheet structure analogous with the immunoglobulin fold (Novotny and Auffray, 1984). In many cases, sequence and/or structural similarities can be shown to be more similar to V domains or C domains based on the presence of the additional V loop sequence (Fig. 1).

3. THE T-LYMPHOCYTE CELL-SURFACE RECEPTOR FOR ANTIGEN

Just as the major function of antibody is the recognition of foreign antigens and their removal from the organism, the major function of cytotoxic T lymphocytes is the recognition and destruction of foreign cells, or cells modified by infection with virus, malignant transformation, or parasitic organisms. This form of antigen recognition is mediated by cytotoxic T lymphocytes with cell-surface receptors for antigen. The recognition of antigen by T lymphocytes is a fundamentally different process from the recognition of antigen by antibody in one important sense. While antibody recognizes antigen alone, in the absence of other molecular structures, the T-cell antigen receptor recognizes antigen only in the context of host-specific, or "self," proteins. The "self" molecules are cell-surface glycoproteins encoded by genes of the major histocompatibility complex. This phenomenon, known as MHC-restrictive recognition, ensures that cytotoxic T lymphocytes destroy only "self" cells rendered abnormal by infection or transformation.

The T-cell antigen receptor is a macromolecular complex composed of a number of proteins. Many of the components of the T-cell antigen receptor have been defined and characterized through the generation of monoclonal antibodies. The complex contains binding components for antigen and MHC as well as components responsible for signal transmembrane and control of ion channels (Meuer et al., 1984). The antigen-binding components, and perhaps MHC-binding components as well, reside in a disulfide-linked dimer with subunits of 40,000 and 45,000 M_r (Fig. 1).

The analysis of the T-cell receptor antigen-binding subunits by recombinant DNA technology represents one of the major accomplishments of the molecular approach to the immune system in recent years. It is not surprising, however, that the structure of T-cell receptor subunits is remarkably similar to that of immunoglobulin. Like antibody, the α- and β-chains of the T-cell receptor are encoded in separate variable and constant region genes that are recombined during T-cell differentiation to produce one of many possible antigen binding sites (Davis, 1985). As with the immunoglobulin heavy chain genes, two diversity elements D and J are included in the recombination mechanism to generate additional diversity. However, immunoglobulin heavy chains, only two C genes are present and these do not appear to be functionally distinct. A number of additional differences between the T-cell receptor genes and immunoglobulins are that recombination may take place in the absence of a J or D gene (Hedrick et al., 1984), and the TCR V genes appear to be somewhat fewer in number (Barth et al., 1985). The detailed structure and rearrangement of the T-cell receptor α-and β-chain genes has been the subject of several recent reviews (Hood et al., 1985; Davis, 1985).

However, in spite of minor differences in the mechanisms of rearrangement, the T-cell receptor and immunoglobulin chains are clearly closely related (Hedrick et al., 1984). Figure 4 shows a sequence comparison of the β-TCR receptor chain with variable and constant region sequences of immunoglobulins. Significant sequence homology is seen particularly in regions surrounding the cystines forming disulfide bridges (Fig. 5). In addition, sequence homology occurs among many of the amino acids forming the β-sheet structures of the immunoglobulin fold. Therefore, the three-dimensional structure of the T-cell receptor subunits will undoubtedly be similar to that of the immunoglobulin light chains.

During the isolation of cDNAs encoding the α- and β-subunits of the T-cell antigen receptor by subtractive hybridization, a third gene was identified that rearranged during T-cell differentiation and showed significant sequence homology to the TCR α- and β-chains (Saito et al., 1984). This molecule, termed the TCR γ-chain, is preferentially expressed in cytotoxic T lymphocytes (Kranz et al., 1985) and may form part of a second type of T-cell receptor complex. This chain also has extensive sequence homology to immunoglobulin V and C genes, although its function has not been established.

4. CLASS I MHC GENES

4.1. H-2 K, D, and L

The transplantation antigens H-2K, D, and L are cell-surface glycoproteins found on virtually all somatic cells of the mouse. These molecules consist of a 44,000-M_r heavy chain in noncovalent association with a 12,000-M_r light chain, β_2-microglobulin. In contrast to the invariant light chain, the heavy chain shows extreme structural polymorphism with more than 100 alleles found at each locus. The polymorphic class I molecule is recognized in context with antigen by the cytotoxic T lymphocyte and serves as a marker for "self." Different allelic forms of the transplantation antigen are preferentially recognized with different viral antigens. For instance, VSV infected cells are killed much more efficiently by stimulated cytotoxic T cells when the H-2L^d is on the surface of the target cell than when the H-2K^d molecule is present (Fig. 6). Rejection of allografts, transplantation between genetically nonidentical members of the same species, is also mediated by cytotoxic T lymphocytes recognizing cell-surface class I MHC antigens.

The transplantation antigen heavy-chain genes are located on chromosome 17 of the mouse in a cluster of immunologically important genes known as the major histocompatibility complex (MHC) (Fig. 7). Many, though not all, of the genes of the MHC are composed of immunoglobulinlike homology units. Transplantation antigens are constructed of

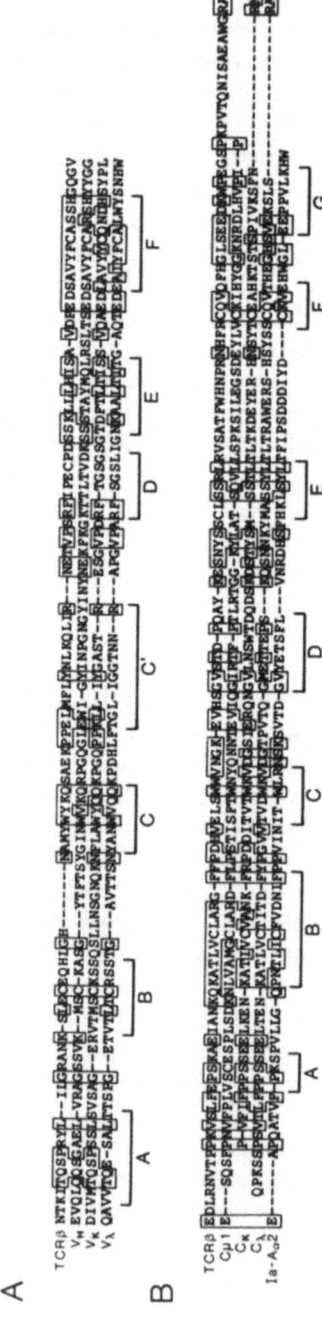

Figure 5. Sequence comparisons of a T-cell receptor β-subunit sequence with immunoglobulin variable and constant genes. The sequence of the T-cell receptor clone TM86 for the β-subunit is shown in alignment with sequences of V_H, V_κ, V_λ, C_μ, C_κ, C_λ, and constant-region sequence of the class II MHC antigen A_{α_2} gene. Boxes indicate conserved amino acid residues and underlined regions indicate the amino acids involved in the β-sheet structures of the immunoglobulin fold (Fig. 4). Sequences are from Hedrick *et al.* (1984) (TCR-β) and Kabat *et al.* (1983).

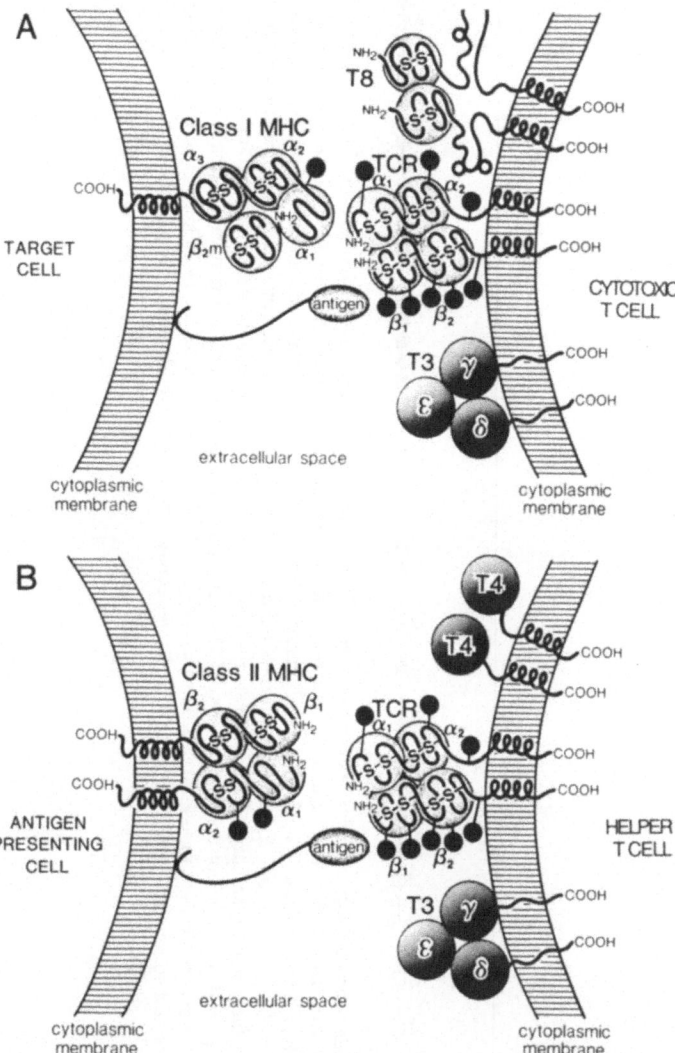

Figure 6. Cell-surface interactions in restrictive recognition. (A) The interaction of cytotoxic T lymphocytes with a target cell involves several molecules constructed of immunoglobulinlike domains, including the T-cell receptor subunits, the class I MHC molecule, β_2-microglobulin, and the T8 accessory molecule. (B) The interaction of a helper T lymphocyte with an antigen presenting cell likewise involves the interaction of the T-cell antigen receptor, the class II MHC molecule, and the T4 accessory molecule. The T3 subunits are not related to the immunoglobulin superfamily. (●) Probable locations of N-linked carbohydrate; (○) O-linked carbohydrate.

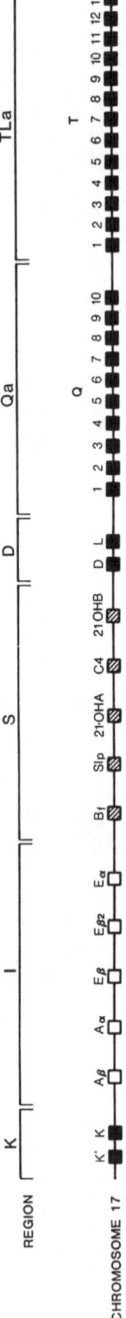

Figure 7. The mouse major histocompatibility complex. The genes of the mouse histocompatibility complex include the class I genes (dark boxes), class II (open boxes), and class III (hatched boxes). Class I genes include the transplantation antigens H-2K, D, and L, as well as genes of the Qa region, Q 1–10, and TLa region, T 1–13. The class II genes, located in the I region, include the subunits of the Ia-A molecule, A_α and A_β, and the subunits of the Ia-E molecule, E_α and E_β. Both class I and class II molecules are related by structure to immunoglobulins. The class III genes, located in the S region, are structurally unrelated to immunoglobulins and encode the complement components B_f and C4, the sex-limited protein Slp, and two forms of the enzyme steroid 21-hydroxylase (White *et al.*, 1984). A general schematic of the mouse MHC is represented that may vary in different mouse strains. For instance, the C57BL10 mouse lacks the H-2L gene. The BALB/c mouse has 8 rather than 10 Qa genes and somewhat more TLa genes. The K′ gene may represent a pseudogene. (From Weiss *et al.*, 1984.)

three external domains of 91 amino acid residues, two of which contain central disulfide-linked cystine residues (see Fig. 1). The protein is anchored to the membrane by virtue of a hydrophobic transmembrane domain and a short carboxy-terminal peptide is present inside the cell membrane. Analysis of the structure of transplantation antigen genes (Steinmetz *et al.*, 1982; Evans *et al.*, 1982*a*; Kvist *et al.*, 1983) demonstrated that each of the three external domains was encoded by a separate exon, and the membrane proximal domain showed sequence homology to immunoglobulin C regions (Fig. 1). Therefore, the class I MHC molecule as well is composed of the same repetitive structure of immunoglobulin homology units.

Gene-transfer studies, coupled with *in vitro* recombination and mutagenesis studies, have been used to correlate the protein structure with functional domains. First, exon shuffling studies have demonstrated that the majority of polymorphic sites recognized by allospecific monoclonal antibodies are present in the outer two domains: α_1 and α_2 (Evans et al., 1982*b*; Arnold *et al.*, 1984). The membrane proximal domain, α_3, is relatively nonantigenic, although several monoclonal antibody-binding sites have been mapped to this domain (Ozato *et al.*, 1984). Likewise, most sites recognized by cytotoxic T lymphocytes are also present in the external two domains, although some determinants may be formed conformationally by the interation of the two domains. The function of the cytoplasmic tail is unclear, as it can be removed and the expressed protein is still functional (Murre *et al.*, 1984). Site-specific mutagenesis has shown that removal of one of the glycosylation sites does not affect surface expression (Shiroishi *et al.*, 1984). However, conversion of one of the cystines to arginine, preventing formation of the disulfide bridge in the α_2, disrupts most monoclonal antibody binding determinants and prevents recognition by cytotoxic T cells (Shiroishi *et al.*, 1985).

4.2. Qa/TLa Genes

One of the surprises to arise from molecular analysis of the class I gene of the major histocompatibility complex was that despite the presence of only three loci for transplantation antigens in the genome, Southern blot analysis demonstrated more than 30 class I genes in the genome (Margulies *et al.*, 1982; Winoto *et al.*, 1983). Transplantation antigen genes are therefore part of a large multigene family (Fig. 7). Further analysis demonstrated that all the additional class I genes mapped to two gene clusters in the Qa and TLa genetic regions, telomeric to the major histocompatibility complex on chromosome 17 of the mouse (Winoto *et al.*, 1983). Several of these genes encode the Qa2,3 antigen and the TLa

antigen, cell-surface molecules of undefined function restricted to sub-
types of lymphoid cells (Goodenow *et al.*, 1982). TLa (thymus-leukemia
antigen) is found on the surfaces of thymocytes and on some T lymphoma cells. Qa2,3 antigens are present on about 25% of thymus cells
and in 60–79% of splenic lymphocytes (Harris *et al.*, 1984). Both molecules consist of 35,000–45,000-M_r heavy chains also associated with β_2-
microglobulin. Other Qa antigenic specificities, Qa-1 through Qa-9, also
map genetically to the Qa2,3 or TLa loci, and it is likely that they represent the products of additional class I genes. However, the relationship
between these antigenic specificities and the large number of class I
genes in the Qa and TLa regions has not yet been established.

The class I genes of the Qa and TL clusters have been isolated and
mapped by cosmid mapping from two mouse strains. The C57B1/10
strain has 10 class I genes in the Qa2,3 cluster and 13 class I genes in
the TL cluster. The number of class I genes in the Qa and TLa clusters
of the BALB/c strain is slightly greater (Weiss *et al.*, 1983; Steinmetz *et
al.*, 1982). Although complete characterization of all the class I genes in
this cluster, and identification of their gene products, has not yet been
accomplished, the analysis of Qa and TL region genes to date has demonstrated several interesting findings.

First, the genes from these clusters that have been analyzed show
a high degree of sequence homology to the class I genes encoding transplantation antigens. At least two of these genes, however, by virtue of
a frame shift in the exon encoding the transmembrane domain, produce
class I molecules that lack a membrane-binding domain and are secreted
(Barra *et al.*, 1985; Devlin *et al.*, 1985; Lalanne *et al.*, 1985). The Q10 gene
(Fig. 7) has been shown to produce a secreted transplantation antigenlike
molecule expressed specifically in the liver (Cosman *et al.*, 1982). A
similar truncated class I molecule has been identified in mutagenized
human lymphoid cells (Krangel, 1985) and in the rat with expression
restricted to the testis (Kastern, 1985). The possible function of secreted
transplantationlike molecules is not clear. However, the Q10 gene has
been postulated to act by blocking class I gene recognition by the T-cell
antigen receptor, thus regulating the cell-mediated immune response
(Cosman *et al.*, 1982).

Second, the large pool of H-2-like genes in the genome has been
suggested as an explanation for the high degree of polymorphism seen
among class I MHC genes. Mutant strains of mice resulting in inappro-
priate graft rejection have been isolated and the sequences of the mutant
transplantation antigen genes determined. In most cases, the mutation
results from the change of several amino acid residues to that seen in
one of the other transplantation antigen genes. Molecular analysis of

several H-2K mutations (Weiss *et al.*, 1983; Mellor *et al.*, 1983) demonstrated that the mutant sequence was also found in one of the Qa regions genes and suggested that the H-2K mutations were the result of gene conversion events. Thus, a second function for the Qa/TLa region genes may be to serve as a pool of genetic information for the rapid generation of polymorphism among class I genes by gene conversion events. This is supported by the finding that some Qa region genes may be true pseudogenes, i.e., genetic remnants of previously active genes that, because of mutational drift, are no longer able to encode proteins.

4.3. β_2-Microglobulin

The light chain of the class I molecule is encoded by a single gene on chromosome 2 of the mouse. This small protein, β_2-microglobulin, is nonpolymorphic, with only two known alleles (Michaelson, 1983). The light chain is required, in most cases, for expression of the class I molecule, as demonstrated by somatic cell mutational analysis of the β_2-microglobulin gene (Hyman and Stallings, 1977). Analysis of the sequence of the β_2-microglobulin gene (Parnes and Seidman, 1982) demonstrated a major exon encoding an immunoglobulinlike domain containing a disulfide loop. The β_2-microglobulin gene also demonstrates mRNA splicing between the second and third base of the junctional codon, suggesting a common evolutionary history with the class I major histocompatibility antigens and immunoglobulins. In addition, significant sequence homology was observed between the β_2-microglobulin gene and the fourth exon (α_3) of the class I genes and immunoglobulin constant region genes. As with other immunoglobulin-related sequences, sequence homology is greatest surrounding the cystine residues.

5. CLASS II MHC GENES

The class II molecules of the major histocompatibility complex are similar in both structure and function to the class I MHC molecules. The class II transplantation antigens, also known as Ia antigens, are encoded by genes in the I subregion of the major histocompatibility complex (Fig. 7). Like the three types of class I molecule—H-2K, D, and L—two types of class II molecule, E and A, are encoded by four genes. The A molecule is a heterodimer composed of a 33,000-M_r heavy chain, A_α, and a 29,000-M_r light chain, A_β. Similarly, the E molecule is composed of E_α and E_β subunits. All four of the class II genes have been cloned and multiple alleles sequenced. Ia molecules are limited in distribution to the surfaces of B lymphocytes and antigen-presenting cells, including macrophages. Like class I genes, the class II genes exhibit extensive polymorphism such that the Ia chains of any two mouse strains can differ by as much as 10% in sequence (Hood *et al.*, 1983).

T cells can recognize foreign antigen only in the context of a class II antigen and respond when a specialized B cell or macrophage cell "presents" antigen in conjunction with a class II molecule of the appropriate allelic type (Fig. 6). Some antigens are preferentially presented with the E molecule; others are presented in the context of the A antigen. Thus, antigens are said to be A or E restricted. T cells activated in response to antigen and the appropriate class II molecule are required to generate a B-cell-mediated immune response and elaborate antibody. The net result is that strains of mice possessing the appropriate Ia molecule will respond to a particular antigen with an intense antibody response, while strains possessing a different Ia allele will respond less well or not at all. The class II genes, then, form the basis of individual variation in immune response and susceptibility to infection. The functional proof of this is that strains that are low responders to synthetic antigens can be converted to high responders by the creation of transgenic mouse strains using Ia genes of foreign haplotypes (Le Meur *et al.*, 1985).

Sequence analysis of class II genes demonstrate that, like the class I genes, they demonstrate structural similarity as well as sequence homology to the immunoglobulin families (Mengle-Gaw and McDevitt, 1985). Moreover, the α- and β-chain of the class II molecule presumably form a tetrameric structure on the cell surface, like the class I molecule in association with β_2-microglobulin (Fig. 1). The polymorphism of the external domains of the class II molecule may also be generated through assortment of class I MHC gene sequences through gene conversion or other forms of genetic exchange.

6. CELL-SURFACE RECEPTORS FOR TRANSEPITHELIAL TRANSPORT OF IMMUNOGLOBULIN

While most antibody remains in the blood, the polymeric antibodies IgM and IgA appear in external secretions. A receptor-mediated mechanism is responsible for the transport of antibodies across epithelial tissues and their secretion into external fluids. IgA and IgM interact with a cell-surface receptor at the sinusoidal surface of gut epithelial cells and are encapsulated with the receptor by means of endocytosis. The vesicles containing the receptor–immunoglobulin complex are transported across the epithelial cell and expelled by means of exocytosis at the luminal surface. The ligand-binding domain of the receptor, with bound immunoglobulin, is then released from the cell surface by proteolytic cleavage near the membrane-insertion site. The extracellular portion of the receptor is secreted with the secretory immunoglobulin. Thus, the immunoglobulin receptor can be used for only a single round of transport. The ligand binding domain of the receptor is found associated with immunoglobulin in external secretions and is known as secretory component (Mostov and Blobel, 1982).

A remarkable result of the application of molecular cloning to the study of secretory component was the finding that the cell-surface receptor for immunoglobulin is structurally similar to its ligand. The cDNA clones for the polymeric immunoglobulin receptor of rabbit liver and lactating mammary gland have been analysed and their sequence determined (Mostov *et al.*, 1984). The immunoglobulin receptor is synthesized as a 773-amino acid precursor containing an 18-amino acid signal peptide, an extracellular portion of 629 amino acids, a transmembrane domain of 23 amino acids, and a cytoplasmic tail of 103 amino acids. Interestingly, the external portion of the receptor consists of five highly conserved immunoglobulinlike domains, of 100 to 155 amino acids each, with centrally placed cystines presumably forming disulfide bonds. These external four domains show sequence homology to immunoglobulin V domains and the Thy-1 glycoprotein (Fig. 8), while the membrane proximal domain is most similar to immunoglobulin C regions. Thus, the receptor for transepithelial transport of immunoglobulins is itself structurally similar, and probably evolutionarily related, to immunoglobulins (Williams, 1984). Additional functionally distinct immunoglobulin receptors are present on other cell types, including the IgE receptor of mast cells, the IgG receptor for transplacental transport, and IgG receptors of macrophages and antibody-dependent cytotoxic cells. It would not be unexpected if many of these receptors were composed of immunoglobulinlike domains.

7. THE T-CELL ACCESSORY MOLECULES T4 AND T8

A major difference between the antibody molecule and the T-cell surface receptor for antigen is that the latter exists as part of a macromolecular complex. Two additional T-cell surface molecules, T4 and T8, may form part of this complex on functionally distinct T lymphocytes (Fig. 6). The T8 antigen, also known as CD8, Leu-2 (man), or Lyt-2

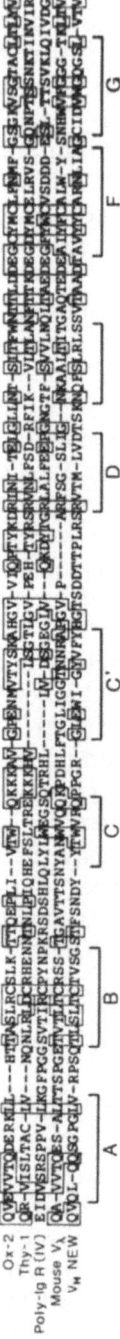

Figure 8. Sequence comparisons of immunoglobulin-related molecules Thy-1 and OX-2. An alignment of portions of the OX-2 antigen, Thy-1 antigen, polyimmunoglobulin receptor (exon IV), and mouse V_λ and V_H sequences are shown. Underlined amino acid residues represent portions of the sequence involved in β-sheet structure of the immunoglobulin fold (see Fig. 4). Sequences are from Clark et al., 1985 (OX-2), Williams and Gagnon, 1982 (Thy-1), Mostov et al., 1984 (poly-Ig receptor), and Kabat et al., 1983 (V_H and V_1).

(mouse), was one of the first mouse T-cell alloantigens described and is expressed predominantly on cytotoxic T lymphocytes. The T4 antigen, also known as Leu-3 (man) and L3/T4 (mouse) is restricted to helper T lymphocytes. Although the function of these molecules is not understood, antibodies to T8 can block T-cell-mediated cytotoxicity and antibodies to T4 block the T-helper response to antigen. It has been suggested that the T4 and T8 molecules play some role in the recognition of antigen and MHC molecule and stabilization of the complex of T-cell antigen receptor, antigen, and MHC molecule on the cell surface (Williams, 1985).

The mouse T8 antigen has been characterized by means of monoclonal antibodies, molecularly cloned, and its sequence determined. Mouse T8 is a heterodimer consisting of a 38,000-M_r α-chain associated with a 35,000-M_r α'-chain. The human T8 molecule is a homodimer with subunits of 34,000 M_r, although it is associated with a 45,000-M_r structure on T cells. The sequence of cDNA clones for the human T8 antigen has been determined (Littman et al., 1985; Sukhatme et al., 1985). Mature T8 consists of 214 amino acids, including a transmembrane region and intracytoplasmic domain. The external portion of the T8 molecule consists of two domains, one of which demonstrates sequence homology to immunoglobulin V region domains. Homology in this external domain is 30–35% with λ and κ V regions, 20–22% with heavy-chain V regions, and 24% to T-cell receptor α- and β-V regions (Sukhatme et al., 1985). The membrane proximal portion of the molecule is not homologous to other known proteins but a 21-amino acid region in the center is homologous to the hinge region of the mouse IgA heavy chain. Therefore, like other members of the immunoglobulin superfamily, T8 contains structural features in common with antibody and T-cell antigen receptors, suggesting evolution from a common primordial gene. Interestingly, the T8 molecule shows the highest degree of homology with immunoglobulin λ and κ V region genes and the single T8 gene located near the κ-chain locus on mouse chromosome 6. It has been suggested that T8 is, in fact, the evolutionary product of a single V region gene that lost the ability to rearrange (Hood et al., 1985).

Recent molecular analysis of a cDNA clone encoding the human T4 antigen demonstrates that the T4 molecule is also structurally related to the immunoglobulin superfamily (Maddon et al., 1985). Similar to the T8 structure, the T4 sequence predicts an integral transmembrane protein with extracellular immunoglobulin V-like and C-like domains. The V-like domain shares an overall 32% sequence identity with immunoglobulin V regions. Computer-assisted secondary structure prediction has suggested that the V-like domain contains seven β-sheets which closely parallel those found in immunoglobulin V domains. However, unlike the T8 structure, a short peptide sequence separates the V- and C-like domains which has considerable structural and sequence similarity to J regions of immunoglobulins and T-cell receptor chains. This J-like sequence is not apparent in the T8 structure. Thus, while T4 and T8 molecules share major structural features, both molecules being integral membrane proteins with extracellular domains homologous to immunoglobulins, at the level of amino acid sequence they are very different. The proteins differ dramatically in size and share a low overall sequence homology. Because T4 contains contiguous V-like and J-like structural elements which are expressed together without DNA recombination, it has been suggested that the T4 represents a more primitive gene than T8, having evolved from the immunoglobulinlike molecules before the emergence of rearrangement mechanisms (Maddon *et al.*, 1985).

8.   RELATED MEMBERS OF THE IMMUNOGLOBULIN SUPERGENE FAMILY EXPRESSED IN THE NERVOUS SYSTEM

Two additional proteins with structural similarities to immunoglobulin domains, Thy-1 and OX-2, have been described and characterized at the molecular level. Both molecules are expressed in the mammalian nervous system as well as on lymphocytes and some fibroblasts. Thy-1 is a mouse T cell and nervous system isoantigen, while OX-2 represents a new class of molecules discovered by means of a monoclonal antibody. The function of these molecules is not understood, though their similarities to immunoglobulins has suggested that they may be involved in cellular interactions.

8.1.   *The Thy-1 Glycoprotein*

Reif and Allen (1966) described two strain-specific isoantigens in mice designated θ-AKR and θ-C3HeB/Fe. These isoantigens were detected with an antiserum made by immunizing mice of the AKR strain with thymocytes of the C3H strain. Interestingly, the supposedly thymus-specific antisera reacted strongly with adult mouse brain tissue but not with immature brain. The θ antigen was present at very low levels during the first 4 days after birth and rose to the adult level at 5–6 weeks of age, while levels of θ antigen in the thymus were essentially at adult levels throughout postnatal life (Reif and Allen, 1966). The appearance of θ antigen paralleled the histological and physiological maturation of the brain; in adult mice, this antigen is the most abundant surface glycoprotein on most neuronal cells (Williams and Gagnon, 1982).

Historically, the θ antigen, renamed Thy-1, has been an important immunological marker for thymus-derived lymphocytes in the mouse. The expression of Thy-1, while restricted to several tissues, is not conserved through evolution. Thy-1 is expressed on neuronal structures, of both neuron and glial origin, and on fibroblasts in mice, rats, man, and most other mammals (Williams and Gagnon, 1982). The expression of Thy-1 antigen on lymphocytes and other tissues varies from species to species. For instance, Thy-1 is present on all T lymphocytes in the mouse. In the rat, the antigen is present on some lymphoid progenitor cells but is not a surface molecule of T lymphocytes. A Thy-1 homologue has been identified in man, but it is present only in nervous tissue and is not found on peripheral lymphocytes (Cotmore *et al.*, 1981). Homologues of Thy-1 have also been identified in the frog (Mansour and Cooper, 1984), chicken (Rostas *et al.*, 1983), dog (McKenzie and Fabre, 1981), and squid (Williams and Gagnon, 1982). Because of its consistent expression in the brain and variable expression on lymphocytes, Thy-1 should probably be considered a neuronal molecule rather than a lymphocyte-surface molecule.

The structure of the Thy-1 glycoprotein has been determined from protein sequencing and recombinant DNA techniques (Williams and Gagnon, 1982; Moriuchi *et al.*, 1983). The Thy-1 molecule is a single polypeptide chain of 111 (rat) or 112 (mouse) amino acids containing two intrachain disulfide bridges. Three N-linked carbohydrates are attached at amino acids 23, 75, and 99. The Thy-1.1 and Thy-1.2 alleles differ by only a single amino acid residue at position 89 (arginine to glutamine), and the sequence homology between the rat and mouse Thy-1 is 85%. The amino acid sequence of Thy-1 is homologous with immunoglobulin V domains and, in fact, resembles a single free immunoglobulin homology unit (Williams and Gagnon, 1982). Similarities between Thy-1 and immunoglobulin heavy-chain and light-chain V domains are shown in Fig. 8.

Although a membrane-bound glycoprotein, the mechanism of association of Thy-1 with the cytoplasmic membrane is unusual. Unlike other members of the immunoglobulin superfamily, Thy-1 is not attached to the cell membrane by a carboxy-terminal hydrophobic peptide. Rather, the carboxy-terminal cystine is linked, through a phosphethanolamine group, to a long-chain fatty acid, probably stearic acid. Although the structure of this phospholipid has yet to be completely determined, it is clear that it is responsible for membrane attachment (Tse, 1985). Thus, Thy-1 is attached to the cell membrane via a lipid, rather than polypeptide, moiety.

Murine Thy-1 is encoded by a single locus on chromosome 9. In light of the unusual mechanism of membrane attachment, the structural analysis of the Thy-1 gene was surprising (Seki *et al.*, 1985; Ingraham *et al.*, 1985; Evans *et al.*, 1984). Both rat and mouse Thy-1 molecules are encoded in four exons (Fig. 2). The first, located 2.2 kb upstream of the remainder of the gene, encodes only a portion of the 5′ untranslated mRNA (Ingraham *et al.*, 1985). The alternate use of two promoters generates two species of mRNA through alternate splicing to the next downstream exon (Ingraham and Evans, 1986). The second exon encodes part of the leader peptide, the third exon encodes the remainder of the leader peptide and the major portion of the Thy-1 glycoprotein. Unexpectedly, a fourth exon encodes an additional carboxy-terminal peptide. This peptide is extremely hydrophobic and is analogous to the transmembrane domains of membrane-bound immunoglobulin, T-cell antigen receptors, and class I and class II MHC antigens. Although pulse-labeling studies in cell culture have as yet failed to detect carboxy-terminal processing (Luescher and Bron, 1985), it is likely that the Thy-1 glycoprotein is initially synthesized as a precursor with a membrane-spanning domain. Presumably, Thy-1 is then modified by proteolytic cleavage and post-translational addition of the carboxy-terminal lipid. The significance is not yet clear, but a similar type of processing and carboxy-terminal modification occurs with the variable surface coat protein of the parasite *Trypanosoma bruceii* (Boothryod *et al.*, 1981) and possibly with the synaptic acetylcholine esterase (Kim and Roseberry, 1985; Futerman *et al.*, 1985).

8.2. The OX-2 Antigen

The MRC OX-2 mouse monoclonal antibody was produced against rat thymocyte membrane glycoproteins (McMaster and Williams, 1979) and shown to precipitate a 47,000-M_r glycoprotein. This molecule is expressed on thymocytes, follicular dendritic cells, vascular endothelium, smooth muscle, some B lymphocytes, and neurons. Because of the expression on neurons as well as thymocytes, similarity to Thy-1 was proposed and the antigen was characterized through protein sequencing and molecular cloning (Clark *et al.*, 1985).

The OX-2 antigen is a glycoprotein of 248 amino acids with a transmembrane domain of 27 amino acids and an intracytoplasmic "tail" of 19 amino acids. The sequence defines six potential glycosylation sites for N-linked carbohydrate and two potential disulfide bridges. The OX-2 sequence shows a striking resemblance to the immunoglobulin superfamily in that the external portion of the molecule defines two immunoglobulinlike domains. The amino-terminal domain shows sequence homology to immunoglobulin V genes, mouse Thy-1, and the poly Ig receptor domain IV (Fig. 8). The membrane proximal domain shows sequence homology with mouse and human immunoglobulin C genes (Clark *et al.*, 1985). The amino-terminal domain of OX-2 shows the greatest degree of sequence homology with the Thy-1 glycoprotein, a sequence that is also similar to immunoglobulin V regions. Therefore, the OX-2 antigen has a structure remarkably similar to that of an immunoglobulin light chain or a T-cell receptor β-chain, with domains similar to variable and constant regions. Although the structure of the OX-2 gene is not yet available, it would be very surprising if it did not have an immunoglobulinlike structure with an intron separating V- and C-like exons.

Like the Thy-1 glycoprotein, the expression of the OX-2 antigen is not restricted to a single cell type but appears on cell surfaces in a variety of tissues. Although no defined function has been found, their similarity to a group of molecules involved in antigen recognition or cell–cell recognition may suggest that they mediate cell–cell interactions on a variety of cell types (Cohen *et al.*, 1981). In any regard, it is interesting that both are found on neurons and, in the case of Thy-1, this is the tissue in which expression is conserved from species to species (Williams, 1985). Since all members of the immunoglobulin superfamily may have derived form a common ancestor, Thy-1 and OX-2 have been proposed as contemporary molecules with the properties of evolutionary precursors. Moreover, because of their common expression in the nervous system, it is possible that Ig-related molecules first evolved to mediate cell–cell interactions on primitive sensory cells. Thus, mechanisms of cellular interactions in the nervous and immune systems may have a common evolutionary history.

9. CONCLUSION: THE EVOLUTION OF THE IMMUNOGLOBULIN SUPERFAMILY

Molecular analysis of the immune response has shown that gene products with unrelated functions are, in fact, composed of similar subunits, immunolglobulinlike homology units. Although this type of analysis suggests a common evolutionary history for the immunoglobulins,

T-cell receptors, and MHC antigens, the possiblity of convergent, rather than divergent, evolution generating similar structures cannot be excluded.

Two main arguments exist against evolutionary convergence. First, amino acid sequence homologies between members of the immunoglobulin superfamily provide strong evidence for a common genetic ancestor. The three-dimensional structure of the enzyme Cu,Zn-superoxide dismutase shows a striking similarity to the three-dimensional structure of immunoglobulins, although there is no detectable amino acid sequence homology (Amzel and Poljak, 1979). Thus, similar structures generated through convergent evolution need not show similarities in primary sequence.

Second, all members of this superfamily demonstrate the same pattern of RNA splicing, between the second and third bases of the junctional codon. Sharp (1981) suggests that RNA splicing patterns are unlikely to be influenced by selective pressure through evolution. Similar RNA splicing, therefore, provides additional evidence of common ancestry. Thus, convergent evolution is an unlikely explanation of similarities between immunoglobulinlike genes, although it cannot be formally excluded.

The three-dimensional structure of the immunoglobulin domain favors several types of structural interactions that are found through the immunoglobulin superfamily. First, immunoglobulin domains interact in cis with similar domains in the same polypeptide, or in trans with other subunits, to form structures with a twofold axis of symmetry. The structures of the immunoglobulins, class I and class II MHC molecules, and T-cell antigen receptors, exemplify these interactions. In the case of the class I, class II molecules and β_2-microglobulin, similar tetrameric surface structures are formed through the interactions of subunits containing different numbers of immunoglobulin homology units (Fig. 1).

Second, molecules composed of immunoglobulinlike domains appear to favor interactions with other molecules composed of similar domains. The T-cell antigen receptor interacts with portions of the class I or class II molecules, the T8 accessory molecule may interact with MHC products (Williams, 1985), and the polyimmunoglobulin receptor binds IgM and IgA. The latter provides one of the clearest examples of a cell-surface receptor and its ligand having similar structures. Finally, the structure of the immunoglobulin and T-cell receptor V regions provides for the creation of a binding site for antigen.

A evolutionary scheme taking into account the known similarities between immunoglobulin-related genes is shown in Fig. 9. The immunoglobulin domain hypothesis (Hill et al., 1966) proposes that immu-

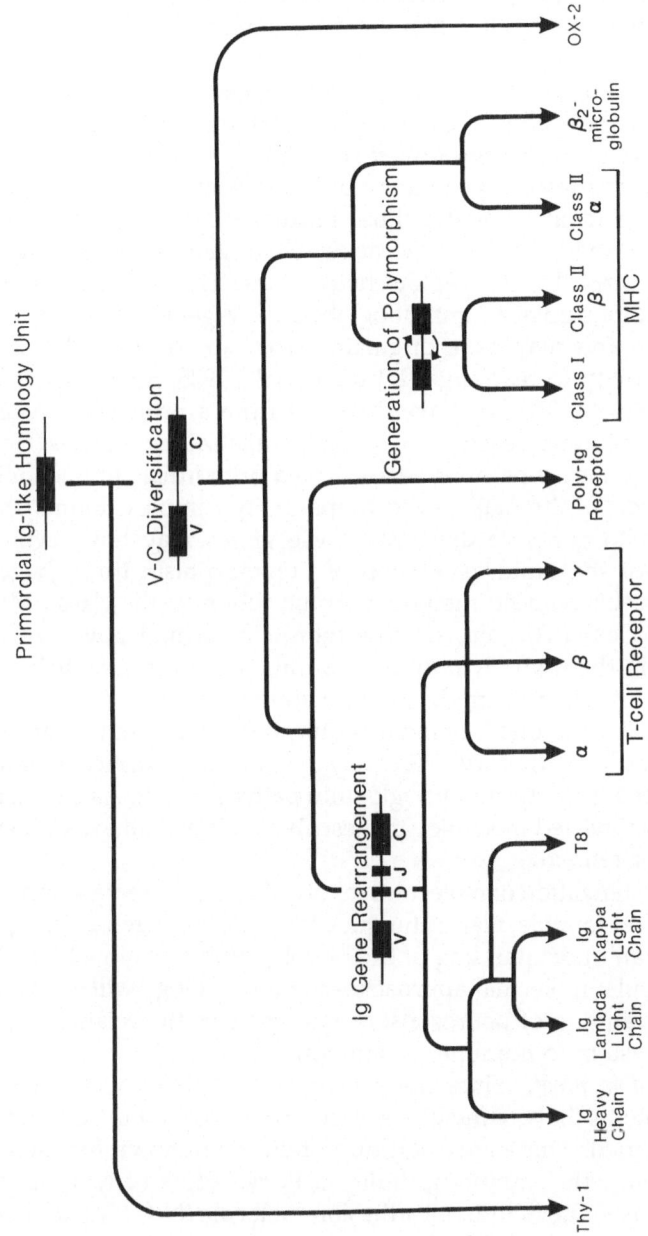

Figure 9. Evolutionary scheme for the immunoglobulin superfamily. A hypothetical scheme based on the immunoglobulin domain hypothesis (Hill *et al.*, 1966) is shown originating from a single primordial immunoglobulinlike homology unit. Thy-1 and OX-2 are similar in structure to the proposed single domain, and free V and C domains postulated as immunoglobulin precursors. The locations of the T8 and poly(Ig) receptor are somewhat arbitrary, based, for the former, on similarities with Ig κ-chain V genes and, for the latter, on its Ig-binding capability.

noglobulins initially derived from a single primordial domain that gave rise, through duplication, to a structure like a cell-surface immunoglobulin light chain on a primitive lymphoid or neurosensory cell. Interestingly, the Thy-1 and OX-2 structures are very like those of these proposed primitive immunoglobulins. The immunoglobulin precursor genes then developed the ability to undergo recombination, perhaps through the "capture" of a transposon (Sui *et al.*, 1984). The evolution of J and D structures permitted the generation of binding site diversity for both T-cell receptor and immunoglobulins. Further duplication and divergence separated the light-chain and heavy-chain genes of immunoglobulins and the three T-cell receptor chain genes. The T8 molecule was found to be more similar to immunoglobulin κ V genes than other immunoglobulins and may have originated from an "orphan" V region that lost the ability to rearrange (Littman *et al.*, 1985; Sukhatme *et al.*, 1985). Alternatively, T8 may have diverged from a primordial V gene and, for some unknown reason, evolved in parallel with κ V genes (Hood *et al.*, 1985). A separate pathway leads from a primordial light-chainlike gene to products of the major histocompatibility complex. Rather than recombination to generate diversity, these genes may have evolved mechanisms for the rapid generation of polymorphism through gene conversion. In this respect, the polymorphic subunits, the class I MHC gene and the class I β-chain, may be more similar and have evolved separately from the nonpolymorphic subunits, the class II α-chain and β_2-microglobulin. The polymeric immunoglobulin receptor contains five V homology units and one C homology unit. It may represent a separate evolutionary pathway that will ultimately include other surface receptors or have diverged from the immunoglobulin pathway itself. As additional immunoglobulin-related molecules are described, this evolutionary scheme may be further refined.

The characterization of molecules involved in the vertebrate immune response has led to molecular techniques for classifying genes and gene products and an understanding of their evolutionary relationships. It is hoped that similar molecular approaches to neurobiology will contribute to our understanding of neurosensory systems and the relationship of the nervous system to immune recognition.

Note added in proof. Since the preparation of this review, several additional proteins that are members of the immunoglobulin superfamily have been defined. These are a serum protein of unknown function (α-1B glycoprotein), the immunoglobulin G F_c receptors of lymphocytes and macrophages, the neural cell adhesion molecule (N-CAM), and neurocytoplasmic protein 3 (NP3). Recent reevaluation of the sequences of the δ- and ε-subunits of the T-cell receptor, which were not originally included in the immunoglobulin superfamily, has led to the suggestion that they may each consist of a single V-like homology unit. It is note-worthy of this expanded list that four members of this superfamily, Thy-1, OX-2, N-CAM, and NP3, are expressed in the nervous system.

ACKNOWLEDGMENTS

I am grateful to H. A. Ingraham, B. E. Rothenberg, and K. C. Evans for their helpful comments and suggestions. Work by the author is supported by grants GM33868 and HD18012 from the National Institutes of Health and by funds from the G. Harold and Leila Y. Mathers Charitable Foundation. The author is a Pew Scholar in the Biomedical Sciences.

REFERENCES

Amzel, L. M., and Poljak, R. J., 1979, Three-dimensional structure of immunoglobulins, *Annu. Rev. Biochem.* **48:**961–997.

Arnold, B., Burget, H. G., Hamman, U., Hammerling, G., Kees, U., and Kvist, S., 1984, Cytolytic T cells recognize the two amino-terminal comains of H-2 antigens in tandem in influenza A infected cells, *Cell* **38:**79–87.

Barra, Y., Tanaka, T., Isselbacher, K., Khoury, G., and Jay, G., 1985, Stable transfer and restricted expression of a cloned class I gene encoding a secreted transplantation-like antigen, *Mol. Cell. Biol.* **5:**1295–1300.

Barth, R. K., Kim, B. S., Lan, N. C., Hunkapiller, T., Sobieck, N., Winoto, A., Gershenfeld, H., Okada, C., Hnasburg, D., Weissman, I. L., and Hood, L., 1985, The murine T-cell receptor uses a limited repertoire of expressed V_β gene segments, *Nature (Lond.)* **316:**517–523.

Boothryod, J. R., Paynter, C. A., Cross, G. A. M., Bernards, A., and Borst, P., 1981, Variant surface glycoproteins of *Trypanosoma brucei* are synthesized with cleavable sequences at the carboxy and amino termini. *Nucl. Acid Res.* **9:**4735–4743.

Clark, M. J., Gagnon, J., Williams, A. F., and Barclay, A. N., 1985, MRC OX-2 antigen: A lymphoid/neuronal membrane glycoprotein with a structure like a single immunoglobulin light chain, *EMBO J.* **4:**113–118.

Cohen, F. E., Novotny, J., Sternberg, M. J. E., Campbell, D. G., and Williams, A. F., 1981, Analysis of structural similarities between brain Thy-1 antigen and immunoglobulin domains, *Biochem. J.* **195:**31–40.

Cosman, D., Khoury, G., and Jay, G., 1982, Three classes of mouse H-2 messenger RNA distinguished by analysis of cDNA clones, *Nature (Lond.)* **295:**73–75.

Cotmore, S. F., Crowhurst, S. A., and Waterfield, M. D., 1981, Purification of Thy-1 related glycoprotein from human brain and fibroblasts: Comparison between these molecules and murine glycoproteins carrying the Thy-1.1 and Thy-1.2 antigens, *Eur J. Immunol.* **11:**597–603.

Davis, M. M., 1985, The murine T cell receptor, 1985, *Annu. Rev. Immunol.* **3:**537–560.

Devlin, J. J., Lew, A. M., Flavell, R. A., and Coligan, J. E., 1985, Secretion of a soluble class I molecule encoded by the Q10 gene of the C57BL/10 mouse, *EMBO J.* **4:**369–374.

Evans, G. A., Margulies, D. H., Ozato, K., Camerini-Otero, R. D., and Seidman, J. G., 1982a, Structure and expression of a mouse major histocompatibility antigen, H-2L^d, *Proc. Natl. Acad. Sci. USA* **79**:1994–1998.

Evans, G. A., Margulies, D. H., Shykind, B., Seidman, J. G., and Ozato, K., 1982b, Exon shuffing: Mapping polymorphic determinants on hybrid mouse transplantation antigens, *Nature (Lond.)* **300**:755–757.

Evans, G. A., Ingraham, H. A., Lewis, K., Cunningham, K., Seki, T., Moriuchi, T., Chang, H. C., Silver, J., and Hyman, R., 1984, Expression of the Thy-1 glycoprotein gene by DNA-mediated gene transfer, *Proc. Natl. Acad. Sci. USA* **81**:5532–5536.

Futerman, A. H., Low, M. G., Ackerman, K. E., Sherman, W. R., and Silman, I., 1985, Identification of covalently bound inositol in the hydrophobic membrane-anchoring domain of *Torpedo* acetylcholinesterase, *Biochem. Biophys. Res. Commun.* **129**:312–317.

Goodenow, R. S., McMillan, M., Nicolson, M., Sher, B. T., Eakle, K., Davidson, N., and Hood, L., 1982, Identification of the class I genes of the mouse major histocompatibility complex by DNA-mediated gene transfer, *Nature (Lond.)* **300**:231–237.

Harris, R. A., Hogarth, P. M., Penington, D. G., and McKenzie, I. F. C., 1984, Qa antigens and their differential distribution of lymphoid, myeloid and stem cells, *J. Immunogenet.* **11**:265–281.

Hedrick, S. M., Nielsen, E. A., Kavaler, J., Cohen, D. I., and Davis, M. M., 1984, Sequence relationships between putative T-cell receptor polypeptides and immunoglobulins, *Nature (Lond.)* **308**:153–158.

Hill, R. L., Delaney, R., Fellowe, R. E., Jr., and Lebovitz, H. E., 1966, The evolutionary history of the immunoglobulins, *Proc. Natl. Acad. Sci. USA* **56**:1762–1769.

Honjo, T., 1983, Immunoglobulin genes, *Annu. Rev. Immunol.* **1**:499–528.

Hood, L., Steinmetz, M., and Malissen, B., 1983, Genes of the major histocompatibility complex of the mouse, *Annu. Rev. Immunol.* **1**:529–568.

Hood, L., Kronenberg, M., and Hunkapiller, T., 1985, T cell antigen receptors and the immunoglobulin supergene family, *Cell* **40**:225–229.

Hyman, R., and Stallings, V., 1977, Analysis of hybrids between an H-2$^+$, TL$^+$ lymphoma and its H-2$^-$, TL$^-$ variant subline, *Immunogenetics* **4**:171–181.

Ingram, M. A., and Evans, G. A., 1986, Characterization of two atypical promoters and alternate mRNA processing in the mouse Thy 1.2 glycoprotein gene, *Mol. Cell. Biol.* **6**:2923–2931.

Ingraham, H. A., Lawless, G. M., and Evans, G. A., 1986, The mouse Thy-1.2 glycoprotein gene: Complete sequence and identification of an unusual promoter, *J. Immunol.* **136**:1482–1489.

Kabat, E. A., Wu, T. T., Bilofsky, H., Reid-Miller, M., and Perry, H., 1983, *Sequences of Proteins of Immunological Interest*, USPHS, Washington, D.C.

Kastern, W., 1985, Characterization of two class I major histocompatibility rat cDNA clones, one of which contains a premature termination codon, *Gene* **34**:227–233.

Kim, B. H., and Roseberry, T. L., 1985, A small hydrophobic domain that localizes human erythrocyte acetylcholinesterase in liposomal membrane is cleaved by papain digestion, *Biochemistry* **24**:3586–3592.

Krangel, M. S., 1985, Unusual RNA splicing generates a secreted form of HLA-A2 in a mutagenized B lymphoblastoid cell line, *EMBO J.* **4**:1205–1210.

Kranz, D. M., Disteche, C. M., Swisshelm, K., Pravtcheva, D., Ruddle, F., Eisen, H. N., and Tonegawa, S., 1985, Chromosomal locations of the murine T-cell receptor alpha chain and the T-cell gamma chain, *Science* **227**:941–945.

Kvist, S., Roberts, S., and Dobberstein, B., 1983, Mouse histocompatibility genes: Structure and organization of a K^d gene, *EMBO J.* **2**:245–254.

Lalanne, J. L., Transey, C., Guerin, S., Darche, S., Meulien, P., and Kourilsky, P., 1985, Expression of class I genes in the major histocompatibility complex: Identification of eight distinct mRNAs in the DBA/2 mouse liver, *Cell* **41**:469–478.

Le Meur, M., Gerlinger, P., Benoist, C., Mathis, D., 1985, Correcting an immune-response deficiency by creating E_{alpha} gene transgenic mice, *Nature (Lond.)* **316**:38–42.

Littman, D. R., Thomas, Y., Maddon, P. J., Chess, L., and Axel, R., 1985, The isolation and sequence of the gene encoding T8: A molecule defining functional classes of T lymphocytes, *Cell* **40**:237–246.

Luescher, B., and Bron, C., 1985, Biosynthesis of mouse Thy-1 antigen, *J. Immunol.* **134**:1084–1089.

Maddon, P. J., Littman, D. R., Godfrey, M., Maddon, D. E., Chess, L., and Axel, R., 1985, The isolation and nucleotide sequence of a cDNA encoding the T cell surface protein T4: A new member of the immunoglobulin gene family, *Cell* **42**:93–104.

Mansour, M. H., and Cooper, E. L., 1984, Purification and characterization of *Rana pipens* brain Thy-1 glycoprotein, *J. Immunol.* **132**:2515–2523.

Margulies, D. H., Evans, G. A., Flaherty, L., and Seidman, J. G., 1982, H-2-like genes in the TLa region of mouse chromosome 17, *Nature (Lond.)* **295**:168–170.

Max, E. E., 1984, Immunoglobulins: Molecular genetics, in: *Fundamental Immunology* (W. E. Paul, ed.), pp. 167–204, Raven Press, New York.

McKenzie, J. L., and Fabre, J. W., 1981, Studies with a monoclonal antibody on the distribution of Thy-1 in the lymphoid and extracellular connective tissues of the dog, *Transplantation* **31**:275–282.

McMaster, W. R., and Williams, A. F., 1979, Identification of Ia glycoproteins in rat thymus and purification from rat spleen, *Eur. J. Immunol.* **9**:426–433.

Mellor, A. L., Weiss, E. H., Ramachandran, K., and Flavell, R. A., 1983, A potential donor gene for the bml gene conversion event in the C57BL mouse, *Nature (Lond.)* **306**:792–795.

Mengle-Gaw, L., and McDevitt, H. O., 1985, The class II genes of the major histocompatibility complex, *Annu. Rev. Immunol.* **3**:369–398.

Meuer, S. C., Acuto, O., Hercend, T., Schlossman, S. F., and Reinherz, E., 1984, The human T-cell receptor, *Annu. Rev. Immunol.* **2**:23–66.

Michaelson, J., 1983, Genetics of beta$_2$-microglobulin in the mouse, *Immunogenetics* **17**:219–259.

Mirsky, R., 1982, The use of antibodies to define and study major cell types in the central and peripheral nervous systems, in: *Neuroimmunology* (J. Brockes, ed.), pp. 141–181, Plenum Press, New York.

Moriuchi, T., Chang, H. C., Denome, R., and Silver, J., 1983, Thy-1 cDNA sequence suggests a novel regulatory mechanism, *Nature (Lond.)* **301**:80–82.

Mostov, K. E., and Blobel, G., 1982, A transmembrane precursor of secretory component, *J. Biol. Chem.* **257**:11816–11821.

Mostov, K. E., Friedlander, M., and Blobel, G., 1984, The receptor for transepithelial transport of IgA and IgM contains multiple immunoglobulin-like domains, *Nature (Lond.)* **308**:37–43.

Murre, C., Reiss, C. S., Bernabeu, C., Chen, L. B., Burakoff, S. J., and Seidman, J. G., 1984, Constructure, expression and recognition of an H-2 molecules lacking its carboxyl terminus, *Nature (Lond.)* **307**:432–436.

Novotny, J., and Auffray, C., 1984, A program for prediction of protein secondary structure from nucleotide sequence data: Application to histocompatibility antigens. *Nucl. Acid Res.* **12**:243–255.

Ozato, K., Evans, G. A., Shykind, B., Margulies, D. H., and Seidman, J. G., 1984, Hybrid H-2 histocompatibility gene products assign domains recognized by alloreactive T cells, *Proc. Natl. Acad. Sci. USA* **80**:2040–2043.

Parnes, J. R., and Seidman, J. G., 1982, Structure of wild-type and mutant mouse beta$_2$-microglobulin genes, *Cell* **29**:661–669.

Reif, A. E., and Allen, J. M., 1966, Mouse thymic isoantigens, *Nature (Lond.)* **209**:521–523.

Rostas, J. A. P., Shevenac, T. A., Sinclair, C. M., and Jeffrey, P. L., 1983, The purification and characterization of a Thy-1-like glycoprotein from chicken brain, *Biochem. J.* **213**:143–152.

Saito, H., Kranz, D. M., Takagaki, Y., Hayday, A. C., Eisen, H. N., and Tonegawa, S., 1984, Complete primary structure of a heterodimeric T-cell receptor deduced from cDNA sequences, *Nature (Lond.)* **309**:757–762.

Schachner, M., Faissner, A., Kruse, J., Linder, J., Meier, D. H., Rathjen, F. G., and Wernecke, H., 1983, Cell-type specificity and developmental expression of neural cell-surface components involved in cell interactions of structurally related molecules, *CSH Symp. Quant. Biol.* **47**:557–568.

Seki, T., Moriuchi, T., Chang, H. C., Denome, R., and Silver, J., 1985, Structural organization of the rat Thy-1 gene, *Nature (Lond.)* **131**:485–487.

Sharp, P. A., 1981, Speculations on RNA splicing, *Cell* **23**:643–646.

Shiroishi, T., Evans, G. A., Appella, E., and Ozato, K., 1984, Role of a disulfide bridge in the immune function of major histocompatibility class I antigen as studied by *in vitro* mutagenesis, *Proc. Natl. Acad. Sci. USA* **81**:7544–7548.

Shiroishi, T., Evans, G. A., Appella, E., and Ozato, K., 1985, *In vitro* mutagenesis of a mouse class I gene for the examination of structure–function relationships, *J. Immunol.* **134**:623–629.

Steinmetz, M., Winoto, A., Minard, K., and Hood, L., 1982, Clusters of genes encoding mouse transplantation antigens, *Cell* **28**:489–498.

Sui, G., Clark, S. P., Yoshikai, Y., Malissen, M., Yanagi, Y., Strauss, E., Mak, T. W., and Hood, L., 1984, The human T cell antigen receptor is encoded by variable, diversity, and joining gene segments that rearrange to generate a complete V gene, *Cell* **37**:393–401.

Sukhatme, V. P., Sizer, K. C., Vololmer, A. C., Hunkapiller, T., and Parnes, J. P., 1985, The T cell differentiation antigen Leu-2/T8 is homologous to immunoglobulin and T cell receptor variable regions, *Cell* **40**:591–597.

Thomas, M. L., Barclay, A. N., Gagnon, J., and Williams, A. F., 1985, Evidence from cDNA clones that the rat leukocyte-common antigen (T200) spans the lipid bilayer and contains a cytoplasmic domain of 80,000 M_r, *Cell* **41**:83–93.

Tse, A. G. D., Barclay, A. N., Watts, A., and Williams, A. F., 1985, A glycophospholipid tail at the carboxyl terminus of the Thy-1 glycoprotein of neurons and thymocytes, *Science* **230**:1003–1008.

Van den Elsen, P., Shepley, B. A., Cho, M., and Terhorst, C., 1985, Isolation and characterization of a cDNA clone encoding the murine homologue of the human 20K T3-T-cell receptor glycoprotein, *Nature (Lond.)* **314**:542–544.

Wall, R., and Kuehl, M., 1983, Biosynthesis and regulation of immunoglobulins, *Annu. Rev. Immunol.* **1**:393–422.

Weiss, E. H., Mellor, A., Golden, L., Fahrner, K., Simpson, E., Hurst, J., and Flavell, R. A., 1983, The structure of a mutant H-2 gene suggests that the generation of polymorphism in H-2 genes may occur by gene conversion-like events, *Nature (Lond.)* **301**:671–674.

White, P. C., Chaplin, D. D., Weis, J. H., Dupont, B., New, M. I., and Seidman, J. G., 1984, Two steroid 21-hydroxylase genes are located in the murine S region, *Nature (Lond.)* **312**:465–467.

Williams, A. F., The immunoglobulin superfamily takes shape, 1984, *Nature (Lond.)* **308**:12–13.

Williams, A. F., 1985, Immunoglobulin-related domains for cell surface recognition, *Nature (Lond.)* **314:**579–580.

Williams, A. F., and Gagnon, J., 1982, Neuronal cell Thy-1 glycoprotein: Homology with immunoglobulin, *Science* **216:**696–703.

Winoto, A., Steinmetz, M., and Hood, L., 1983, Genetic mapping in the major histocompatibility complex by restriction enzyme site polymorphisms: Most mouse class I genes map to the Tla complex, *Proc. Natl. Acad. Sci. USA* **80:**3425–3429.

Zipursky, S. L., Venkatesh, T. R., Teplow, D. B., and Benzer, S., 1984, Neuronal development in the drosophila retina: Monoclonal antibodies as molecular probes, *Cell* **36:**15–26.

8

Specificity of Prohormone Processing
The Promise of Molecular Biology

LLOYD D. FRICKER, DANE LISTON, MARK GRIMES, and EDWARD HERBERT

1. INTRODUCTION

Cells that use peptides as intercellular messengers employ a biosynthetic strategy that differs substantially from the biosynthesis of other chemical messengers. Most neuropeptides are initially synthesized as larger precursor proteins (prohormones) that are cleaved enzymatically to produce the bioactive peptides. The primary structures of many prohormones have been established in recent years, largely due to the advent of recombinant DNA technique (for review, see Douglas *et al.*, 1984). In many prohormones, two or more bioactive domains will overlap, with a potential proteolytic processing site occurring within the sequence of a bioactive peptide. Differential cleavage of such internal sites can give rise to peptides with marked differences in both potency at a particular receptor and/or selectivity for various receptors, yielding products with substantial differences in biological function. Thus, a fundamental ques-

LLOYD D. FRICKER • Molecular Pharmacology Department, Albert Einstein College of Medicine, Bronx, New York 10461.   DANE LISTON, MARK GRIMES, and EDWARD HERBERT • Institute for Advanced Biomedical Research, The Oregon Health Sciences University, Portland, Oregon 97201.

tion of neuropeptide biosynthesis concerns the control of the specificity of cleavage of the precursor.

Several structural features of prohormones appear to be quite general. For example, most of the prohormones possess a hydrophobic signal sequence of 15–30 amino acids at the amino terminal of the protein. This sequence seems to be important for targeting the protein to the secretory pathways of the cell (Blobel and Dobberstein, 1975). Many prohormones contain either multiple copies of a similar bioactive peptide or several distinct peptide sequences and thus have been called polyfunctional proteins, or polyproteins. The bioactive domains are typically flanked by pairs of basic amino acid residues (lysine, arginine) that appear to act as signals for enzymatic cleavage of the precursor. Initial action of an endopeptidase followed by a carboxypeptidase B-like exopeptidase would liberate the active form of the peptide.

While pairs of basic amino acid residues seem to be the most common sites for endoproteolytic cleavage of the prohormone, other processing sites are utilized. These include cleavage at single arginine residues (prodynorphin), four adjacent basic residues (pro-opiomelanocortin) and acidic or hydrophobic amino acids (atrial pronatriodilatin). Numerous nonproteolytic modifications have been observed as well, such as the addition of oligosaccharides, sulfate, or phosphate to various residues of the protein backbone. Two forms of modification, amidation at the carboxy terminus and acetylation of the amino terminus, seem to be especially important and can yield drastic alterations in the affinity of a peptide for a particular receptor.

Given the complexity of the neuropeptide precursors and the potential for differential processing, it is not surprising that significant tissue-specific differences in the final peptide products are observed. This is certainly the case for the biosynthesis of the opioid peptides, a large family of neuropeptides derived from three distinct precursors, pro-opiomelanocortin (POMC), prodynorphin, and proenkephalin. All three of these precursors have been shown to yield varied end products in different tissues. These products exhibit large differences in affinity for the three types of opiate receptors: μ, δ, and κ (Corbett et al., 1982; Chavkin and Goldstein, 1981). Thus, through regulation of prohormone processing, a wide range of physiological responses may be affected.

1.1. Tissue-Specific Processing of Opioid Peptides

Early studies on the processing of POMC found substantial differences between various regions of brain and pituitary (for review, see

Liotta and Krieger, 1983). Within the anterior lobe of the pituitary, POMC is largely processed into ACTH and both β-lipotropin and β-endorphin. In the intermediate lobe, ACTH is rapidly converted to α-MSH, and β-lipotropin is converted to β-endorphin (Scott et al., 1974; Crine et al., 1979; Gianoulakis et al., 1979). While β-endorphin is present in both lobes of the pituitary, the form (and biological activity) of this neuropeptide differs in the two lobes (Smyth and Zakarian, 1980; Liotta et al., 1981, Eipper and Mains, 1981; Weber et al., 1982b). In the anterior pituitary, nearly all the β-endorphin immunoreactivity is associated with β-endorphin 1–31. In the intermediate lobe, β-endorphin 1–31 represents a minor component, and the shorter β-endorphin 1–26 and 1–27 are the predominant forms. A large fraction of all forms of β-endorphin in the intermediate pituitary are α,N-acetylated, which greatly reduces the affinity for the opiate receptors and eliminates the analgesic activity of the peptides (Smyth et al., 1979; Deakin et al., 1980). The function of these N-acetylated forms of β-endorphin is unknown.

Similar differences in the processing of POMC are also found in various brain regions. In the hypothalamus, β-endorphin 1–31 is the predominant form of this neuropeptide, whereas α,N-acetyl-β-endorphin 1–27 and 1–26 have been reported to be present in hippocampus and brain stem (Zakarian and Smyth, 1982). However, α, N-acetylated forms of β-endorphin account for a small fraction of total brain β-endorphin (Weber et al., 1981). ACTH (1–39), ACTH (1–13) amide (desacetyl-α-MSH), and a α-MSH are all reported to be present in various brain regions (Orwoll et al., 1979; Gramsch et al., 1980; Loh et al., 1980; Barnea et al., 1982; Evans et al., 1982; Geis et al., 1984). The distinction between α-MSH and the unacetylated ACTH (1–13) amide is important, since these two peptides exhibit significant differences in several behavioral assays (O'Donohue et al., 1981, 1982).

The processing of prodynorphin also shows significant differences in various brain and pituitary tissues. Prodynorphin contains three different bioactive domains: dynorphin A, dynorphin B (rimorphin), and α/β-neo-endorphin (Kakidani et al., 1982). In the posterior pituitary, both dynorphin A 1–17 and dynorphin 1–8 are present in comparable amounts, whereas levels of the shorter dynorphin 1–8 are much higher than dynorphin 1–17 levels in most brain regions (Weber et al., 1982a; Seizinger et al., 1984). Similar variations are found with α- and β-neo-endorphin, which differ in size by a single amino acid (Kangawa et al., 1981; Minamino et al., 1981). High levels of both α- and β-neo-endorphin are present in the posterior pituitary, while in other tissues the ratio of α- to β-neo-endorphin ranges from three in the pons/medulla, hypothalamus, and spinal cord, to 30 in the striatum (Weber *et al.*, 1982c). As with POMC-derived peptides, the various forms of prodynorphin-derived peptides have significant differences in potency toward the three types of opiate receptors (Chavkin and Goldstein, 1981; Corbett *et al.*, 1982).

The most complicated opioid peptide precursor in terms of the number of potential processing sites is proenkephalin. This prohormone contains six copies of the pentapeptide Met-enkephalin and one of Leu-enkephalin (Comb *et al.*, 1982; Gubler *et al.*, 1982; Noda *et al.*, 1982). Numerous alternate processing sites are present in the protein, with the potential for the production of an octapeptide (YGGFMRGL) and a heptapeptide (YGGFMRF) displaying opioid activity. The earliest studies on proenkephalin processing were performed in the bovine adrenal medulla, in which numerous enkephalin-containing peptides ranging from 600 M_r to as large as 23,400 M_r accumulate. These peptides represent as much as 90% of the total enkephalin immunoreactivity (for review, see Udenfriend and Kilpatrick, 1983). These larger peptides are thought to represent intermediates in the processing of proenkephalin. Studies of brain tissue demonstrated that most brain regions do not accumulate these high-molecular-weight intermediates to the same degree as does the bovine adrenal medulla (Liston *et al.*, 1983). However, the bovine hypothalamus is an exception, and as much as 75% of Met-enkephalin immunoreactivity is present in peptides larger than the enkephalins (Liston *et al.*, 1984; Liston and Rossier, 1984). Subsequently, Met-enkephalin was localized to those magnocellular neurons of the bovine hypothalamus that contain the neurohypophyseal hormone, oxytocin (Vanderhaeghen *et al.*, 1983). These cells possess a characteristic anatomy, with cell bodies in the supraoptic and paraventricular nuclei and projections to the posterior lobe of the pituitary. Examination of the proenkephalin-derived peptides in the cell body, axon, and terminals revealed that proteolytic processing of the prohormone was occurring concurrently with axonal transport of the secretory granules down the axon (Liston *et al.*, 1984). This was analogous to the processing of other neurohypophyseal hormones, oxytocin, and vasopressin (Gainer *et al.*, 1977). Thus, the proteolytic enzymes that process proenkephalin to the active peptides are contained and function within the secretory granules. Comparison of the enkephalin-containing peptides in the bovine adrenal medulla and supraoptic nucleus also revealed that the processing intermediates in these two tissues were not the same, demonstrating that distinct, tissue-specific processing pathways exist for proenkephalin (Liston *et al.*, 1984).

1.2. Limitations of Classic Approaches to the Study of Prohormone Processing

The observation of tissue-specific processing suggests that differences in the endoproteolytic cleavage of the precursors is occurring, due either to distinct endoproteases or to unique environments that affect the specificity of an endoprotease. What is known about the various processing enzymes is directly related to the ease of assaying the enzymatic activities. An exopeptidase, carboxypeptidase E, has been very well characterized, largely a result of an extremely simple assay to detect this enzyme (for review, see Fricker, 1985). Simple assays for both the acetylating and amidating enzymes exist, and much has been learned recently about these enzymes (Bradbury et al., 1982; Eipper et al., 1983; Glembotski, 1982; Mains and Eipper, 1984). The least well characterized enzymes are the trypsinlike endopeptidases, which initially cleave the prohormone at pairs of basic amino acids. The difficulties in obtaining substrates for these enzymes as well as characterizing the reaction products have slowed the isolation and characterization of these enzymatic activities.

Direct studies on prohormone processing enzymes are also complicated due to the low abundance of these enzymes relative to other cellular proteolytic activities. One strategy used to overcome this problem is to isolate secretory granules by subcellular fractionation of the tissue and to characterize the peptidases in the secretory granules. However, unless lysosomes are completely removed from the secretory granules, the results can be confusing because lysosomes contain very high levels of peptidases. As a result of these problems, the classic approach to this isolation and characterization of peptidases has met with limited success. Other approaches are currently being tried, some of which have yielded exciting results.

2. GENE-TRANSFER SYSTEMS

In the absence of both purified enzymes and natural substrates, it is difficult to evaluate the relative importance of precursor structure and enzyme specificity in determining where endoproteolytic cleavage occurs in a large, complex prohormone. To circumvent some of these limitations, approaches that employ recombinant DNA are being applied to study prohormone processing. To date such experiments have involved the transfer of DNA encoding a neuropeptide precursor, either as cDNA or gene, into heterologous cells. These recombinant cells can be used to examine prohormone biosynthesis at several levels. At the genetic level, events involved in the transcription of a prohormone gene and processing of the transcript to mature mRNA can be explored. At the level of proteolytic processing of the prohormones, gene transfer can help determine the relative involvement of precursor structure and protease specificity in the production of diverse peptide products from a single precursor. Subsequent modification of the products by glycosylation, acetylation, amidation, and phosphorylation may be examined in heterologous cells. Factors that alter the secretion of the final products may be explored in cells that possess defined secretory pathways. Ultimately, the reconstitution of a prohormone-processing system by transfer of both prohormone and processing enzymes into a model cell should allow for the characterization of isolated processing events.

One of the most exciting potential applications of gene transfer experiments is the use of these techniques to reduce the level of putative processing enzymes. This can be done by transferring expression vectors that produce RNA complementary to the mRNA for a particular precursor or processing enzyme. The antisense RNA hybridizes with the target mRNA, forming a duplex that is poorly translated. The physiological consequences of selective removal of a peptide or processing event may then be studied, perhaps providing new insight into the physiological functions of these molecules.

2.1. Transfer of Proenkephalin into Mouse Pituitary Cells

The gene-transfer approach that has been most widely used to study processing specificity has involved the insertion of a gene coding for a prohormone into a cell type that normally does not express the prohormone. The system we have chosen to study uses gene transfer into a mouse cell line, AtT-20$_{D16V}$. These cells are derived from anterior pituitary corticotrophs: they share many properties with the original tissue with respect to hormone production. For example, AtT-20 cells produce large quantities of the prohormone POMC, the biosynthetic progenitor of adrenocorticotropic hormone (ACTH), β-endorphin, and the melanocyte-stimulating hormones (MSH). The processing of POMC to the active daughter peptides is well characterized in these cells (Roberts *et al.*, 1978; Mains *et al.*, 1977). Through gene transfer, we have presented these cells with the gene for human proenkephalin, a protein that shares many properties with POMC in terms of gene and protein structure.

A plasmid was constructed by subcloning the human proenkephalin gene into pBR322 (Comb *et al.*, 1983). For convenience in plasmid constructions a 2.5-kb deletion was introduced in intron III. This construction, pHENK 5.5, contained 200 bp of 5′ flanking DNA and 2.66 kb of 3′ flanking DNA. AtT-20 cells were cotransformed with pRSVneo (Gorman *et al.*, 1983) by the calcium phosphate precipitation technique (Graham and Van der Eb, 1973). The plasmid pRSVneo carries the bacterial neo gene, which confers resistance to the aminoglycoside drug, G418. After 2 weeks in selective media, 25 colonies were grown in culture for further analysis.

Genomic DNA analysis (Southern, 1975) indicated that 14 of the 25 clones contained one or more copies of the hENK gene. The number of gene copies present in the AtT-20/hENK clones ranged from one to as many as 20. Northern blot analysis (Thomas, 1980) showed that several of these clones expressed relatively high levels of a 1.4-kb mRNA that hybridized to human proenkephalin cDNA (Fig. 1). This mRNA is iden-

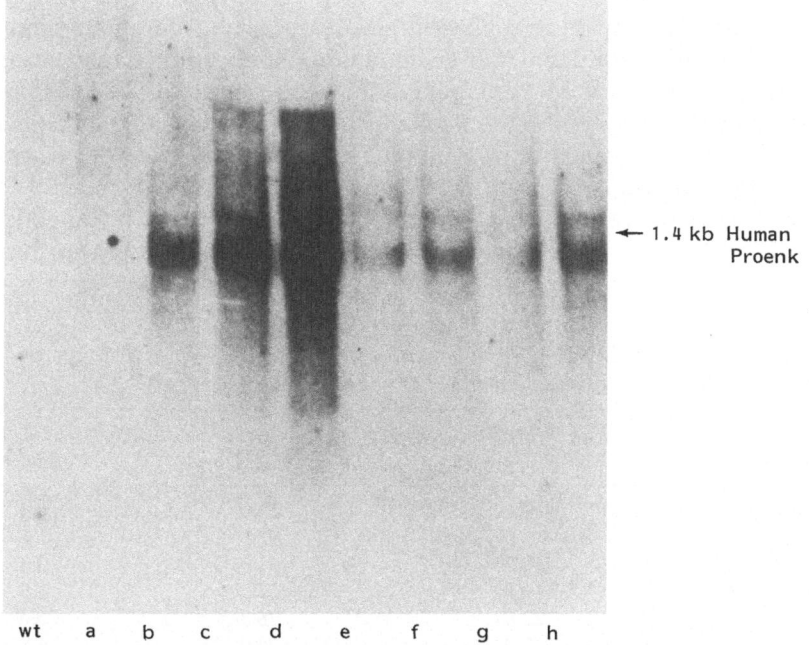

Figure 1. Northern blot analysis of RNA isolated from AtT-20 clones transformed with the human proenkephalin gene. Total RNA was isolated, size-fractionated by electrophoresis through a 1.7% agarose gel, transferred to nitrocellulose, and hybridized with a labeled 0.918-kb HincII human proenkephalin cDNA probe. All lanes contained 25 µg total RNA except lane f (19 µg) and lane h (17 µg). Control AtT-20 RNA is designated wt. Autoradiography was performed for 4 days at −70°C in the presence of an intensifying screen.

tical in size to the mature proenkephalin message in human pheochro-mocytoma. Thus, the AtT-20/hENK cells were able to use the splice junctions in the primary transcript of the human proenkephalin gene and produce mature mRNA of the correct size.

To determine whether proenkephalin protein was produced by the transformants, peptides were extracted and assayed for Met-enkephalin immunoreactivity. Some clones were found to contain relatively high levels of Met-enkephalin-IR, in some cases matching or exceeding the levels of the endogenous peptide, ACTH (Table I). In the clone exhibiting the highest expression of Met-enkephalin-IR (AtT-20/hENK-d) proen-kephalin reached 26% of the expression of POMC.

Given the high level of proenkephalin expression in these cells, it was of interest to determine whether processing of the precursor was taking place. Gel exclusion chromatography of cellular extracts indicated that most of the Met-enkephalin-IR material was present as low-molec-ular-weight peptides ($<3000\ M_r$) (Fig. 2A). These small peptides were further analyzed by reverse-phase high-pressure liquid chromatography (HPLC). The predominant Met-enkephalin-IR species eluted in the same position as authentic Met-enkephalin (Fig. 2B). Thus, AtT-20 cells pro-cess proenkephalin to Met-enkephalin with high efficiency, as shown by the relatively small accumulation of high-molecular-weight process-ing intermediates.

It was interesting that only a small amount of heptapeptide (YGGFMRF) and no octapeptide (YGGFMRGL) were present in these

Table I. Cellular Levels of Met-Enkephalin-IR and ACTH-IR in AtT-20 Clones Containing the Human Proenkephalin Gene[a,b]

| Clone | pmole IR/mg protein | | % |
	Met-enkephalin	ACTH	PE/POMC
WT	< 1.4	53.5	< 0.4
a	5.0	73.5	1.1
b	19.3	96.0	3.3
c	11.8	127.5	1.5
d	122.6	78.5	26.0
f	10.0	88.0	1.9
g	16.0	26.5	10.1
h	69.6	135.0	8.6

[a] Values are the average of two separate cultures, each assayed in duplicate.
[b] The molar ratio of proenkephalin to proopiomelanocortin (PE/POMC) was estimated by first dividing the moles of Met-enkephalin by six (for the number of copies of Met-enkephalin in proenkephalin). This value was divided by the moles of ACTH (one copy per mole of POMC) and multiplied by 100 to arrive at the % ratio PE/POMC.

extracts. It appears that these two peptides, which are flanked by pairs of basic amino acids in the sequence of proenkephalin, are further processed into Met-enkephalin. This may involve endoproteolytic cleavage at the single arginine residue, possibly between the methionine and arginine residues.

2.2. Other Gene-Transfer Studies

Several studies are now in the literature that have employed recombinant DNA to express peptide precursor proteins in heterologous cells. These studies vary in the use of cDNA or genes, DNA- or RNA-mediated gene transfer, and in the host cell used as the test system. An examination of these studies reveals much about the ability of cells to process and secrete heterologous peptides.

Proinsulin has been used in several studies to examine precursor processing. Using the SV40 virus, Gruss and Khoury (1981) inserted the rat preproinsulin gene I behind the late viral promotor. The construct included the single intron (119 bp) of the proinsulin gene, as well as putative regulatory signals at its 5' and 3' ends. Following infection of African green monkey kidney cells, polyadenylated mRNA was found that corresponded in size to insulin mRNA from rat insulinoma cells, indicating that proper splicing of the primary transcript was occurring. Furthermore, the preproinsulin mRNA possessed a 5' terminus identical to authentic rat preproinsulin I mRNA, suggesting that the transcription initiators of the proinsulin gene had been used. Immunoreactive insulin was present in both cell extracts and in the tissue culture medium. Immunoprecipitation of [^{35}S]cysteine-labeled proteins from infected cells yielded a protein that comigrated with authentic proinsulin. Thus, these cells possess the capacity to remove the signal peptide from preproinsulin but do not process the prohormone further.

Plasmids constructed from proinsulin genes or cDNAs and SV40 promoters have also been used to drive expression of proinsulin in heterologous cells. Lomedico (1982) examined the expression of the rat insulin II gene in Cos cells. This gene contains two introns (Lomedico et al., 1979) that were accurately removed during the processing of the primary transcript. When transfected Cos cells were labeled with [^{35}S]cysteine and the solubilized proteins subjected to immunoprecipitation with insulin antiserum, a single labeled protein was found that comigrated with authentic proinsulin, demonstrating that the signal peptide of preproinsulin had been removed. In related experiments, Laub and Rutter (1983) compared the expression of human insulin gene and cDNA in Cos cells, again under the control of SV40 promoters. They

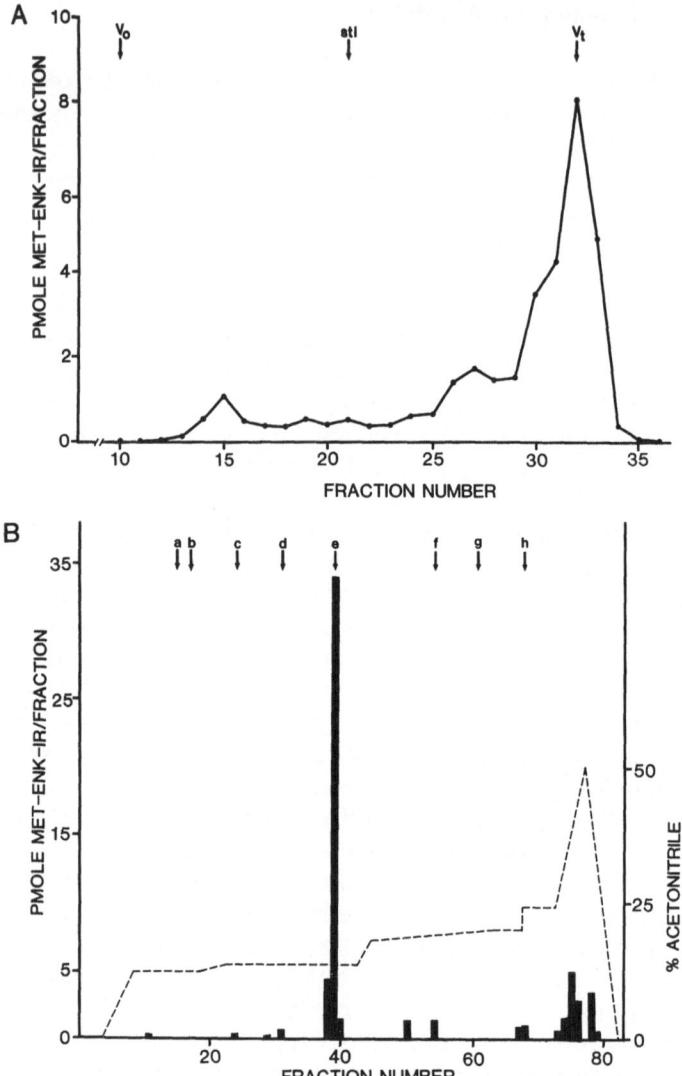

Figure 2. (A) Gel exclusion chromatography of extracts from AtT-20 cells transformed with human proenkephalin gene. Cellular protein was extracted in 1 M acetic acid, the high-speed supernatant was applied to a column (1.0 × 55 cm) of Sephadex G-100 equilibrated with 1 M acetic acid. Aliquots of fractions were dried and digested sequentially with trypsin (10 μg/ml) and carboxypeptidase B (0.1 μg/ml) before radioimmunoassay for Met-enkephalin-IR. The major peak of Met-enkephalin-IR eluted in the total volume of the column in the same position as authentic Met-enkephalin. However, approximately 50% of the total Met-enkephalin-IR eluted as larger peptides. These peptides represent the accumulation of intermediates in the processing of proenkephalin to low-molecular-weight peptides. V_0, void volume; sti, elution volume of soybean trypsin inhibitor (20,100 M_r); V_t, total volume. (B) Reverse-phase HPLC of low-molecular-weight peptides present in

found that intronic regions were not necessary for expression of the heterologous protein that accumulated in the cell culture medium. Interestingly, alternate pathways for splicing of the insulin gene were observed, resulting in the formation of a novel chimeric peptide via a codon frame shift, as well as authentic proinsulin produced by normal splicing.

All three of these studies used monkey kidney cells as hosts for gene transfer; in all cases it was found that the signal sequence of pre-proinsulin was removed without further proteolytic processing of the precursor. However, when a cell line derived from mouse anterior pituitary (AtT-20) was used as a host for a preproinsulin cDNA-SV40 recombinant plasmid, Moore et al. (1983) observed that proinsulin appeared to be proteolytically processed to insulin-immunoreactive material of the same molecular weight as authentic insulin. Upon stimulation with secretagogues, only the insulinlike material, and not proinsulin, was released into the tissue culture medium along with the endogenous POMC-derived peptides. Interestingly, transformed mouse fibroblast L cells were found to secrete only proinsulin, which was not stored and whose secretion rate did not depend on secretagogues. These results indicate that not all cells possess the capacity to recognize the processing and secretion signals present in a peptide precursor protein. Rather, the proteolytic processing machinery and secretion mechanisms may only be present in particular cells, in this case those derived from neuroendocrine tissue.

Another neuroendocrine peptide, growth hormone, has been the target of numerous gene-transfer experiments. Several different promoter systems have been used to drive the transcription of the growth hormone gene or cDNA, including SV40 promoters (Pavlakis et al., 1981) and retroviral promoters (Doehmer et al., 1982; Miller et al., 1984), as well as the endogenous growth hormone promoter (Robins et al., 1982) and a promotor from another neuroendocrine precursor, prolactin (Su-

←
_____

AtT-20/hENK cells. Fractions 28–34 from the G-100 column chromatography were pooled (before digestion with trypsin and carboxypeptidase B) and injected onto an Altex octyl column eluted with a discontinuous gradient of acetonitrile (dashed line). Met-enkephalin-IR was determined in the resulting fractions by radioimmunoassay following digestion with trypsin and carboxypeptidase B. The major peak of immunoreactivity (70% of total) eluted with the same retention time as authentic Met-enkephalin. The recovery of Met-enkephalin-IR from the column was 88%. Calibration standards: a, Met-enkephalin sulfoxide; b, Met-enkephalin-Arg^6Arg^7; c, Met-enkephalin-Arg^6; d, Arg^0-Met-enkephalin; e, Met-enkephalin; f, Leu-enkephalin; g, Met-enkephalin-Arg^6-Gly^7-Leu^8; h, Met-enkephalin-Arg^6-Phe^7.

powit *et al.*, 1984). These studies have focused on genetic elements that control the regulation of expression of growth hormone in heterologous cells; they have not addressed the specificity of proteolytic processing. All the cell types used in these studies (Cos, African green monkey kidney, mouse L cells, mouse NIH/3T3, rat 208F, human A431) appeared to remove the signal sequence, however.

Proteolytic processing of heterologous peptide precursors has been addressed in some studies. Hellerman *et al.* (1984) used a recombinant retrovirus encoding human preproparathyroid hormone (preproPTH) to infect a rat pituitary cell line, GH4. These cells, which synthesize and secrete prolactin, were shown by pulse-chase experiments and protein sequencing to synthesize preproPTH, rapidly remove the 25 amino acid signal sequence, and cleave the 6 amino-acid pro sequence to yield authentic PTH. In addition, the transformed cells released mature PTH into the medium in response to the secretagogue thyrotropin-releasing hormone. Thus, these neuroendocrine-derived cells exhibited efficient and accurate proteolytic processing of a simple, heterologous peptide precursor the secretion of which was regulated in parallel with the endogenous peptide, prolactin.

In contrast to the results reported for preproinsulin, accurate proteolytic processing of preprosomatostatin has been reported to occur in Cos cells (Warren and Shields, 1984). Following transfer of a SV-40 preprosomatostatin recombinant plasmid to Cos cells, small amounts of both preprosomatostatin and prosomatostatin were detected in cellular extracts, and low levels of the mature peptide somatostatin were secreted into the medium. It is possible that Cos cells possess proteases that recognize the processing signals within the sequence of prosomatostatin, and not those within proinsulin. This suggests rather strict sequence requirements of the endoproteolytic processing enzymes involved in neuroendocrine peptide biosynthesis as well as the fortuitous presence of one or a restricted number of these enzymes in Cos cells. Another possibility is that most cells possess proteases capable of processing peptide precursors correctly, but at a much lower activity than is present in specialized neurosecretory cells. In this case, the use of more sensitive methods for the detection of processed peptides may uncover previously unnoticed proteolytic cleavages (Warren and Shields, 1984).

Thus, it appears from these studies that removal of the hydrophobic signal sequence of the neuropeptide precursors is a function present in perhaps all cell types. However, the discrete proteolytic processing of the prohormone to yield the final peptide products may be normally restricted to cells of neuroendocrine origin that express particular endoproteolytic activities. Determination of the relative specificity of these special processing endoproteases with respect to the range of precursors that are cleaved must await additional gene-transfer experiments and ultimately purification of the enzymes themselves.

2.3. Reduction of Enzyme Activity with Antisense RNA

One very exciting application of gene-transfer techniques to the question of processing enzyme specificity involves the selective reduction of a putative processing enzyme from the cellular environment by blocking translation of the mRNA that codes for the enzyme. This may be achieved by introducing a cDNA for the enzyme under study in an expression vector similar to those used for transfer of the precursor proteins. During construction of the expression vector, however, the cDNA is oriented with respect to the promoter in such a way as to drive the transcription of the antisense strand of the cDNA, rather than the sense strand which codes for the protein. The presence of the antisense RNA leads to a decrease in the activity of the corresponding enzyme (Izant and Weintraub, 1984, 1985), permitting examination of the phenotypic consequences of loss of the target protein. This approach has been used successfully to inhibit thymidine kinase gene expression in mouse fibroblasts following direct microinjection of vector DNA into the cellular nucleus (Isant and Weintraub, 1984). Subsequently, it has been shown that calcium phosphate-mediated gene transfer was also effective in reducing enzyme activity in both transient expression systems and stably transformed cell lines (Izant and Weintraub, 1985).

The mechanisms by which antisense RNA inhibits enzymatic activity are unclear but almost certainly involve the formation of double-stranded RNA hybrid molecules *in vivo*. Indeed, such an antisense RNA : mRNA duplex has been observed in *Xenopus* oocytes following microinjection of antisense β-globin RNA (Melton, 1985), and RNA duplexes were found in cultured L cells following transfection with antisense thymidine kinase DNA (Kim and Wold, 1985). Such duplex formation would presumably block translation of the sense mRNA. The crucial subcellular site of action of the antisense RNA is unknown. In *Xenopus* oocytes, the site of the translational inhibition seems to be cytoplasmic, but events in the nucleus cannot be ruled out. In transformed L cells, greater than 95% of the sense message is localized in the nucleus when antisense RNA is present. This suggests that the antisense RNA may sequester the mRNA to the nucleus. In any event, the final result is to remove the target mRNA from the translational machinery, leading to a specific decrease in protein level. In some cases, it appears that the antisense RNA must cover the 5' end of the mRNA, particularly the

AUG codon, to block translation efficiently (Melton, 1985), although this is not necessary in all systems (Kim and Wold, 1985). Izant and Weintraub (1985) found that fragments as short as 52 bp complementary to 5' untranslated sequences can be sufficient for inhibiting the activity of the target protein. This approach suggests the feasibility of gene replacement experiments, wherein translation of an endogenous protein (prohormone or processing enzyme) is inhibited specifically with a short 5' untranslated antisense sequence and replaced by a modified protein under the transcriptional control of an unrelated promoter. Such experiments could greatly facilitate the study of specific mutations in eukaryotic systems.

3. ENZYMATIC STUDIES

The results of the gene-transfer experiments as well as analysis of tissue-specific differences in peptide processing suggest that despite the specificity of some processing pathways for a particular prohormone, many pathways appear to be common to the various peptide systems. Most of the differences in prohormone processing could be explained by selective endopeptidases that cleave the precursors only at specific sites. However, some of the endopeptidases that recognize pairs of basic amino acids appear to be nonselective for any one precursor. Other processing events, such as acetylation, amidation, and carboxypeptidase B-like activity, also appear to be common to many different peptide precursors.

It is not surprising that *in vitro* experiments with partially purified processing enzymes have yielded results similar to those produced in *in vivo* experiments. That is, while some of trypsinlike endopeptidases have been reported to be specific for individual prohormones, many of the other reported processing enzymes are capable of processing a wide variety of peptide precursors. However, many of these studies are only preliminary results, since no enzyme has yet been proved to be a processing enzyme.

An enzyme must satisfy several commonly accepted criteria to be designated a prohormone-processing enzyme (Docherty and Steiner, 1982). First, the cellular and subcellular localization of the enzyme must be consistent with the proposed function of the enzyme. Since peptide processing is thought to take place in secretory granules, the processing enzymes should be present in these granules. Second, the enzyme must be active under conditions (such as pH and ion concentrations) that simulate the internal environment of secretory granules. Of crucial importance is the ability of the enzyme to process the prohormone correctly, without further degradation of the products. Finally, reduction of the enzyme activity, such as with highly specific inhibitors, mutagenesis, or antisense mRNA experiments, must have an effect on processing in an *in vivo* system. While no enzyme has met all of the criteria, many enzymes have been proposed to be processing enzymes, and for some the evidence is rather compelling.

3.1. Trypsinlike Processing Enzymes

Trypsinlike endopeptidases that process nerve growth factor and epidermal growth factor have been identified as kallikreins (Bothwell *et al.*, 1979). These enzymes cleave the precursors at specific sites containing basic amino acids, and then bind the products (Taylor *et al.*, 1970). The resulting complex may serve to protect the growth factors from further degradation. Interestingly, although both enzymes have many properties in common, they appear to be very specific for their own growth factor precursor (Server and Shooter, 1976). The nerve growth factor converting enzyme will not process the precursor for epidermal growth factor (Frey *et al.*, 1979). Other kallikreins have been proposed to be processing enzymes for other precursors, such as POMC (Powers and Nasjletti, 1982, 1983). However, none of these studies has shown that the kallikreins present in the pituitary are capable of processing POMC into the expected products.

An enzyme with many properties in common with cathepsins B and D has also been proposed to be a POMC-processing enzyme. This enzyme is present in highly purified secretory granules from the anterior (Chang and Loh, 1983), intermediate (Loh and Gainer, 1982), and neural lobes of pituitary (Chang *et al.*, 1982). All enzymes show an acidic pH optimum (pH 4–6) and a sensitivity to thiol protease inhibitors. The purified enzyme from bovine intermediate lobe pituitary secretory granules is capable of processing POMC into products of the proper size and immunological properties (Loh *et al.*, 1985).

A similar enzymatic activity has been found in secretory granules isolated from anglerfish pancreatis islets (Fletcher *et al.*, 1980, 1981). This activity processes prosomatostatin, proglucagon, and proinsulin into the expected peptides, although it is unclear as to whether more than one enzyme is responsible for the observed activity. Within the secretory granules, both soluble and membrane-bound converting activities were detected. The two forms of enzyme activity showed similar substrate and inhibitor specificities and were present in comparable amounts.

Several trypsinlike endopeptidases have been reported to be present in bovine adrenal chromaffin granules (Lindberg *et al.*, 1982, 1984; Evangelista *et al.*, 1982; Mizuno *et al.*, 1982, 1985). One of these enzyme activities is a serine protease with a neutral pH optimum (Lindberg *et al.*, 1982). This enzyme generates enkephalin from higher-molecular-weight precursors and does not cleave β-lipotropin (Lindberg *et al.*, 1984). The enzyme activity hydrolyzed enkephalin-containing peptides (such as peptide E and peptide F) at pairs of basic amino acids, although peptides with single basic amino acids (Met-enkephalin-Arg^6-Phe^7) were hydrolyzed slowly. The enzyme was inhibited by soybean trypsin inhibitor and aprotinin, suggesting some similarities with other serine proteases (trypsin and kallikrein). However, the apparent substrate specificity of this chromaffin granule endopeptidase differs considerably from either trypsin or the kallikreins.

Recently, Mizuno *et al.* (1985) described a similar endopeptidase in adrenal chromaffin granules. As reported by Lindberg *et al.* (1984), the enzyme cleaves between pairs of basic amino acids producing two peptides, each with a basic amino acid attached. While the carboxyl-terminal basic amino acid could be removed by the carboxypeptidase B-like enzyme (Fricker, 1985), the amino-terminal basic amino acid would require aminopeptidase activity. An enzyme that removes an amino-terminal arginine has been reported to be present in bovine pituitary secretory vesicles (Gainer *et al.*, 1984).

A trypsinlike endopeptidase that cleaves opioid peptide precursors at single basic residues has also been described. This enzymatic activity shows a marked specificity for leumorphin (dynorphin B-29), which is processed into dynorphin B (Devi and Goldstein, 1983). Both dynorphin A and peptide E inhibit this enzyme with high affinities, whereas a variety of other peptides do not influence enzymatic activity (Devi and Goldstein, 1986). Interestingly, ACTH also potently inhibited the dynorphin-converting activity (Devi and Goldstein, 1986), although until this enzymatic activity has been purified, it is impossible to determine whether these other peptides also serve as substrates. This thiol protease is maximally active at pH 8, and shows only 10% maximal activity at pH 6 (Devi and Goldstein, 1983).

Most of these trypsinlike endopeptidases are reported to be very specific for particular precursors, although the results are quite preliminary. The emerging trend is that there are specific endopeptidase-processing enzymes, and it is unlikely that a single enzyme is responsible for processing many prohormones. Several endopeptidases may even be required to process a single prohormone into the various fragments. These fragments, which still contain the basic amino acids on their C-terminus, must be further processed by a carboxypeptidase B-like exopeptidase. Recent studies suggest that a single carboxypeptidase B-like enzyme is responsible for processing many different peptide hormones.

3.2. Carboxypeptidase E (Enkephalin Convertase)

Early studies on the carboxypeptidase B-like processing enzyme detected a unique enzymatic activity in bovine adrenal chromaffin granules (Fricker and Snyder, 1982). This activity was shown to produce both [Met] and [Leu] enkephalin from either Arg[6] or Lys[6] enkephalin, and so was designated both enkephalin convertase and carboxypeptidase E (CPE). The properties of this enzyme were distinct from other carboxypeptidases, such as carboxypeptidase B (Folk, 1971), carboxypeptidase N (Plummer and Erdos, 1981), and a carboxypeptidase reported to be present in partially purified chromaffin granules (Hook et al., 1982), which was suggested to be a lysosomal enzyme (Fricker and Snyder, 1982). CPE is significantly stimulated by $CoCl_2$ and is inhibited by the chelating agents, 1,10-phenanthroline and EDTA (Fricker and Snyder, 1982) as well as by some thiol inhibitors (Fricker et al., 1982; Hook and Eiden, 1984). Carboxypeptidase E is maximally active at pH 5.6, with very little detectable activity above pH 7.0.

Further studies on CPE in different tissues found high levels of enzymatic activity in brain and pituitary (Fricker et al., 1982). Within the pituitary, this enzymatic activity was found to be associated with the secretory granules (Hook and Loh, 1984). In the rat, CPE levels in the anterior pituitary were 20-fold higher than the average brain levels (Fricker et al., 1982). Within the brain, CPE activity showed significant variations, with highest levels in the hippocampus and hypothalamus and lowest levels in the cerebellum (Fricker and Snyder, 1982).

Upon fractionation of the tissues, it became apparent that two forms of CPE were present in the secretory granules: a soluble and a membrane-bound form (Fricker and Snyder, 1982). The soluble CPE was purified to homogeneity from either bovine brain, pituitary, or adrenal medulla extracts using ion exchange, gel filtration, and affinity chromatography (Fricker and Snyder, 1983). After solubilization of the membrane-bound enzyme with 1% Triton X-100 and 1 M NaCl, purification to homogeneity was achieved using an identical purification scheme (Supattapone et al., 1984). The only detectable difference between the two forms of CPE was a slight difference in size, with the membrane-bound form (52,000 M_r) larger than the soluble form (50,000 M_r). Both forms of CPE possess the same amino acid sequence at their N-termini (Fricker et al., 1986). Purified CPE from either soluble or membrane fractions of brain, pituitary, and adrenal medulla showed similar enzymatic properties (Fricker and

Snyder, 1983; Supattapone *et al.*, 1984). The effect of ions and inhibitors on enzymatic activity was comparable, as originally observed with crude tissue homogenates. A series of synthetic substrates were hydrolyzed with similar kinetic constants for CPE from the different tissues. The membrane-bound form gave results similar to the soluble enzyme form with respect to the different ions, inhibitors, and substrates. Thus, no difference in CPE from the various tissues could be detected, and the only measurable difference between the soluble and membrane-bound forms was the apparent molecular weight.

The similarity of CPE from the brain, pituitary, and adrenal medulla suggests that this enzyme is involved in the production of a wide variety of neuropeptides. In the adrenal medulla chromaffin granules, CPE is probably involved in the biosynthesis of the enkephalin peptides. Since the anterior and intermediate lobes of the pituitary contain high levels of CPE and pro-opiomelanocortin peptides but low levels of enkephalin, it is likely that CPE is involved in the production of pro-opiomelano-cortin-derived peptides. The distribution of CPE in rat brain, determined either by autoradiography with a specific inhibitor (Lynch *et al.*, 1984) or by immunostaining (Hook *et al.*, 1985) shows a correlation with many neuropeptides. An enzyme with many properties similar to brain, pituitary, and adrenal CPE has been identified in rat insulinoma secretory granules (Docherty and Hutton, 1983). Although complete characterization of this enzyme has not been reported, these preliminary results suggest that the same carboxypeptidase B-like enzyme is involved in the biosynthesis of peptide hormones as well as peptide neurotransmitters.

The similarity of CPE in the different neuropeptide systems predicts that this enzyme has been evolutionarily stable and probably has been involved in peptide biosynthesis for a long time. To test this, we examined tissue extracts from several diverse species for CPE-like carboxypeptidase activity. Human, bovine, mouse, frog, shark, and *Aplysia* neural tissues contained a carboxypeptidase with similar properties (Fricker and Herbert, in preparation). Several active site-directed inhibitors of carboxypeptidase B-like enzymes inhibited human, bovine, mouse, frog, and shark CPE with similar inhibition constants. Interestingly, the *Aplysia* CPE displayed a considerably lower affinity for lysine-derived inhibitors than for arginine-derived ones. However, the arginine-based inhibitors showed a similar potency for CPE from *Aplysia* with CPE from the other species. Thus, while some difference could be found, there are many similarities among CPE from the different species.

As with bovine tissues, CPE from all species examined existed in both a soluble and a membrane-bound form (Fricker and Herbert, in preparation). The fraction of total CPE activity that was soluble ranged from 50 to 80% for the mammalian tissues. However, soluble CPE accounted for only 10 to 20% of total CPE activity in frog, shark, and *Aplysia* tissues. Both the soluble and membrane-bound forms of CPE showed similar tissue distributions and enzymatic properties. The similarities of the two forms of CPE raise some questions concerning the origin of the soluble and membrane-bound CPE. One possibility is that the membrane-bound form served as the precursor to the smaller soluble enzyme. Alternately, both forms could arise from different mRNAs without any interconversion of the two forms.

Recently, a clone that encodes CPE has been isolated from a bovine pituitary cDNA library (Fricker *et al.*, 1986). The predicted amino acid sequence from this clone matches the known amino acid sequence of several proteolytic fragments of CPE. Interestingly, the amino acid sequence of CPE shows some homology with carboxypeptidase A and B, suggesting a common evolutionary origin. All the amino acids thought to be essential for catalytic activity of carboxypeptidase A and B (Plummer, 1969; Schmid and Herriott, 1976; Lipscomb, 1983) are present in CPE in the corresponding position. These amino acids include the zinc-binding histidines and glutamate and the tyrosine and glutamate thought to participate in the hydrolysis of the peptide bond.

Other features, such as the region that binds the free carboxy-terminus of the peptide substrate (Schmid and Herriott, 1976) also show a high degree of homology. Of the 90 amino acids that have been conserved among bovine CPB, crayfish CPB, bovine CPA, and rat CPA (Titani *et al.*, 1984), 35 are present in bovine CPE (39% homology). The overall homologies of CPE with bovine CPB (17.3% amino acid identity) and of CPE with bovine CPA (20%) are considerably less than the homology of CPA with CPB (47.8%). This suggests that CPE diverged from the other carboxypeptidases before CPA and CPB diverged.

Analysis of bovine genomic DNA (Southern blots) with CPE cDNA probes reveals a single gene for this enzyme (Fig. 3). When the genomic DNA is digested with Bam HI or Eco RI, CPE cDNA probes hybridize to a single band (Fig. 3, lanes 1 and 2). Digestion of genomic DNA with Hind III produces several fragments that hybridize to the CPE cDNA probes (Fig. 3, lane 3). Since there is a single Hind III site within the 350 base-pair (bp) region of the cDNA probes, these probes should hybridize to two Hind III fragments.

In addition to the major bands on the Southern blot, other faint bands appear in all the lanes. Since the homology between CPE and CPA or CPB at the nucleotide level is not sufficient for hybridization to occur under the stringent conditions used, these minor bands on the

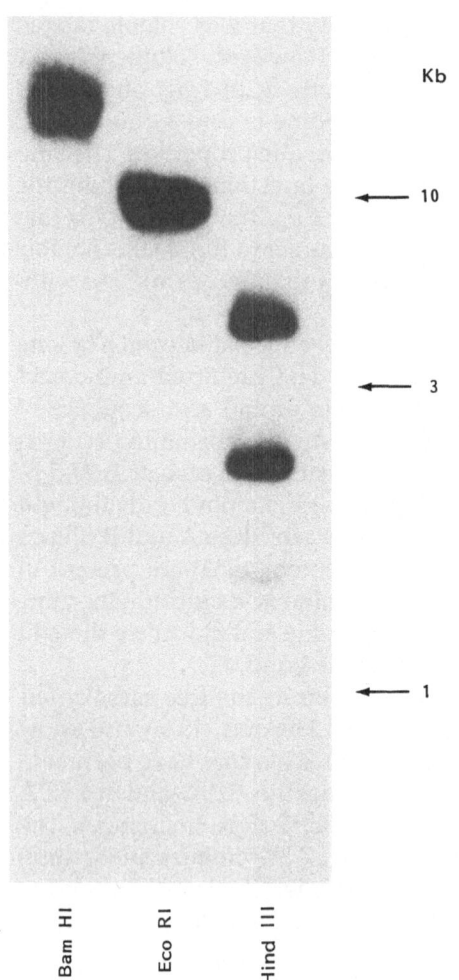

Figure 3. Restriction enzyme analysis of the bovine carboxypeptidase E gene. Bovine genomic DNA (20 μg) was digested with Bam HI, Eco RI, or Hind III, electrophoresed on a 0.8% agarose gel, and transferred to a nitrocellulose filter (Southern, 1975). A 350-bp piece of cDNA corresponding to a region coding for CPE was labeled with ^{32}P by nick translation and hybridized overnight at 55°C. (From Fricker *et al.*, 1986). A 1-kb ladder (Bethesda Research Laboratories, Gaithersburg, Maryland) was used as DNA size standards.

Southern blot are probably not CPA or CPB genes. It is more likely that there are other members of the carboxypeptidase family with greater homology to CPE. One possible candidate is the plasma enzyme, carboxypeptidase N (Plummer and Erdos, 1981). This enzyme shares many properties in common with CPE, such as a specificity for basic amino acids, a significant stimulation by cobalt chloride, inhibition by chelating agents, and similar molecular weight. The major differences between CPE and CPN are the binding affinities of several active-site-directed inhibitors (Fricker *et al.*, 1983) and the optimum pH for catalytic activity (pH 5–6 for CPE, pH 7–8 for CPN). These differences could result from minor changes in the amino acid sequences of the two enzymes, which evolved to function in different environments. It will be of interest to compare the amino acid sequence of CPE with CPN, once known.

In summary, the neuropeptide-synthesizing carboxypeptidase E is a member of a family of carboxypeptidases, including carboxypeptidase A, carboxypeptidase B, and possibly carboxypeptidase N. Although related to other enzymes, CPE appears to be specific for neuropeptide biosynthesis, based on its cellular and subcellular distribution. Since CPE is the only carboxypeptidase that has been detected in highly purified secretory granules from several different tissues, it is likely that this enzyme is involved in the biosynthesis of a variety of peptide hormones and neurotransmitters.

Other enzymatic activities that are often involved in neuropeptide biosynthesis, such as the acetylating and the amidating activities, may also be general processing enzymes (Mains *et al.*, 1983). By contrast, many studies suggest that some of the trypsinlike processing enzymes are very specific for certain cleavage sites within a neuropeptide precursor. However, this apparent specificity is not necessarily a direct result of different trypsinlike processing enzymes. It is possible that the secretory granule environment influences enzyme specificity and that tissue-specific prohormone processing is a result of differences within secretory granules. Thus, a discussion of the specificity of prohormone processing should include a comparison of the secretory granule environments in which processing occurs.

4. THE INTERNAL ENVIRONMENT OF SECRETORY GRANULES

Neuropeptide-containing secretory granules from different tissues have many properties in common. The most obvious property that could affect enzyme activity (or specificity) is the internal pH of the granules. The internal pH of secretory granules from bovine adrenal medulla (Casey *et al.*, 1977; Johnson and Scarpa, 1976) and bovine posterior pituitary (Russell, 1984) are comparable with most estimates close to pH 5.6. While some of the putative processing enzymes are active at this pH, other enzymes have been reported to have neutral pH optima. These enzymes could still function as processing enzymes, since it is believed that the secretory granules initially have a neutral internal pH. An ATP-dependent proton pump on the secretory granule membrane acidifies the internal environment as the granule matures. Thus, enzymes with a neutral pH optimum would be active in young granules and would become less active as the granule matures. Enzymes with an acidic pH optimum would initially be less active, becoming activated as the granule is acidified (Gainer *et al.*, 1985). This may be a mechanism by which the cell controls the specificity of prohormone processing. However, since different peptidergic secretory granules are thought to undergo similar changes in their internal pH, it is unclear as to how differential processing could result from this.

Secretory granules from different tissues also have other properties in common. The most widely studied peptide-containing secretory granules are the chromaffin granules of the bovine adrenal medulla. These granules contain relatively high levels of catecholamines, nucleotides, magnesium, calcium, ascorbate, and a class of proteins designated *chromogranins* (Smith and Winkler, 1967; Kirshner, 1974; Winkler and Westhead, 1980). Although catecholamines are not always found in peptidergic secretory granules, some of the other components are thought to be present in various types of secretory granules (Whittaker, 1974; Howell, 1974; Pletscher *et al.*, 1974; Lagercrantz, 1976). Since chromogranins are a major class of proteins in the chromaffin granules, it is of interest that similar acidic proteins are found in a wide variety of secretory granules. Such proteins include the neurophysins of the neurohypophyseal vesicles (Hope and Pickup, 1974), vesiculin of the cholinergic vesicle from *Torpedo* (Whittaker, 1974), soluble acidic lipoproteins (Koenig, 1974), secretory protein-1 of the parathyroid gland (Morrissely *et al.*, 1980), and chromogranin A of the adrenal chromaffin granules (Smith and Winkler, 1967; Houge-Angelletti, 1977). These latter two proteins share many properties, such as physical characteristics, amino acid composition, and amino-terminal sequence, and may represent the same protein (Cohn *et al.*, 1982).

Chromogranin immunoreactivity is found in a variety of tissues, where it is often localized in neuropeptide producing cells (O'Connor *et al.*, 1983; Cohn *et al.*, 1984; Somogyi *et al.*, 1984; O'Connor, 1983; O'Connor and Frigan, 1984; Lloyd and Wilson, 1983). Within the central nervous system (CNS), chromogranin immunoreactivity is widely distributed and is not limited to catecholaminergic pathways. High levels of chromogranin immunoreactivity are found in bovine hippocampus and pituitary (Somogyi *et al.*, 1984). Interestingly, these regions also contain the highest levels of carboxypeptidase E activity (Fricker *et al.*, 1982). The subcellular distribution of these two proteins shows many similarities in some brain regions, such as the hippocampus and hypothalamus (Somogyi *et al.*, 1984; Lynch *et al.*, 1984). The distribution of both chromogranin and CPE activity correlates better with peptidergic systems in general than with any specific neurotransmitter.

The function of chromogranins is not clear, and several plausible ideas have been proposed. These proteins could function to lower the osmotic pressure inside the vesicle by transiently binding solutes (Sen and Sharp, 1982). Since the secretory vesicles contain such a high concentration of ions and other solutes, a complex among adenosine triphosphate (ATP), catecholamines, and chromogranins would lower the osmotic pressure inside the vesicle (Helle et al., 1985). However, the stability of the chromaffin granule can be accounted for, perhaps entirely, by spontaneous interactions between catecholamines and nucleotides (Kopnell and Westhead, 1982). Chromogranins are also found in much lower quantities in sympathetic nerve vesicles (Lagercrantz, 1976), and they are not always colocalized with catecholamines in the CNS (Somogyi et al., 1984). Alternatively, chromogranins, or smaller peptides processed from the larger proteins, may have some unindentified hormonal activity since they are secreted into the bloodstream upon stimulation of the splanchnic nerve (Blaschko et al., 1967).

Chromogranins could also be involved in the control of prohormone processing. While there is no evidence for such a role, recent findings make this idea worth investigating. The amino acid sequence of chromogranin A, deduced from the nucleotide sequence of the cDNA clones (Iacangelo et al., 1986) reveals several pairs of basic amino acids within the protein. Three of these pairs are Lys-Arg residues, which are often found in prohormones. It is possible that these pairs of basic amino acids within chromogranin A are recognized by the endopeptidases that cleave prohormones at pairs of basic amino acids. This could account for the smaller forms of chromogranins observed in secretory granules. However, the abundance of the single-chain 70,000-M_r chromogranin A in adrenal chromaffin granules suggests that significant cleavage at the pairs of basic amino acids does not occur. These regions of chromogranin might still interact with the substrate-binding site of an endopeptidase without serving as substrates themselves. Many examples exist where a peptide inhibitor binds to an enzyme to form an inactive enzyme–inhibitor complex, with only limited proteolysis of the inhibitor protein (for review, see Fritz et al., 1974).

Since chromogranin A does not possess much secondary structure in solution (Daniels et al., 1978; Sen and Sharp, 1982), these pairs of basic amino acids are probably accessible to the prohormone-processing endopeptidases. Even if the affinity of the enzyme for chromogranin is low, the relative abundance of chromogranin in the secretory granules is an important factor. The level of chromogranin in the adrenal chromaffin granules is much higher than the level of proenkephalin (Winkler and Westhead, 1980). Thus, if chromogranins interact even weakly with the prohormone processing enzymes, this could influence the biosynthesis of neuropeptides.

In summary, the environment inside the secretory vesicle can influence the activity of polypeptide hormone-processing enzymes. The activation or inactivation of the enzymes during maturation of the vesicle may be dependent on the change in pH. The chromogranins, which are major soluble protein components of many peptidergic secretory vesicles, may play a role in processing. However, it is unclear as to how chromogranins could control activation or specificity of processing enzymes. Further studies using both *in vitro* and *in vivo* approaches are necessary to discern how the environment inside the secretory vesicle affects prohormone processing.

5. PERSPECTIVES

As more is learned about the structure of prohormones and processing enzyme, a general scheme of prohormone biosynthesis is emerging. The synthesis of bioactive peptides from their precursor proteins requires the sequential action of endopeptidase(s) and exopeptidase(s). The endoproteolytic cleavage most frequently occurs at pairs of basic amino acid residues. However, evidence is mounting that not all these sites in a precursor protein are equivalent. For example, the precursors to the opioid peptides have been shown to undergo selective endoproteolytic cleavage in different tissues to produce different sets of peptides. It is not known whether this form of tissue specific processing is a result of distinct endoproteases in the tissues or differences in local environments that alter the structure of the substrate or specificity of the enzyme. Nevertheless, these studies clearly indicate that endoproteases play a key role in the regulation of neuropeptide production.

Using current techniques of molecular biology, the gene for a prohormone can be transferred to a cell that does not normally express that prohormone. This permits comparison of the processing of two different prohormones within an identical environment. The gene-transfer approach has been applied to the study of the specificity of endopeptidases in neuroendocrine cells. The expression of the proenkephalin gene in pituitary cells (AtT-20 cells) that normally express POMC, but not proenkephalin, has enabled us to compare processing of two different precursors in the same cellular environment. The finding that AtT-20 cells appear to cleave all 10 pairs of basic amino acids in proenkephalin is of particular interest in view of the fact that only four of eight pairs of basic amino acids in POMC are cleaved in the same cells. There are no obvious features in the primary structure of the two precursors that would explain why so many potential cleavage sites in POMC are not cut while apparently all these sites are cleaved in proenkephalin by AtT-20 cells. Thus, features of secondary and tertiary structure of the precursor appear to be critical in determining which cleavage sites in a precursor are actually cut.

The use of classic biochemical approaches in studying processing enzymes has resulted in the identification of a cobalt activated carboxypeptidase B-like enzyme in the secretory granules of neuroendocrine cells. This enzyme, designated both enkephalin convertase and carboxypeptidase E (CPE), is present in neuroendocrine tissues of a wide variety of animals. Through the use of recombinant DNA methods, it has been possible to clone cDNA that codes for this enzyme and to determine its primary amino acid sequence from the nucleotide sequence of the cDNA. Although this carboxypeptidase is related to other enzymes, including carboxypeptidase A, B, and N, its cellular and subcellular distribution suggest that it is specific for processing neuroendocrine peptides. Furthermore, since CPE is the only carboxypeptidase found in purified secretory granules from several neuroendocrine tissues, it is likely that this enzyme is a general one involved in the processing of a wide variety of peptides. Nucleic acid homology studies with the cDNA probes indicate that the mRNA that codes for this enzyme is present in many different neuroendocrine tissues, strongly supporting the idea that it is involved in the processing of many kinds of neuropeptides.

Other enzymes involved in neuropeptide biosynthesis, such as amidating and acetylating enzymes, also appear to be general processing enzymes. By contrast, some of the endopeptidases appear to be specific for a given type of precursor. However, this specificity could be related to differences in the intracellular environment in which these enzymes operate.

The internal environment of secretory vesicles may play an important role in the regulation of prohormone processing. Changes in the internal pH could have a significant effect on the activity of the various processing enzymes. One of the major proteins within secretory vesicles, the chromogranins, might also play a role in prohormone processing by competing with the prohormones for the enzymes. While there is no direct evidence for such a role, the presence of Lys-Arg sequences within chromogranin make this idea worth investigating.

Several very powerful gene-transfer approaches are now being developed to answer questions raised in this chapter about specificity of prohormone processing and the role various enzymes or environmental factors play in processing peptides in the cell. The availability of cloned genes that code for prohormones and for candidate-processing enzymes makes it possible to analyze the processing capability of these enzymes in different cellular environments. One can transfect and express these genes in a wide variety of secretory and nonsecretory cells and assess the ability of these enzymes to function in a variety of cell types. One can also mutate sequences in the coding regions of the cloned genes and analyze the effect of specific amino acid substitutions in the enzymes and precursors on processing and secretion in transfected cells. Finally, it is possible to reduce cellular levels of specific proteins through the production of antisense RNA *in vivo* and ask what effect this reduction has on the processing of a neuroendocrine peptide. With this technique it should be possible to evaluate the role of a specific enzyme in the cell with the same degree of confidence attained in genetic experiments with yeast or bacteria. Thus, the future holds promise of spectacular advances in our knowledge of the processes that control the production of neuropeptides.

REFERENCES

Barnea, A., Cho, G., and Porter, J. C., 1982, Molecular-weight profiles of immunoreactive corticotropin in the hypothalamus of the aging rat, *Brain Res.* **232**:355–363.

Blaschko, H., Comline, R. S., Schneider, F. H., Silver, M., and Smith, A. D., 1967, Secretion of a chromaffin granule protein, chromogranin, from the adrenal gland after splanchnic stimulation, *Nature (Lond.)* **215**:58–59.

Blobel, G., and Dobberstein, B., 1975, Transfer of proteins across membranes. I. Presense of proteolytically processed and unprocessed nascent immunoglobulin light chains on membrane-bound ribosomes of murine myeloma, *J. Cell Biol.* **67**:835–851.

Bothwell, M. A., Wilson, W. H., and Shooter, E. M., 1979, The relationship between glandular kallikrein and growth factor-processing proteases of mouse submaxillary gland, *J. Biol. Chem.* **254**:7287–7294.

Bradbury, A. F., Finnie, M. D. A., and Smyth, D. B., 1982, Mechanisms of C-terminal amide formation by pituitary enzymes, *Nature (Lond.)* **298**:686–688.

Casey, R. P., Njus, D., Radda, G. K., and Sehr, P. A., 1977, Active proton uptake by chromaffin granules: Observation by amino distribution and phosphorus-31 nuclear magnetic resonance techniques, *Biochemistry* **16**:972–977.

Chang, T. L., and Loh, Y. P., 1983, Characterization of proopiocortin converting activity in rat anterior pituitary secretory granules, *Endocrinology* **112**:1832–1838.

Chang, T. L., Gainer, H., Russell, J. T., and Loh, Y. P., 1982, Proopiocortin-converting enzyme activity in bovine neurosecretory granules, *Endocrinology* **111**:1607–1614.

Chavkin, C., and Goldstein, A., 1981, Specific receptor for the opioid peptide dynorphin: Structure–activity relationships, *Proc. Natl. Acad. Sci. USA* **78**:6543–6547.

Cohn, D. V., Zangerle, R., Fischer-Colbrie, R., Chu, L-L-H., Elting, J. J., Hamiton, J. W., and Winkler, H., 1982, Similarity of secretory protein I from parathyroid gland to chromogranin A from adrenal medulla, *Proc. Natl. Acad. Sci. USA* **79:**6056–6059.

Cohn, D. V., Elting, J. J., Frick, M., and Elde, R., 1984, Selective localisation of the parathyroid secretory protein-I/adrenal medulla chromogranin A protein family in a wide variety of endocrine cells of the rat, *Endocrinology* **114:**1963–1974.

Comb, M., Seeburg, P. H., Adelman, J., Eiden, L., and Herbert, E., 1982, Primary structure of the human met- and leu-enkephalin precursor and its mRNA, *Nature (Lond.)* **295:**663–666.

Comb, M., Rosen, H., Seeburg, P., Adelman, J., and Herbert, E., 1983, Primary structure of the human proenkephalin gene, *DNA* **2:**213–229.

Corbett, A. D., Paterson, S. J., McKnight, A. T., Magnan, J., and Kosterlitz, H. W., 1982, Dynorphin 1-8 and dynorphin 1-9 are ligands for the Kappa-subtype of opiate receptor, *Nature (Lond.)* **299:**79–81.

Crine, P., Seidah, N. G., Gossard, F., Lis, M., and Chretien, M., 1979, Processing of the two forms of the common precursor for a α-melanotropin and β-endorphin in the rat pars intermedia, *Biol. Cell.* **36:**119–125.

Daniels, A. J., Williams, R. J. P., and Wright, P. E., 1978, The character of the stored molecules in chromaffin granules of the adrenal medulla: A nuclear magnetic resonance study, *Neuroscience* **3:**573–585.

Deakin, J. F. W., Dostrousky, O., and Smyth, D. B., 1980, Influence of N-terminal acetylation and C-terminal proteolysis on the analgesic activity of β-endorphins, *Biochem. J.* **189:**501–506.

Devi, L., and Goldstein, A., 1983, Dynorphin converting enzyme with unusual specificity from rat brain, *Proc. Natl. Acad. Sci. USA* **81:**1892–1896.

Devi, L., and Goldstein, A., 1986, Opioid and other peptides as inhibitors of leumorphin (dynorphin B-29) converting activity, *Peptides* **7:**87–90.

Docherty, K., and Steiner, D. F., 1982, Post-translational proteolysis in polypeptide hormone biosynthesis, *Annu. Rev. Physiol.* **44:**625–638.

Docherty, K., and Hutton, J. C., 1983, Carboxypeptidase activity in the insulin secretory granule, *FEBS Lett.* **162:**137–141.

Doehmer, J., Barinaga, M., Vale, W., Rosenfeld, M. G., Verman, I. M., and Evans, R. M., 1982, Introduction of rat growth hormone gene into mouse fibroblasts via a retroviral DNA vector, expression and regulation, *Proc. Natl. Acad. Sci. USA* **79:**2268–2272.

Douglass, J., Civelli, O., and Herbert E., 1984, Polyprotein gene expression: Generation of diversity of neuroendocrine peptides, *Annu. Rev. Biochem.* **53:**665–715.

Eipper, B. A., and Mains, R. E., 1981, Further analysis of post-translational processing of β-endorphin in rat intermediate pituitary, *J. Biol. Chem.* **256:**5689–5695.

Eipper, B. A., Mains, R. E., and Glembotski, C. C., 1983, Identification in pituitary tissue of a peptide α-amidating activity, *Proc. Natl. Acad. Sci. USA* **80:**5144–5149.

Evangelista, R., Ray, P., and Lewis, R. V., 1982, A "trypsin-like" enzyme in adrenal chromaffin granules: A proenkephalin processing enzyme, *Biochem. Biophys. Res. Commun.* **106:**895–902.

Evans, C. J., Lorenz, R., Weber, E., and Barchas, J. D., 1982, Variants of α-melanocyte stimulating hormone in rat brain and pituitary: Evidence that acetylated α-MSH exists only in the intermediate lobe of the pituitary, *Biochem. Biophys. Res. Commun.* **106:**910–919.

Fletcher, D. J., Noe, B. D., Bauer, G. E., and Quigley, J. P., 1980, Characterization of the conversion of a somatostatin precursor to somatostatin by islet secretory granules, *Diabetes* **29:**593–599.

Fletcher, D. J., Quigley, J. P., Bauer, G. E., and Noe, B. D., 1981, Characterization of proinsulin and proglucagon converting activities in isolated islet secretory granules, *J. Cell Biol.* **90**:312–322.

Folk, J. E., 1971, Carboxypeptidase B. in: *Enzymes*, 3rd ed. (P. D. Boyer, ed.), pp. 57–59, Academic Press, New York.

Frey, P., Forand, R., Maciag, T., and Shooter, E. M., 1979, The biosynthetic precursor of epidermal growth factor and the mechanism of its processing, *Proc. Natl. Acad. Sci. USA* **76**:6294–6298.

Fricker, L. D., and Snyder, S. H., 1982, Enkephalin convertase: Purification and characterization of a specific enkephalin-synthesizing carboxypeptidase localized to adrenal chromaffin granules, *Proc. Natl. Acad. Sci. USA* **79**:3886–3890.

Fricker, L. D., Supattapone, S., and Snyder, S. H., 1982, Enkephalin convertase: A specific enkephalin synthesizing carboxypeptidase in adrenal chromaffin granules, brain, and pituitary gland, *Life Sci.* **31**:1841–1844.

Fricker, L. D., Plummer, T. H. Jr., and Snyder, S. H., 1983, Enkephalin convertase: Potent, selective, and irreversible inhibitors, *Biochem. Biophys. Res. Commun.* **111**:994–1000.

Fricker, L. D., and Snyder, S. H., 1983, Purification and characterization of enkephalin convertase, an enkephalin-synthesizing carboxypeptidase, *J. Biol. Chem.* **258**:10950–10955.

Fricker, L. D., 1985, Neuropeptide biosynthesis: Focus on the carboxypeptidase processing enzyme, *Trends Neurosci.* **8**:210–214.

Fricker, L. D., Evans, C. J., Esch, F. S., and Herbert, E., 1986, Cloning and sequence analysis of cDNA for bovine carboxypeptidase E, *Nature (Lond.)* **323**:461–464.

Fritz, H., Tschesche, H., Greene, L. J., and Truscheit, E. (eds.), 1974, *Proteinase Inhibitors*, Springer-Verlag, New York.

Gainer, H., Sarne, Y., and Brownstein, M. J., 1977, Biosynthesis and axonal transport of rat neurohypophysial proteins and peptides, *J. Cell Biol.* **73**:366–381.

Gainer, H., Russell, J. T., and Loh, Y. P., 1984, An aminopeptidase activity in bovine pituitary secretory vesicles that cleaves the N-terminal arginine from β-lipotropin 60-65, *FEBS Lett.* **175**:135–139.

Gainer, H., Russell, J. T., and Loh, Y. P., 1985, The enzymology and intracellular organization of peptide precursor processing: The secretory vesicle hypothesis, *Neuroendocrinology* **40**:171–184.

Geis, R., Martin, R., and Voight, K. H., 1984, α-MSH-like peptides from the rat hypothalamus and pituitary: Differences in the degree of N-acetylation, *Horm. Metab. Res.* **16**:266–267.

Gianoulakis, C., Sidah, N. G., Routhier, R., and Chretien, M., 1979, Biosynthesis and characterization of adrenocorticotropic hormone, α-melanocyte-stimulating hormone, and an NH_2-terminal fragment of the adrenocorticotropic hormone/β-lipotropin precursor from rat pars intermedia, *J. Biol. Chem.* **254**:11903–11906.

Glembotski, D. B., 1982, Characterization of the peptide acetyltransferase activity in bovine and rat intermediate pituitaries responsible for the acetylation of β-endorphin and α-melanotropin, *J. Biol. Chem.* **257**:10501–10509.

Gorman, B., Padmanabhan, R., and Howard, B., 1983, High efficiency DNA-mediated transformation of primate cells, *Science* **221**:551–553.

Graham, F., and Van der Eb, A. J., 1973, A new technique for the assay of infectivity of human adenovirus DNA, *Virology* **52**:456–467.

Gramsch, C., Kleber, G., Hollt, V., Pasi, A., Merhaiein, P., and Herz, A., 1980, Proopiocortin fragments in human and rat brain: β-Endorphin and α-melanotropin are the predominant peptides, *Brain Res.* **192**:109–119.

Gruss, P., and Khoury, G., 1981, Expression of simian virus 40-rat preproinsulin recombinants in monkey kidney cells: Use of preproinsulin RNA processing signals, *Proc. Natl. Acad. Sci. USA* **78**:133–137.

Gubler, U., Seeburg, P., Hoffman, B. J., Gage, L. P., and Udenfriend, S., 1982, Molecular cloning establishes proenkephalin as precursor of enkephalin-containing peptides, *Nature (Lond.)* **295**:206–208.

Helle, K. B., Read, R. K., Pihl, K. E., and Serech-Hanssen, G., 1985, Osmotic properties of the chromogranins and relation to osmotic pressure in catecholamine storage granules, *Acta Physiol. Scand.* **123**:21–33.

Hellerman, J. G., Cone, R. C., Potts, J. T., Rich, A., Mulligan, R. C., and Kronenberg, H. M., 1984, Secretion of human parathyroid hormone from rat pituitary cells infected with a recombinant retrovirus encoding preproparathyroid hormone, *Proc. Natl. Acad. Sci. USA* **81**:5340–5344.

Hogue-Angeletti, R. A., 1977, Nonidentity of chromogranin A and dopamine β-mono oxygenase, *Arch. Biochem. Biophys.* **184**:364–372.

Hook, V. Y. H., and Eiden, L. E., 1984, Two peptidases that convert ^{125}I-Lys-Arg-(Met) enkephalin and ^{125}I-(Met) enkephalin-Arg^6, respectively, to ^{125}I-(Met) enkephalin in bovine adrenal medullary chromaffin granules, *FEBS Lett.* **172**:212–218.

Hook, V. Y. H., and Loh, Y. P., 1984, Carboxypeptidase B-like converting enzyme activity in secretory granules of rat pituitary, *Proc. Natl. Acad. Sci. USA* **81**:2776–2780.

Hook, V. Y.H., Eiden, L. E., and Brownstein, M. J., 1982, A carboxypeptidase processing enzyme for enkephalin precursors, *Nature (Lond.)* **295**:341–342.

Hook, V. Y. H., Mezey, E., Fricker, L. D., Pruss, R. M., Siegel, R. E., and Brownstein, M. J., 1985, Immunochemical characterization of carboxypeptidase B-like peptide-hormone-processing enzyme, *Proc. Natl. Acad. Sci. USA* **82**:4745–4749.

Hope, D. B., and Pickup, J. C., 1974, Neurophysins, in: *Handbook of Physiology*, Section 7: *Endocrinology*, Vol. IV (E. Knobil and W. H. Sawyer, eds.), pp. 173–189, American Physiology Society, Washington, D.C.

Howell, S. L., 1974, The molecular organization of the β-granule of the islets of Langerhans, *Adv. Cytopharmacol.* **2**:319–327.

Iacangelo, A., Affholter, H.-U., Eiden, L. E., Herbert, E., and Grimes, M., 1986, Bovine chromogranin A sequence and the distribution of its messenger RNA in endocrine tissues, *Nature (Lond.)* **323**:82–86.

Izant, J. G., and Weintraub, H., 1984, Inhibition of thymidine kinase gene expression by anti-sense RNA: A molecular approach to genetic analysis, *Cell* **36**:1007–1015.

Izant, J. G., and Weintraub, H., 1985, Constitutive and conditional suppression of exogenous and endogenous genes by anti-sense RNA, *Science* **229**:346–352.

Johnson, R. G., and Scarpa, A., 1976, Internal pH of isolated chromaffin vesicles, *J. Biol. Chem.* **251**:2189–2191.

Kakidani, H., Furatani, Y., Takahashi, H., Noda, M., Morimoto, Y., Hirose, T., Asai, M., Inayama, S., Nakanishi, S., and Numa, S., 1982, Cloning and sequence analysis of cDNA for porcine α-neo-endorphin/dynorphin precursor, *Nature (Lond.)* **298**:245–249.

Kangawa, K., Minamino, N., Chino, N., Sakokibara, S., and Matsuo, H., 1981, The complete amino acid sequence of α-neo-endorphin, *Biochem. Biophys. Res. Commun.* **99**:871–878.

Kim, S. K., and Wold, B. J., 1985, Stable reduction of thymidine kinase activity in cells expressing high levels of anti-sense RNA, *Cell* **42**:129–138.

Kirshner, N., 1974, Molecular organization of the chromaffin vesicles of the adrenal medulla, *Adv. Cytopharmacol.* **2**:265–272.

Koenig, H., 1974, The soluble acidic lipoproteins (SALPs) of storage granules: Matrix constituents which may bind stored molecules, *Adv. Cytopharmacol.* **2**:273–301.

Kopnell, W. N., and Westhead, E. W., 1982, Osmotic pressures of solutions of ATP and catecholamines relating to storage in chromaffin granules, *J. Biol. Chem.* **257**:5707–5710.

Lagercrantz, H., 1976, On the composition and function of large dense cored vesicles in sympathetic nerves, *Neuroscience* **1**:81–92.

Laub, O., and Rutter, W. J., 1983, Expression of the human insulin gene and cDNA in a heterologous mammalian system, *J. Biol. Chem.* **258**:6043–6050.

Lindberg, I., Yang, H. Y. T., and Costa, E., 1982, An enkephalin-generating enzyme in bovine adrenal medulla, *Biochem. Biophys. Res. Commun.* **106**:186–193.

Lindberg, I., Yang, H. Y. T., and Costa, E., 1984, Further characterization of an enkephalin-generating enzyme from adrenal medullary chromaffin granules, *J. Neurochem.* **42**:1411–1419.

Liotta, A. S., and Krieger, D. T., 1983, Pro-opiomelanocortin-related and other pituitary hormones in the central nervous system, in *Brain Peptides* (D. T. Krieger, M. J. Brownstein, and J. B. Martin, eds.), pp. 613–660, Wiley, New York.

Liotta, A. S., Yamaguchi, H., and Krieger, D. T., 1981, Biosynthesis and release of β-endorphin-, N-acetyl β-endorphin-, β-endorphin-(1-27)-, and N-acetyl β-endorphin-(1-27)-like peptides by rat pituitary neurointermediate lobe: β-Endorphin is not further processed by anterior lobe, *J. Neurosci.* **1**:585–595.

Lipscomb, W. N., 1983, Structure and catalysis of enzymes, *Annu. Rev. Biochem.* **52**:17–34.

Liston, D., and Rossier, J., 1984, Distribution and characterization of synenkephalin immunoreactivity in the bovine brain and pituitary, *Reg. Peptides* **8**:79–87.

Liston, D., Vanderhaegen, J-J., and Rossier, J., 1983, Presence in brain of synenkephalin, a proenkephalin-immunoreactive protein which does not contain enkephalin, *Nature (Lond.)* **302**:62–65.

Liston, D., Patey, G., Rossier, J., Verbanck, P., and Vanderhaegen, J-J., 1984, Processing of proenkephalin is tissue-specific, *Science* **225**:734–737.

Lloyd, R. V., and Wilson, B. S., 1983, Specific endocrine tissue marker defined by a monoclonal antibody, *Science* **222**:628–630.

Loh, Y. P., and Gainer, H., 1982, Characterization of pro-opicortin-converting activity in purified secretory granules from rat pituitary neurointermediate lobe, *Proc. Natl. Acad. Sci. USA* **79**:108–112.

Loh, Y. P., Eskay, R. L., and Brownstein, M. J., 1980, α-MSH-like peptides in rat brain: Identification and changes in level during development, *Biochem. Biophys. Res. Commun.* **94**:916–923.

Loh, Y. P., Parish, D. C., and Tuteja, R., 1985, Purification and characterization of a paired basic residue-specific proopiomelanocortin converting enzyme from bovine pituitary intermediate lobe secretory vesicles, *J. Biol. Chem.* **260**:7194–7205.

Lomedico, P. T., 1982, Use of recombinant DNA technology to program eukaryotic cells to synthesize rat proinsulin: A rapid expression assay for cloned genes, *Proc. Natl. Acad. Sci. USA* **79**:5798–5802.

Lomedico, P., Rosenthal, N., Efstratiadis, A., Gilbert, W., Kolodner, R., and Tizard, R., 1979, The structure and evolution of the two nonallelic rat preproinsulin genes, *Cell* **18**:545–558.

Lynch, D. R., Strittmatter, S. M., and Snyder, S. H., 1984, Enkephalin convertase localization by [^{3}H]guanidinoethylmercaptosuccinic acid autoradiography: Selective association with enkephalin-containing neurons, *Proc. Natl. Acad. Sci. USA* **81**:6543–6547.

Mains, R. E., and Eipper, B. A., 1984, Secretion and regulation of two biosynthetic enzyme activities, peptidyl-glycine α-amidating monooxygenase and a carboxypeptidase, by mouse pituitary corticotropic tumor cells, *Endocrinology* 115:1683–1690.

Mains, R. E., Eipper, B. A., and Ling, N., 1977, Common precursor to corticotropins and endorphins, *Proc. Natl. Acad. Sci. USA* 74:3014–3018.

Mains, R. E., Eipper, B. A., Glembotski, C. C., and Dores, R. M., 1983, Strategies for the biosynthesis of bioactive peptides, *Trends Neurosci.* 6:229–235.

Melton, D. A., 1985, Injected anti-sense RNAs specifically block messenger RNA translation in vivo, *Proc. Natl. Acad. Sci. USA* 82:144–148.

Miller, A. D., Ong, E. S., Rosenfeld, M. G., Verma, I. M., and Evans, R. M., 1984, Infections and selectable retrovirus containing an inducible rat growth hormone minigene, *Science* 225:993–998.

Minamino, N., Kangawa, K., Chino, N., Sakakibana, S., and Matsuo, H., 1981, β-Neoendorphin, a new hypothalamic "big" Leu-enkephalin of porcine origin: Its purification and the complete amino acid sequence, *Biochem. Biophys. Res. Commun.* 99:864–870.

Mizuno, K., Miyata, A., Kangawa, K., and Matsuo, H., 1982, A unique proenkephalin-converting enzyme purified from bovine adrenal chromaffin granules, *Biochem. Biophys. Res. Commun.* 108:1235–1242.

Mizuno, K., Kojima, M., and Matsuo, H., 1985, A putative prohormone processing protease in bovine adrenal medulla specifically cleaving in between Lys-Arg sequences, *Biochem. Biophys. Res. Commun.* 128:884–891.

Moore, H. P. H., Walker, M. D., Lee, F., and Kelly, R. B., 1983, Expressing a human proinsulin cDNA in a mouse ACTH-secreting cell: Intracellular storage, proteolytic processing, and secretion on stimulation, *Cell* 35:531–538.

Morrissely, J. J., Shoststall, R. E., Hamilton, J. W., and Cohn, D. W., 1980, Synthesis, intracellular distribution and secretion of multiple forms of parathyroid secretory protein I, *Proc. Natl. Acad. Sci. USA* 77:6406–6410.

Noda, M., Furutani, Y., Takahashi, H., Toyosato, M., Hirose, T., Inayama, S., Nakanishi, S., and Numa, S., 1982, Cloning and sequence analysis of cDNA for bovine adrenal preproenkephalin, *Nature (Lond.)* 295:202–206.

O'Connor, D. T., 1983, Chromogranin: Widespread immunoreactivity in polypeptide hormone producing tissues and in serum, *Reg. Peptides* 6:263–280.

O'Connor, D. T., and Frigon, R. P., 1984, Chromogranin A: The major catecholamine storage vesicle soluble protein, *J. Biol. Chem.* 259:3237–3247.

O'Connor, D. T., Burton, D., and Deftos, L. J., 1983, Chromogranin A: Immunohistology reveals its universal occurrence in normal polypeptide hormone producing endocrine glands, *Life Sci.* 33:1657–1663.

O'Donohue, T. L., Handelmann, G. E., Chaconas, T., Miller, R. L., and Jacobowitz, D. M., 1981, Evidence that N-acetylation regulates the behavioral activity of α-MSH in the rat and human central nervous system, *Peptides* 2:333–344.

O'Donohue, D. L., Handelmann, G. E., Miller, R. L., and Jacobowitz, D. M., 1982, N-Acetylation regulates the behavioral activity of α-melanotropin in a multineurotransmitter neuron, *Science* 215:1125–1127.

Orwoll, E., Kendall, J. W., Lamorena, L., and McGilvra, R., 1979, Adrenocorticotropin and melanocyte-stimulating hormone in the brain, *Endocrinology* 104:1845–1852.

Pavlakis, G. N., Hizuka, N., Gorden, P., Seeburg, P., and Hamer, D. H., 1981, Expression of two human growth hormone genes in monkey cells infected by simian virus 40 recombinants, *Proc. Natl. Acad. Sci. USA* 78:7398–7402.

Pletscher, A., DaPrada, M., Berneis, K. H., Steffen, H., Lutold, B., and Weder, H. G., 1974, Molecular organization of amine storage organelles of blood platelets and adrenal medulla, *Adv. Cytopharmacol.* **2**:257–264.

Plummer, T. H. Jr., 1969, Isolation and sequence of peptides at the active center of bovine carboxypeptidase B, *J. Biol. Chem.* **244**:5246–5253.

Plummer, T. H. Jr., and Erdos, E. G., 1981, Human plasma carboxypeptidase N, *Methods Enzymol.* **80**:442–449.

Powers, C. A., and Nasjletti, A., 1982, A novel kinin-generating protease (kininogenase) in the porcine anterior pituitary, *J. Biol. Chem.* **257**:5594–5600.

Powers, C. A., and Nasjletti, A., 1983, A kininogenase resembling glandular kallikrein in the rat pituitary pars intermedia, *Endocrinology* **112**:1194–1200.

Roberts, J. L., Phillips, M., Rosa, P. A., and Herbert, E., 1978, Steps involved in the processing of common precursor forms of adrenocorticotropin and endorphin in cultures of mouse pituitary cells, *Biochemistry* **17**:3609–3618.

Robins, D. M., Pack, I., Seeburg, P. H., and Axel, R., 1982, Regulated expression of human growth hormone genes in mouse cells, *Cell* **29**:623–631.

Russell, J. T., 1984, ΔpH, H^+ diffusion potentials, and Mg^+ATPase in neurosecretory vesicles isolated from bovine neurohypophyses, *J. Biol. Chem.* **259**:9496–9507.

Schmid, M. F., and Herriott, J. R., 1976, Structure of carboxypeptidase B at 2.8 Å resolution, *J. Biol. Chem.* **103**:175–190.

Scott, A. P., Lowry, P. J., Ratcliffe, J. G., Rees, L. H., and Landon, J., 1974, Corticotrophin-like peptides in the rat pituitary, *J. Endocrinol.* **61**:355–367.

Seizinger, B. R., Grimm, C., Hollt, V., and Herz, A., 1984, Evidence for a selective processing of proenkephalin B into different opioid peptide forms in particular regions of rat brain and pituitary, *J. Neurochem.* **42**:447–457.

Sen, R., and Sharp, R. R., 1982, Molecular mobilities and the lowered osmolarity of the chromaffin granule aqueous phase, *Biochem. Biophys. Acta* **721**:70–82.

Server, A. C., and Shooter, E. M., 1976, Comparison of the arginine esteropeptidases associated with the nerve and epidermal growth factors, *J. Biol. Chem.* **251**:165–173.

Smith, A. D., and Winkler, H., 1967, Purification and properties of an acidic protein from chromaffin granules of bovine adrenal medulla, *Biochem. J.* **103**:483–492.

Smyth, D. G., and Zakarian, S., 1980, Selective processing of β-endorphin in regions of porcine pituitary, *Nature (Lond.)* **288**:613–615.

Smyth,, D. G., Massey, D. E., Zakarian, S., and Finnie, M. D. A., 1979, Endorphins are stored in biologically active and inactive forms: Isolation of α-N-acetyl peptides, *Nature (Lond.)* **279**:252–254.

Somogyi, P., Hodgson, A. J., DePotter, R. W., Fischer-Colbrie, R., Schober, M., Winkler, H., and Chubb, I. W., 1984, Chromogranin immunoreactivity in the central nervous system: Immunochemical characterization, distribution, and relationship to catecholamine and enkephalin pathways, *Brain Res. Rev.* **8**:193–230.

Southern, E. M., 1975, Detection of specific sequences among DNA fragments separated by gel electrophoresis, *J. Mol. Biol.* **98**:503–517.

Supattapone, S., Fricker, L. D., and Snyder, S. H., 1984, Purification and characterization of a membrane-bound enkephalin-forming carboxypeptidase, "enkephalin convertase," *J. Neurochem.* **42**:1017–1023.

Supowit, S. C., Potter, E., Evans, R. M., and Rosenfeld, M. G., 1984, Polypeptide hormone regulation of gene transcription: Specific 5' genomic sequences are required for epidermal growth factor and phorbol ester regulation of prolactin gene expression, *Proc. Natl. Acad. Sci. USA* **81**:2975–2979.

Taylor, J. M., Cohen, S., and Mitchell, W. M., 1970, Epidermal growth factor: High and low molecular weight forms, *Proc. Natl. Acad. Sci. USA* **67**:164–171.

Thomas, P. S., 1980, Hybridization of denatured RNA and small DNA fragments transferred to nitrocellulose, *Proc. Natl. Acad. Sci. USA* **77**:5201–5205.

Titani, K., Ericsson, L. H., Kumar, S., Jakob, F., Neurath, H., and Zwillig, R., 1984, Amino acid sequence of crayfish (*Astacus fluviatilis*) carboxypeptidase B, *Biochemistry* **23**:1245–1250.

Udenfriend, S., and Kilpatrick, D. L., 1983, Biochemistry of the enkephalin and enkephalin-containing peptides, *Arch. Biochem. Biophys.* **221**:309–323.

Vanderhaegen, J-J., Lotstra, F., Liston, D. R., and Rossier, J., 1983, Proenkephalin, [Met] enkephalin, and oxytocin immunoreactivities are colocalized in bovine hypothalamic magnocellular neurons, *Proc. Natl. Acad. Sci. USA* **80**:5139–5143.

Warren, T. G., and Shields, D., 1984, Expression of preprosomatostatin in heterologous cells: Biosynthesis, post-translational processing, and secretion of mature somatostatin, *Cell* **39**:547–555.

Weber, E., Evans, C. J., and Barchas, J. D., 1981, Acetylated and nonacetylated forms of β-endorphin in rat brain and pituitary, *Biochem. Biophys. Res. Commun.* **103**:982–989.

Weber, E., Evans, C. J., and Barchas, J. D., 1982a, Predominance of the amino-terminal octapeptide fragment of dynorphin in rat brain regions, *Nature (Lond.)* **299**:77–79.

Weber, E., Evans, C. J., Chang, J. K., and Barchas, J. D., 1982b, Antibodies specific for α-N-acetyl-β-endorphins: Radioimmunoassays and detection of acetylated β-endorphins in pituitary extracts, *J. Neurochem.* **38**:436–447.

Weber, E., Evans, C. J., Chang, J. K., and Barchas, J. D., 1982c, Brain distributions of α-neo-endorphin and β-neo-endorphin: Evidence for regional processing differences, *Biochem. Biophys. Res. Commun.* **108**:81–88.

Whittaker, V. P., 1974, Molecular organization of the cholinergic vesicle, in: *Advances in Cytopharmacology*, Vol. 2 (B. Ceccarelli, F. Clement, and J. Meldoles, eds.), pp. 311–317, Raven Press, New York.

Winkler, H., 1976, The composition of adrenal chromaffin granules: An assessment of controversial results, *Neuroscience* **1**:65–80.

Winkler, H., and Westhead, E., 1980, The molecular organization of adrenal chromaffin granules, *Neuroscience* **5**:1803–1823.

Zakarian, S., and Smyth, D. G., 1982, β-endorphin is processed differently in specific regions of rat pituitary and brain, *Nature (Lond.)* **296**:250–253.

Index